ANALYSIS OF BIOLOGICAL NETWORKS

Wiley Series on

Bioinformatics: Computational Techniques and Engineering

Bioinformatics and computational biology involve the comprehensive application of mathematics, statistics, science, and computer science to the understanding of living systems. Research and development in these areas require cooperation among specialists from the fields of biology, computer science, mathematics, statistics, physics, and related sciences. The objective of this book series is to provide timely treatments of the different aspects of bioinformatics spanning theory, new and established techniques, technologies and tools, and application domains. This series emphasizes algorithmic, mathematical, statistical, and computational methods that are central in bioinformatics and computational biology.

Series Editors: **Professor Yi Pan** and **Professor Albert Y. Zomaya**
pan@cs.gsu.edu   zomaya@it.usyd.edu.au

Knowledge Discovery in Bioinformatics: Techniques, Methods, and Applications
Xiaohua Hu and Yi Pan

Grid Computing for Bioinformatics and Computational Biology
Edited by El-Ghazali Talbi and Albert Y. Zomaya

Bioinformatics Algorithms: Techniques and Applications
Ion Mandiou and Alexander Zelikovsky

Analysis of Biological Networks
Edited by Björn H. Junker and Falk Schreiber

ANALYSIS OF BIOLOGICAL NETWORKS

EDITED BY

BJÖRN H. JUNKER and FALK SCHREIBER

WILEY-INTERSCIENCE

A JOHN WILEY & SONS, INC., PUBLICATION

Library of Congress Cataloging-in-Publication Data:

Analysis of biological networks / edited by Björn H. Junker and Falk Schreiber.
    p. ; cm.
    "A WileyInterscience Publication."
    Includes bibliographical references and index.
    ISBN 978-0-470-04144-4 (cloth : alk. paper)
  1. Biological models. 2.   Computational biology.   I. Junker, Björn H. II. Schreiber, Falk.
    [DNLM: 1.   Models, Biological. 2.   Algorithms. 3.   Computational Biology.
    QH 324.8 A532 2008]
    QH 324.8.A532 2008
    570.285–dc22

                                                                    2007034278

Printed in Mexico
10 9 8 7 6 5 4 3

CONTENTS

4 Network Centralities   65

Dirk Koschützki

5 Network Motifs   85

Henning Schwöbbermeyer

FOREWORD

Among the sciences that showed the most rapid growth rate during the second half of the twentieth century, biochemistry occupies one of the foremost positions. Up to this time, the isolation of enzymes and the investigation of their reactions were the main focus of studies. But in 1953, the structure of DNA was published and a complete new field of research was entered. The function of genetic information storage and recovery could be elucidated. Consecutive studies in this direction dealt with the regulation of protein synthesis, which determines the quantity of proteins present and thus, in the case of enzymes, their activity. However, the turnover rate of enzymatic reactions is also controlled by activators, inhibitors, conformation changes, degradation or chemical modification of enzymes, and by other means. A whole network of regulatory mechanisms could be elucidated, which became an overlay over the network of enzymatic pathways.

All these activities resulted in a tremendous flood of data. Only the application of computers and advancement in informatics permitted the completion of many tasks in biochemistry, for example, the sequencing of the human genome with its billions of base pairs. Data management is developing in two directions: on the one hand to achieve the solution of detailed problems, and on the other hand to obtain a general survey of metabolism and regulation in a larger unit, up to a complete living being. The latter aspect includes the presentation of the network of metabolic pathways, the mechanisms of their regulation and the compounds involved in it, the flow rates of metabolites under various external conditions, etc. Analogously to the term "genome" known for a long time, their various components are named "transcriptome," "proteome," "regulome," "metabolome", etc.

The presentation of these aspects has a qualitative and a quantitative side. The latter, with the final target of modeling the complex interconnections within a whole organism, is a task reaching far into the future. It requires a wealth of measurements under various conditions. However, much progress has been achieved already by the simultaneous measurement of metabolite concentrations or of gene expression using microarrays. Evaluation of these data and studying their interconnections by network analysis is strictly a task of computer-supported bioinformatics. The methods and tools required for this research are presented in this book.

A slightly different picture exists for the qualitative aspect, especially if an initial survey is intended. Here, didactic considerations also play a role. Although the facts can be correctly presented by adequate computer programs, to some extent a manual "curating" of the networks may be helpful to give a more clear impression, especially when a large network with multiple interconnections is presented. This may add a "human touch" to bioinformatics. I followed this line when I wrote the book "*Biochemical Pathways*" and constructed the wallchart with the same title.

Dealing with the highly complex functions of living beings requires the cooperation of many fields of science. Their contacts generate as offspring new sciences. As biochemistry developed some time ago from activities in chemistry, biology, physics, medicine, and other areas, later bioinformatics combined biochemistry with informatics and mathematics. Recently, there have been exiting developments in the topological analysis of biological networks, a field that requires the interdisciplinary cooperation mentioned above. The present speed of development allows the prediction of a large growth of this field in the future, and this book may help the reader to obtain an overview of the topic.

GERHARD MICHAL

Tutzing, Germany

PREFACE

Network analysis is a research area that has been employed in specific fields such as social sciences for a long time. However, only since the late 1990s, when several papers on the fundamental design principles of various kinds of large-scale networks were published, has this research area gained wider publicity. It was then found that different kinds of biological networks, and also the World Wide Web, power grids, and other networks, share common properties. These studies led to a tremendous expansion of the research field of network analysis, especially in biology. Nearly one decade later, this field is still growing rapidly. Nonetheless, the analysis of biological networks is still hampered by the interdisciplinary nature of this special research area, which involves concepts and ideas from different sciences such as biology, biochemistry, physics, mathematics, and computer science.

AIMS OF THIS BOOK

This book intends to give a comprehensive overview of the structural analysis of biological networks, located at the interface of biology and computer science. Biological networks represent processes in cells, organisms, or entire ecosystems. Large amounts of data that represent (or are related to) biological networks have been gathered in the past, not least with the help of the latest technological advances. Thus, analysis of these networks is an important research topic in modern bioinformatics, and the analysis of biological networks is gaining more and more attention in the life sciences and particular in the growing field of systems biology.

This edited book is an introduction to the analysis of biological networks for students and scientists from both computer science and biology. It is intended for researchers who want an overview of the field, and who want information about the possibilities (and limits) of network analysis in life sciences, and how it can be applied to their data. It is also intended for graduate students who specialize in the field of biological network analysis. The book presents an overview of biological networks, methods for their analysis, and a summary of important insights obtained through the analysis of biological networks. Each specific area of the field is comprehensively presented and discussed by experts.

HOW TO READ THIS BOOK

In order to reach a broad spectrum of readers—biologists, biochemists, computer scientists, bioinformaticians, and so on—the book does not require deeper knowledge of computer science or biology. Instead, the reader will learn about fundamental techniques from computer science, in particular graph theory, graph algorithms, and network analysis, as well as from biology, in particular biochemistry, molecular biology, ecology, and evolution.

This book consist of three parts: (I) an introduction that gives a brief overview of biological networks and graph theory/graph algorithms; (II) chapters discussing network analysis methods; and (III) chapters dealing with biological networks, the application of appropriate network analysis methods and initial insights gained in this relatively new field of research.

Each chapter can be studied independently; however, we recommend following the chapter structure. Both introductory chapters present some background from biology and computer science, respectively. They are especially intended for readers not familiar with these specific areas and present basic concepts from molecular biology and biological networks in Chapter 1, as well as graph theory and algorithms in Chapter 2. These chapters also introduce basic terminology and definitions used in the other chapters of the book. Readers familiar with these topics may skip these chapters.

In Part II, network analysis methods are presented and discussed. Chapter 3 gives an insight into global network properties and discusses network models. In Chapter 4, different centrality concepts are described. Centrality analysis helps in ranking network elements. Network motifs, small recurring patterns in networks, which can represent potentially important network parts are discussed in Chapter 5. In Chapter 6, clustering methods are presented. Such methods divide networks into parts or modules. Finally, Chapter 7 discusses Petri nets, which are a special representation often used to model biological networks and which offer specific analysis capabilities. Each chapter presents the specific topic from a computer science point of view, introduces concepts and algorithms, and discusses the application of the particular network analysis method to biological networks.

In Part III, networks that can be found in biology are presented. On the molecular level, signal transduction and gene regulation networks are required to regulate various processes in a cell. These networks are presented in Chapter 8. Chapter 9 deals with protein interaction networks, which have their seeds in chemical interactions between proteins. In Chapter 10, the structure of metabolism is described, which connects enzymes and metabolites to metabolic networks. Coming to a more macroscopic level of biological networks and also to much larger timescales, Chapter 11 introduces phylogenetic networks, which describe the evolutionary relationship among different biological species. In Chapter 12, ecological networks are studied. These networks depict the interaction between biological species on smaller timescales. Finally, Chapter 13 deals with correlation networks, which are in the strict sense not intrinsically biological networks, but help to discover causalities and regulatory events from large-scale data.

In all parts, each chapter finishes with a summary concluding the main points of the chapter and some exercises that may help the reader to test their understanding of the information presented in the chapter.

ABOUT THE COVER

The image on the front cover shows the largest connected component of the protein interaction network of *Arabidopsis thaliana*. The objects (vertices) represent proteins, and their connections (edges) show known interactions (usually binding) between them. The vertices of the network are colored according to their degree centrality. The degree of a vertex is the number of its neighbors, and the more neighbors a vertex has, the more central it is. Degree and other centralities are discussed in detail in Chapter 4, and details about protein interaction networks and their analysis can be found in Chapter 9. The image has been produced with the programs Vanted (http://vanted.ipk-gatersleben.de) and POV-Ray (http://www. povray.org/).

ACKNOWLEDGMENTS

First and foremost we thank the contributors of this book for their hard work and especially for their help in fitting each chapter into the structure and concept of this book. Thanks to the editors of Wiley's Bioinformatics series, Yi Pan and Albert Y. Zomaya, for giving us the opportunity to present our research interests as a book in their series. We acknowledge the support, guidance, and patience of Whitney Lesch, Paul Petralia, and the team from Wiley's production department during the production of this book. We thank Gerhard Michal for his support and for writing the foreword, and Henning Schwöbbermeyer, Christian Klukas, and Dirk

Koschützki for helping during the editing process. Finally, we thank our families for their help, support, and patience.

OFFICIAL FTP SITE

Results for all exercises found in the book and supplementary information for Chapter 7 can be found at the following address:

ftp://ftp.wiley.com/public/sci_tech_med/biological_networks/

BJÖRN H. JUNKER
FALK SCHREIBER

Gatersleben, Germany
January 2008

CONTRIBUTORS

Balabhaskar Balasundaram, Industrial and Systems Engineering, College Station, Texas A&M University, Texas

Frederik Börnke, Friedrich-Alexander University, Department of Biochemistry, Erlangen, Germany

Dirk Büssis, Institute for Biochemistry and Biology, University of Potsdam, Potsdam-Golm, Germany

Sergiy Butenko, Industrial and Systems Engineering, College Station, Texas A&M University, Texas

Ursula Gaedke, Department of Ecology/Ecosystem Modeling, Institute for Biochemistry and Biology, University of Potsdam, Potsdam, Germany

Birgit Gemeinholzer, Botanic Garden and Botanical Museum, Berlin, Germany

Feng He, Systems Biology Group, Helmholtz Center for Infection Research (former GBF), Braunschweig, Germany

Monika Heiner, Brandenburg University of Technology at Cottbus, Cottbus, Germany

Björn H. Junker, Leibniz Institute of Plant Genetics and Crop Plant Research (IPK), Gatersleben, Germany and Brookhaven National Laboratory, Upton, New York

Ina Koch, Technical University of Applied Sciences, Berlin, Germany and Max Planck Institute for Molecular Genetics, Berlin, Germany

Dirk Koschützki, Leibniz Institute of Plant Genetics and Crop Plant Research (IPK), Gatersleben, Germany

Leonard Krall, Max Planck Institute of Molecular Plant Physiology, Potsdam-Golm, Germany

Gorka Zamora López, Nonlinear Dynamics Group, Institute for Physics, University of Potsdam, Potsdam, Germany

Hongwu Ma, Helmholtz Center for Infection Research (former GBF), Systems Biology Group, Braunschweig, Germany

Gerhard Michal, Tutzing, Germany

Carsten Müssig, Institute for Biochemistry and Biology, University of Potsdam, Potsdam-Golm, Germany

Anatolij P. Potapov, Department of Bioinformatics, University of Göttingen, UKG, Göttingen, Germany

Falk Schreiber, Leibniz Institute of Plant Genetics and Crop Plant Research (IPK), Gatersleben, Germany

Henning Schwöbbermeyer, Leibniz Institute of Plant Genetics and Crop Plant Research (IPK), Gatersleben, Germany

Márcio Rosa da Silva, Systems Biology Group, Helmholtz Center for Infection Research (former GBF), Braunschweig, Germany

Dirk Steinhauser, Max Planck Institute of Molecular Plant Physiology, Potsdam-Golm, Germany

Ralf Steuer, Institute for Theoretical Biology, Humboldt University, Berlin, Germany

Jibin Sun, Systems Biology Group, Helmholtz Center for Infection Research (former GBF), Braunschweig, Germany

Björn Usadel, Max Planck Institute of Molecular Plant Physiology, Potsdam-Golm, Germany

An-Ping Zeng, Institute of Bioprocess and Biosystems Engineering, Hamburg University of Technology, Hamburg, Germany, Systems Biology Group, and Helmholtz Center for Infection Research (former GBF), Braunschweig, Germany

PART I

INTRODUCTION

1

NETWORKS IN BIOLOGY

Björn H. Junker

1.1 INTRODUCTION

Our environment is a combination of tightly interlinked complex systems at various levels of magnitude. While the exact sciences of physics and chemistry describe our environment from subatomic level up to the molecular level, biology is carrying the burden to deal with an inexact and extremely complex universe that sometimes even seems lawless. Yet biological systems follow "laws" that physicists would rather refer to as "probabilities." By these laws, it is possible to describe biology at different detail levels with a certain precision.

The smallest biological detail level is the molecular level of DNA, RNA, proteins, and metabolites. All these molecules are ingredients of a cell, which in turn is a part of a tissue. Different tissues constitute the organs of an organism. Many organisms together form the ecosystem. Additionally, over time these organisms are subjected to evolution, which results in a certain phylogenetic relationship between them. At all these levels of detail, the relationships between the elements are of great interest. These relationships can be described as networks, in which the elements are the vertices (nodes, points) and the relationships are the edges (arcs, lines; see Chapter 2). Typical biological networks at the molecular level are gene regulation networks, signal transduction networks, protein interaction networks, and metabolic networks. An example of a biological network is given in Fig. 1.1. While parts of all these networks have been modeled since a long time, recent technological

Analysis of Biological Networks, Edited by Björn H. Junker and Falk Schreiber
Copyright © 2008 John Wiley & Sons, Inc.

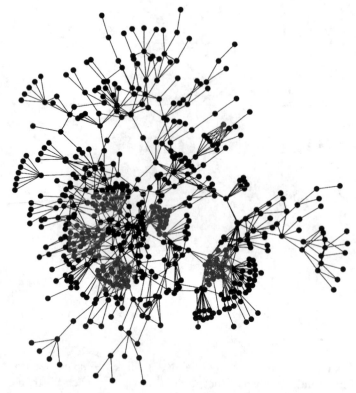

FIGURE 1.1   Example of a biological network. The largest strongly connected component (see Chapter 2) of the human protein interaction network is shown. The network is based on the complete data set for interaction of human proteins downloaded from the Database of Interacting Proteins (DIP, [35]) in January 2005.

advances have made it possible to elicit entire networks, or at least large proportions of them.

The next section contains a concise overview of basic biology and is especially aimed at readers who would like to refresh their knowledge of biology. Section 1.3 introduces the concept of systems biology. In Section 1.4, an overview is given about what findings have been made about different biological networks with modern network analysis methods.

1.2   BIOLOGY 101

1.2.1   Biochemistry and Molecular Biology

The information about the assembly of an organism is stored in the desoxyribonucleic acid (DNA, see Fig. 1.2). DNA is a coiled ladder (*helix*) consisting of two sugar phosphate backbones enclosing pairs of the nucleotide bases adenine, cytosine, guanine, and thymine (A,C,G,T). The nucleotide A pairs only with T, whereas C pairs only with

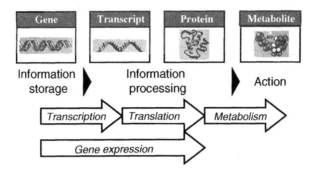

FIGURE 1.2   Information flow from genes to metabolites in cells.

G. While DNA constitutes the passive part of the cell's biochemistry, the active part is contributed by the proteins, as they catalyze reactions and are responsible for many other mechanisms in the cell. The process of information transmission from DNA to proteins is called *gene expression* (Fig. 1.2). This process can be divided into two main parts, *transcription* and *translation*. Transcription is a complicated, highly regulated process, in which a protein complex containing the RNA polymerase opens the DNA helix, reads one strand, and synthesizes a corresponding ribonucleic acid (RNA) like a blueprint. Transcription is initiated and terminated at certain signal sequences, which are called *promoter* and *terminator*, respectively. The corresponding RNA to a certain gene is called *transcript* (Fig. 1.2). In *eukaryotes* (see next section), the RNA then undergoes a process called *splicing*, in which the *introns* (noncoding regions) are excised so that only the *exons* (coding regions) remain. During translation, amino acid chains are synthesized from the (spliced) RNA by the ribosomes. The information of the RNA is read in triplets (*codons*), for which there are $4^3 = 64$ combinations. These are used to code both for 20 amino acids (sometimes more than one codon stands for one amino acid), as well as one start codon and three stop codons.

The structure of a protein is important for its functionality. The primary protein structure is simply the amino acid sequence, where as the secondary structure consists of regular three-dimensional patterns such as loops, helices, or sheets. Furthermore, the tertiary structure describes how these patterns are arranged in space to form a protein or a subunit thereof. Finally, the quaternary structure depicts how the different amino acid chains of the subunits are arranged to form an active protein complex. Proteins can play many different roles in the cell, for example, structural proteins that stabilize the cell's structure, transcription factors that regulate the process of transcription, or enzymes enzyme that catalyze reaction in which one metabolite is converted into another.

Metabolite is a term for all molecules of low molecular weight, such as sugars or amino acids. All the processes mentioned above are subject to tight regulation, which can take place at different levels. An environmental or internal signal (e.g., light, hormone) can be at first multiplied and processed through signal transduction chains. Then, a regulatory action can take place, for example, at the transcriptional level through activation or repression of gene expression, or at the protein level through posttranslational modification.

1.2.2 Cell Biology

Depending on the domain of life, the cells of an organism are organized in different ways. *Prokaryotes* such as bacteria are single cells that are not further subdivided. Their *genome*, the totality of the genes, is organized in one single circular chromosome. In contrast, the cells of *eukaryotes* are structured much more complex (see Fig. 1.3). Like prokaryotes, the cells are filled with the *cytoplasm*, but contrary to prokaryotes, additional *organelles* are separated from the cytoplasm through membranes. Organelles are, for example, *mitochondria* that produce chemical energy, the *endoplasmatic reticulum* that plays a role in protein synthesis, and the *nucleus* (Fig. 1.3). Plant cells are equipped with additional organelles, the *plastids*, which is an umbrella term for *chloroplasts* (responsible for photosynthesis), *chromoplasts* (pigment synthesis and storage), *amyloplasts* (starch synthesis and storage), and *vacuoles* that serve as storage organelle for metabolites.

Inside the nucleus, the genome is organized in several chromosomes, each of which is consisting of two *chromatides*, parallel coils that are connected near the middle to form an x-like structure. On the gene level, this means that a eukaryotic cell generally has at least two copies of every gene. Further on, most cells in most organisms are equipped with two sets of chromosomes, one from each parent.

In living organisms, there is a variety of different cell types responsible for various functions. A number of cells that perform a similar function constitute a *tissue*, examples for animal cells are epithelium and connective tissue, for plant cells epidermis or vascular tissue. A group of tissues that perform a specific function or a set of functions form an *organ*. Typical organs in animals are brain, lung, and liver. Typical organs in plants are leafs, stem, and seeds. All organs together constitute the entire organism.

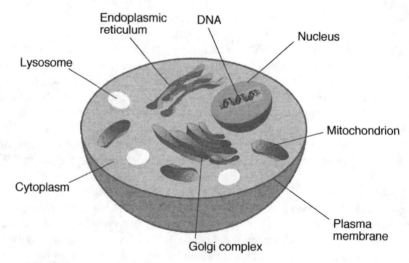

FIGURE 1.3   Schematic illustration of an animal cell with some organelles.

1.2.3   Ecology and Evolution

In the previous two sections, an overview was given about the biochemical and cellular composition of a single organism. In our environment, the organisms constantly interact with each other and are integrated components of the ecosystem. The influence of one organism on another is called a *biotic factor*. This influence might be the predator–prey relationship between two animals, or the relationship between a plant and an insect pollinating this plant. Further on, organisms are influenced also by *abiotic factors* such as climate and geology.

Organisms are subjected to evolution over large timescales. Evolution is the process by which populations of organisms acquire and pass on novel traits from generation to generation. The modern theory of evolution is based on the concept of natural selection, as first outlined in Darwin's 1859 book *"The Origin of Species"* [9]. Individual organisms that possess advantageous traits will be more likely to pass on their genes. In the 1930s, Darwin's theory was combined with Mendel's heredity laws to create the *modern synthesis*, which explains evolution as a change in frequency of alleles within a population between two generations. In modern times, sequence information from certain genes is used to derive evolutionary relationships between different organisms. From this data, phylogenetic trees can be constructed at different detail levels of the taxonomy (Fig. 1.4).

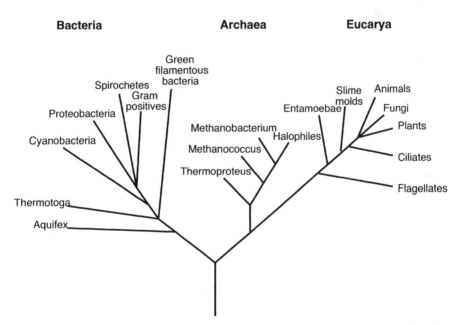

FIGURE 1.4   A speculative phylogenetic tree showing the separation of the three domains of life. Exemplary groups are shown, which represent different detail levels of the phylogeny.

1.3 SYSTEMS BIOLOGY

Biology is currently in the starting phase of a shift that will ultimately transform it into a precise science similar to physics and chemistry. The term "systems biology" is drawing more and more attention. While the origin of systems biology dates back to at least 1969 when Ludwig von Bertalanffy described his systems theory [37], it faced an explosion of interest in the new millennium. Hiroaki Kitano defined systems biology in his book "*Foundations of System Biology*" as "systems biology is a new field in biology that aims at system-level understanding of biological systems" [21]. That means the ultimate goal of systems biology is to understand entire biological systems by elucidating, modeling, and predicting the behavior of all components and interactions.

The central step toward a systems-level understanding of biology was to move away from *reductionist* to *wholist* approaches, sometimes also called bottom-up and top-down approaches, respectively [20]. Traditionally, reductionists look at *one* element of the system to find out the connections to neighbors, roles in all processes that the element is involved in, and mechanisms of action. In contrast, the wholist approach is to first make a snapshot of *all* elements at a certain level (genes, transcripts, proteins, and/or metabolites; see also Fig. 1.2). For this task, since the 1990s, many massively parallel experimental techniques have been developed. The entire set of components of one kind is described with terms ending *-ome* (genome, proteome), whereas the techniques to identify this set ends with *-omics* (genomics, proteomics). To date, more than hundred of these -omics technologies have been defined [1]. While some of them are just new words for old things, some others open an entirely new view on biological systems. The genomes of many organisms were sequenced, starting with *Escherichia coli* in 1997 [8], to reach 680 complete published genomes in November 2007 [2]. Recent technological developments will likely result in an exponential increase of this number [26]. Snapshots of the transcriptome (set of all RNA molecules of one biological sample, [10]) are routinely measured in laboratories all over the world. By the help of experimental techniques such as two-dimensional gels and mass spectrometry, the proteomes of several organisms can be determined [31]. Another recent development is metabolomics, in which a large number of metabolites are measured simultaneously in one sample [13]. With other "-omics" technologies, other "-omes" have been measured, such as the fluxome (the fluxes through metabolic pathways) or the interactome (the interactions between proteins and small molecules). Having established these high-throughput experimental techniques, scientists were confronted with the problem of how to make sense out of the wealth of generated data. One possible solution will be presented in the next section.

1.4 PROPERTIES OF BIOLOGICAL NETWORKS

Just as it is impossible to assemble an airplane by using a list of all parts, it seems impossible to gain any useful information out of the wealth of data generated with the

-omics methods detailed above. One particularly promising approach for the generation of hypothesis out of this data is network analysis, such promising that this entire book is dedicated to this area of research. While network analysis is not a new research field, it is noticeable that some fundamental properties of networks have been elucidated just at the change of the millennium. In 1998, Watts and Strogatz published a paper in which they illustrate that the neural network of the worm *Caenorhabditis elegans*, the power grid of the Western United States, and the collaboration graph of film actors have similar properties: they are highly clustered (densely connected subgraphs, see Chapter 2), yet they have small characteristic path lengths (see Chapter 2) [39]. The authors created the term *small world networks* for this phenomenon, by analogy with the popular small-world phenomenon [27], which states that any person on our planet links to any other person by a chain of on average six acquaintances. One year later, Barabási and Albert created a simple model for these networks, which they found to follow a scale-free power-law distribution and thus named them *scale-free networks* [5]. The consequence of this connectivity distribution is that many vertices have few links, while there are some that are highly connected. As a result, scale-free networks are very robust against failure, such as removal of arbitrary network elements [3]. To date it has been found that power grids, the Internet (routers and cables), the World Wide Web (webpages and links), protein interaction networks, metabolic networks, and many other networks follow these general rules [4]. However, the first obstacle for the application of these methods in biological research is the generation of networks out of the data sets determined with the -omics technologies. Because it is not possible to directly infer any networks from sequences, or from transcript, protein, or metabolite concentrations, additional information is needed, such as information about interactions. In the following sections, it will be briefly discussed which sources are available to derive biological networks, and which novel findings have been made investigating these networks.

1.4.1 Networks on a Microscopic Scale

Biochemical networks have been under investigation for many decades. However, the efforts were until recently limited to the determination of the components of the networks, rather than addressing the design principles of its structure. The fundamental finding about all kinds of networks (as mentioned above) have also been investigated in biological networks, such as regulation networks, protein interaction networks, and metabolic networks.

Transcriptional regulation networks (or gene regulation networks) are controlling gene expression in cells. The expression of one gene can be controlled by the gene product of another gene. Thus, a directed graph (see Chapter 2) in which the vertices are genes and the directed edges represent control can be used to model these networks. Until recently, only fragments of these networks have been modeled, usually quantitatively, by assigning rate laws to every step. For example, quantitative models containing selected genes have greatly improved the understanding of morphogenesis of early embryos of the fruit fly *Drosophila melanogaster* [16]. Recent advances in data collection and analysis made it possible to elucidate large-scale gene regulation networks [23]. It has been found that in this network type, certain motifs (small recurring patterns, see Chapter 5) such as feed-forward loops or single input modules are overrepresented when compared with randomly generated networks [23,36]. Through these investigations it was possible to define the "basic computational elements" of biological networks.

Signal transduction networks can be understood as gene regulation networks extended by signaling chains that contain different kinds of vertices and edges such as protein–protein interaction and phosphorylation. By quantitative modeling, emergent properties have been found in these networks such as integration of signals across multiple timescales, generation of distinct outputs depending on input strength and duration, and self-sustaining feedback-loops [7]. A more detailed explanation of gene regulation and signal transduction networks together with scientific results is given in Chapter 8.

Protein interaction networks are generated out of different types of large-scale experimental and computational approaches [38]. The different methods are resulting in significantly different networks, so that we can speak only of a network for a certain organism determined by using a certain method. The protein interaction network of the baker's yeast (*Saccharomyces cerevisiae*) as determined by systematic two-hybrid analyses was found to follow the laws of scale-free networks [17]. Furthermore, it has been shown that the most highly connected proteins in the cell are the most important for its survival [17]. In the network, this corresponds to the vertices with the highest number of connections (high degree centrality, see Chapter 4). In the same network, it has been shown that certain motifs are overrepresented [41] (see Chapter 5). Through comparison with orthologous networks from other higher eukaryotes, the authors found that these motifs are evolutionarily conserved. More details on protein interaction networks are given in Chapter 9.

Metabolic networks consist of metabolites that are converted into each other by enzymes. These networks have been determined through biochemical experiments over the last few decades, and they can be found in various kinds of biochemistry textbooks. A summary of biochemical pathways is given in the well-known Boehringer map [12]. Since few years, metabolic pathways have also been predicted from the genome of fully sequenced organisms. The KEGG database [19] is a public resource for these predicted pathways. In an early study, it was found that the large-scale structure of the core metabolic network from 43 organisms is identical, being dominated by the same highly connected substrates [18]. For the same set of metabolic networks, it has been stated later that they are organized into many small, highly connected topologic modules that combine in a hierarchical manner into larger, less cohesive units [34]. Several other studies have compared the structure of the metabolic networks of several organisms in order to derive information about their phylogenetic relationship [14,25,32]. While these first studies could not replicate the detail of phylogenetic studies based on sequence information, it was at least possible to deduce from the network whether an organism belongs to the domains of Archaea, Bacteria, or Eukaryotes (see also Fig. 1.4). A more detailed discussion of metabolic networks can be found in Chapter 10.

1.4.2   Networks on a Macroscopic Scale

As stated before, networks are also present in the areas of biology dealing with larger space- or timescales. The interactions of different organisms can be depicted as ecological networks. Food webs have been under investigation since a long time. Qualitative food webs, which contain information only about predator–prey relationship, but no quantities, can be modeled as directed graphs (see Chapter 2). In this context, qualitative food webs are often called *static models*. However, they have not been the subject of many studies [11]. This is probably due to the fact that the available food webs are relatively small compared with biochemical networks, and thus not much new information can be gained out of the structure alone. Nevertheless, through comparison of 50 food webs of lakes, it was found that a relation exists between the number of species in a food web and the links per species [15]. Instead of investigating the structure of food webs alone, they are often modeled quantitatively with rate laws for every step (*dynamic models* [11]). Ecological networks other than food webs can be, for example, plant–pollinator interaction networks, which were found to exhibit an increased number of interactions per species upon increased diversity [28], analogous to the food webs mentioned above. More details on ecological networks are given in Chapter 12.

   Phylogenetic networks describe the evolutionary relationships between organisms. Traditionally, they were presented as bifurcate or binary trees (see Chapter 2) [29]. The branchpoints of the tree represent points of separation of two species during evolution. However, recent studies suggest that population genealogies are often multifurcated (trees, see Chapter 2), or even containing reticulate relationships due to recombination events, which turns them into phylogenetic networks [33]. Recently, a network for the phylogenetic relationships between all groups of prokaryotes has been presented and termed the " net of life" [22]. A more detailed discussion of phylogenetic trees and networks can be found in Chapter 11. As mentioned in the previous section, this topic is linked to several biochemical networks through many studies that have been made to infer phylogeny especially from metabolic networks. Recently, it has been shown that bacterial metabolic networks evolve adaptively by horizontal gene transfer [30].

1.4.3   Other Biological Networks

Correlation networks have only been investigated for a relatively short time, and they represent an exception among biological networks. Their special feature is that these networks are not a direct result of experimental data, but they are determined by collecting large amounts of high-throughput data and calculating the correlations between all elements. So far this has been done for transcripts and metabolites. Barkai and coworkers compared large-scale gene expression data sets of six evolutionarily distant organisms [6]. They found that for all organisms the connectivity of the correlation network follows a power-law, highly connected genes tend to be essential and conserved, and the expression program is highly modular. Furthermore, transcript correlation networks have been used to identify hormone-related genes in plants [24]. Metabolite correlation networks have been constructed from pair-wise analysis of linear correlations between metabolites from profiling data [40]. It was found that the connectivity distribution in these networks also follows the typical power-law for scale-free networks. More examples of correlation networks and their analysis are given in Chapter 13.

1.5   SUMMARY

Biology describes the processes of our environment from the molecular level to the level of the ecosystem. At all levels of detail, many of the respective processes can be modeled by networks. At the microscopic levels, these are gene regulation networks, signal transduction networks, protein interaction networks, and metabolic networks. At the macroscopic level, these are ecological and phylogenetic networks. All these networks have some special characteristics and are quite distinct from each other, but they also share common properties. Although the analysis of large-scale biological networks with modern tools has made significant progress in the last decade, this branch of science is still in its infancy.

1.6   EXERCISES

1. Describe the information flow within a cell, from DNA to metabolism. Name the processes.
2. What are the four levels of protein structure?
3. Describe the organization of a cell.
4. In a regular cell of most organisms, how many copies of each gene are present? Why?
5. Describe the term "systems biology" in your own words.
6. What are -omes and -omics?
7. Why is the measurement of a complete transcriptome not yielding a network?
8. Name at least four microscopic and two macroscopic networks in biology.
9. Why are correlation networks not intrinsic biological networks?

REFERENCES

1. -omes and -omics glossary and taxonomy. http://www.genomicglossaries.com/content/omes.asp.
2. Gold—genomes online database. http://www.genomesonline.org/.
3. R. Albert, H. Jeong, and A.-L. Barabási. Error and attack tolerance of complex networks. *Nature*, 406:378–381, 2000.
4. A.-L. Barabási. Emergence of scaling in complex networks. In S. Bornholdt and H. G. Schuster, editors, *Handbook of Graphs and Networks*, pp. 69–84. Wiley-VCH, Weinheim (Germany) 2003.
5. A.-L. Barabási and R. Albert. Emergence of scaling in random networks. *Science*, 286:509–512, 1999.

6. S. Bergmann, J. Ihmels, and N. Barkai. Similarities and differences in genomewide expression data of six organisms. *PLoS Biology*, 2:85–93, 2004.

7. U. S. Bhalla and R. Iyengar. Emergent properties of networks of biological pathways. *Science*, 283:381–387, 1999.

8. F. R. Blattner, G. Plunket, III, C. A. Bloch, N. T. Perna, V. Burland, M. Riley, J. Collado-Vides, J. D. Glasner, C. K. Rode, G. F. Mayhew, J. Gregor, N. W. Davis, H. A. Kirkpatrick, M. A. Goeden, D. J. Rose, B. Mau, and Y. Shao. The complete genome sequence of *Escherichia coli* k-12. *Science*, 277:1453–1462, 1997.

9. C. Darwin. *On the Origin of Species by Means of Natural Selection*. John Murray, London, 1859.

10. J. L. DeRisi, V. R. Iyer, and P. O. Brown. Exploring the metabolic and genetic control of gene expression on a genomic scale. *Science*, 278:680–686, 1997.

11. B. Drossel and A. J. McKane. Modelling food webs. In S. Bornholdt and H. G. Schuster, editors, *Handbook of Graphs and Networks*, pp. 218–247. Wiley-VCH, Weinheim (Germany) 2003.

12. G. Michal. *Biochemical Pathways (wall charts)*. Boehringer, Mannheim, Basle Switzerland, 1993.

13. O. Fiehn, J. Kopka, P. Dormann, T. Altmann, R. N. Trethewey, and L. Willmitzer. Metabolite profiling for plant functional genomics. *Nature Biotechnology*, 18:1157–1161, 2000.

14. C. V. Forst and K. Schulten. Phylogenetic analysis of metabolic pathways. *Journal of Molecular Evolution*, 52:471–489, 2001.

15. K. Havens. Scale and structure in natural food webs. *Science*, 257:1107–1109, 1992.

16. J. Jaeger, S. Surkova, M. Blagov, H. Janssens, D. Kosman, K. N. Kozlov, Manu, E. Myasnikova, C. E. Vanario-Alonso, M. Samsonova, D. H. Sharp, and J. Reinitz. Dynamic control of positional information in the early *Drosophila* embryo. *Nature*, 430:368–371, 2004.

17. H. Jeong, S. P. Mason, A.-L. Barabási, and Z. N. Oltvai. Lethality and centrality in protein networks. *Nature*, 411:41–42, 2001.

18. H. Jeong, B. Tombor, R. Albert, Z. N. Oltvai, and A.-L. Barabási. The large-scale organization of metabolic networks. *Nature*, 107:651–654, 2000.

19. M. Kanehisa and S. Goto. KEGG: Kyoto encyclopedia of genes and genomes. *Nucleic Acids Research*, 28:27–30, 2000.

20. F. Katagiri. Attacking complex problems with the power of systems biology. *Plant Physiology*, 132:417–419, 2003.

21. H. Kitano. *Foundations of Systems Biology*. The MIT Press, Cambridge, MA, 2001.

22. V. Kunin, L. Goldovsky, N. Darzentas, and C. A. Ouzounis. The net of life: Reconstructing the microbial phylogenetic network. *Genome Research*, 15:954–959, 2005.

23. T. I. Lee, N. J. Rinaldi, F. Robert, D. T. Odom, Z. Bar-Joseph, G. K. Gerber, N. M. Hannett, C. T. Harbison, C. M. Thompson, I. Simon, J. Zeitlinger, E. G. Jennings, H. L. Murray, D. B. Gordon, B. Ren, J. J. Wyrick, J.-B. Tagne, T. L. Volkert, W. Fraenkel, D. K. Gifford, and R. A. Young. Transcriptional regulatory networks in *Saccharomyces cerivisiae*. *Science*, 298:799–804, 2002.

24. J. Lisso, D. Steinhauser, T. Altmann, J. Kopka, and C. Müssig. Identification of brassinoid-related genes by means of transcript co-response analyses. *Nucleic Acids Research*, 33:2685–2696, 2005.

25. H.-W. Ma and A.-P. Zeng. Phylogenetic comparison of metabolic capacities of organisms at genome level. *Molecular Phylogenetics and Evolution*, 31:204–213, 2004.

26. M. Margulies, M. Egholm, W. E. Altman, S. Attiya, J. S. Bader, L. A. Bemben, J. Berka, M. S. Braverman, Y. J. Chen, Z. Chen, S. B. Dewell, L. Du, J. M. Fierro, X. V. Gomes, B. C. Godwin, W. He, S. Helgesen, C. H. Ho, G. P. Irzyk, S. C. Jando, M. L. Alenquer, T. P. Jarvie, K. B. Jirage, J. B. Kim, J. R. Knight, J. R. Lanza, J. H. Leamon, S. M. Lefkowitz, M. Lei, J. Li, K. L. Lohman, H. Lu, V. B. Makhijani, K. E. McDade, M. P. McKenna, E. W. Myers, E. Nickerson, J. R. Nobile, R. Plant, B. P. Puc, M. T. Ronan, G. T. Roth, G. J. Sarkis, J. F. Simons, J. W. Simpson, M. Srinivasan, K. R. Tartaro, A. Tomasz, K. A. Vogt, G. A. Volkmer, S. H. Wang, Y. Wang, M. P., Weiner, P. Yu, R. F. Begley, and J. M. Rothberg. Genome sequencing in microfabricated high-density picolitre reactors. *Nature*, 437:376–380, 2005.

27. S. Milgram. The small-world problem. *Psychology Today*, 1:61–67, 1967.

28. J. M. Olesen and P. Jordano. Geographic patterns in plant-pollinator mutualistic networks. *Ecology*, 83:2416–2424, 2002.

29. N. R. Pace. A molecular view of microbial diversity and the biosphere. *Science*, 276:734–740, 1997.

30. C. Pál, B. Papp, and M. J. Lercher. Adaptive evolution of bacterial metabolic networks by horizontal gene transfer. *Nature Genetics*, 37:1372–1375, 2005.

31. A. Pandey and M. Mann. Proteomics to study genes and genomes. *Nature*, 405:837–846, 2000.

32. J. Podani, Z. N. Oltvai, H. Jeong, B. Tombor, A.-L. Barabási, and E. Szathmáry. Comparable systems-level organization of archaea and eukaryotes. *Nature Genetics*, 29:54–56, 2001.

33. D. Posada and K. A. Crandall. Intraspecific gene genealogies: trees grafting into networks. *Trends in Ecology and Evolution*, 16:37–45, 2001.

34. E. Ravasz, A. L. Somera, D. A. Mongru, Z. N. Oltvai, and A.-L. Barabási. Hierarchical organization of modularity in metabolic networks. *Science*, 297:1551–1555, 2002.

35. L. Salwinski, C. S. Miller, A. J. Smith, F. K. Pettit, J. U. Bowie, and D. Eisenberg. The database of interacting proteins: 2004 update. *Nucleic Acids Research*, 32:D449–D451, 2004.

36. S. S. Shen-Orr, R. Milo, S. Mangan, and U. Alon. Network motifs in the transcriptional regulation network of *Escherichia coli*. *Nature Genetics*, 31:64–68, 2002.

37. L. von Bertalanffy. *General Systems Theory: Foundations, Development, Applications*. George Braziller, New York (NY, USA) 1969.

38. C. von Mering, R. Krause, B. Snel, M. Cornell, S. G. Oliver, S. Fields, and P. Bork. Comparative assessment of large-scale data sets of protein-protein interactions. *Nature*, 417:309–403, 2002.

39. D. J. Watts and S. H. Strogatz. Collective dynamics of 'small-world' networks. *Nature*, 393:440–442, 1998.

40. W. Weckwerth, M. E. Loureiro, K. Wenzel, and O. Fiehn. Differential metabolic networks unravel the effect of silent plant phenotypes. *Proceedings of the National Academy of Sciences USA*, 101:7809–7814, 2004.

41. S. Wuchty, Z. N. Oltvai, and A.-L. Barabási. Evolutionary conservation of motif constituents in the yeast protein interaction network. *Nature Genetics*, 35:176–179, 2003.

2

GRAPH THEORY

Falk Schreiber

2.1 INTRODUCTION

The term *network* is an informal description for a set of elements with connections or interactions between them. A typical example from biology is a protein interaction network. It consists of a set of proteins (elements) and a set of interactions between them (connections). The previous chapter introduced many different biological networks. Given such networks, we could be interested in questions such as Which protein has the highest number of interactions with other proteins in a protein interaction network? Are there clusters of proteins where every protein interacts with every other? Or, in a metabolic network, we might like to study the shortest path of reactions that transform one metabolite into another. Such questions can be answered if we analyze the structure of the network, that is, the way the elements are connected.

To deal with networks in a formal way they are modeled as *graphs*. A graph is a mathematical object consisting of vertices and edges representing elements and connections, respectively. This usage of the term "graph" should not be confused with another meaning often used in biology: the graphical representation of a function in the form of a curve or surface. The theory of graphs reaches back to Leonard Euler and his "Königsberg bridge problem" in 1736. The problem is as follows: In Königsberg (today Kaliningrad), the river Pregel runs through the town as shown in Fig. 2.1. Seven bridges were built over the river. The question is whether it is possible to walk around the town in a way that would involve crossing each bridge exactly once. By

Analysis of Biological Networks, Edited by Björn H. Junker and Falk Schreiber
Copyright © 2008 John Wiley & Sons, Inc.

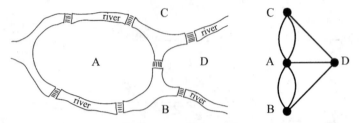

FIGURE 2.1 A map of Königsberg with the river Pregel and the representation of the "Königsberg bridge problem" problem as a graph.

analyzing the structure of the graph representing the problem, as shown in Fig. 2.1, Euler proved that this is not possible.

This chapter gives an introduction to most of the mathematical and computer science terminologies used later in the book. It is aimed at readers not familiar with these topics, and formal concepts are restricted to a minimum. Readers with prior knowledge may wish to skip this chapter. More detailed presentations can be found in many good textbooks about graph theory, network analysis, and algorithms, for example [2–4,6]. Here, we discuss basis terminology and notation for graphs in Section 2.2, special graphs used in modeling biological networks in Section 2.3, typical representations of graphs in Section 2.4, and some fundamental algorithms for the analysis of graphs in Section 2.5.

2.2 BASIC NOTATION

2.2.1 Sets

A *set* $A = \{a_1, a_2, \ldots, a_n\}$ is a collection of distinct objects $a_1, a_2, \ldots, a_n$ considered as a whole, and can be defined by listing its elements between braces. An example is a set $A = \{6, 3, 4, 2, 1\}$ of numbers. The objects a_i of a set A are called its *members*. In case an object is a member of a set this is symbolized by $\in$. For example, in the set defined above 2 is a member of A, written $2 \in A$. Two sets A_1 and A_2 are said to be equal (written $A_1 = A_2$) if every member of A_1 is a member of A_2, and every member of A_2 is a member of A_1. If every member of set A_1 is a member of set A_2 (but not necessarily every member of A_2 is a member of A_1) then the set A_1 is a subset of set A_2, written $A_1 \subseteq A_2$. Two sets A_1 and A_2 can be combined into a new set. The *union* of the sets A_1 and A_2 is the set of all objects that are members of either A_1 or A_2 and is denoted by $A_1 \cup A_2$. The *intersection* of the sets A_1 and A_2 is the set of all objects that are members of both A_1 and A_2 (denoted by $A_1 \cap A_2$). An empty set is denoted by $\emptyset$. Special sets used in this book are the set of natural numbers including zero ($\mathbb{N}_0$), the set of integers ($\mathbb{Z}$), and the set of real numbers ($\mathbb{R}$).

2.2.2 Graphs

A *graph* $G = (V, E)$ consists of a set of *vertices* (also called nodes or points) V and a set of *edges* (arcs, links) E, where each edge is assigned to two (not necessarily

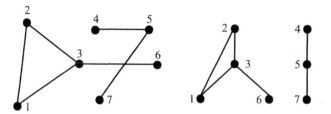

FIGURE 2.2   Two graphical representations of the graph $G = (V, E)$ with vertex set $V = \{1, 2, 3, 4, 5, 6, 7\}$ and edge set $E = \{\{1, 2\}, \{2, 3\}, \{1, 3\}, \{3, 6\}, \{4, 5\}, \{5, 7\}\}$.

disjunct) vertices. An edge e connecting the vertices u, v is denoted by $\{u, v\}$, we say u and v are *incident* with e and *adjacent* (or *neighbors*) to each other. The vertices incident to an edge are called its *end-vertices*. The *degree* of a vertex v is the number of edges that have v as end-vertex. An edge where the two end-vertices are the same vertex is called a *loop*. A *loop-free* graph does not contain loops.

This definition describes undirected graphs, that is, graphs where connections between vertices are without a direction. Undirected graphs are used, for example, to model protein interaction networks (see Chapter 9), phylogenetic networks (see Chapter 11), and correlation networks (see Chapter 13). In the following, we describe general graph concepts based on undirected graphs. Section 2.3 deals with other types of graphs, especially directed graphs that are used, for example, to model gene regulation networks (see Chapter 8) and ecological networks (see Chapter 12).

The usual way to visualize a graph is by drawing a point for each vertex and a line for each edge that connects the corresponding points of its end-vertices, see Fig. 2.2. It is not important how a graph is drawn, as long as it is clearly visible which pairs of vertices are connected by edges and which not. The positions of the vertices and the drawing of the lines are called the *layout* of the graph.

A *subgraph* $G' = (V', E')$ of the graph $G = (V, E)$ is a graph where V' is a subset of V and E' is a subset of E. This implies that E' contains only edges with end-vertices in V'. If graph G' is a subgraph of graph G and the edge set E' contains all edges of E that connect vertices of V', the subgraph is called an *induced subgraph* of G. See Fig. 2.3 for subgraph and induced subgraph.

One graph can have many different graphical representations, see Fig. 2.2. But two graphs can also be the same, see Fig. 2.4. Both graphs $G_1 = (V_1, E_1)$ and $G_2 = (V_2, E_2)$ have different vertex and edge sets. Graph G_1 consists of

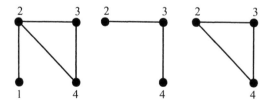

FIGURE 2.3   A graph G, a subgraph of G, and an induced subgraph of G (from left to right).

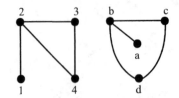

FIGURE 2.4   Two isomorphic graphs.

$V_1 = \{1, 2, 3, 4\}$ and $E_1 = \{\{1, 2\}, \{2, 3\}, \{3, 4\}, \{2, 4\}\}$;   graph G_2 of $V_2 = \{a, b, c, d\}$ and $E_2 = \{\{a, b\}, \{b, c\}, \{b, d\}, \{c, d\}\}$. However, even though both graphs appear to be different, they contain the same number of vertices connected in the same way and are therefore considered as the same or *isomorphic* graphs. Formally, two graphs G_1 and G_2 are isomorphic, if there exists a bijective mapping between the vertices in V_1 and V_2 with the property that any two vertices $u, v \in V_1$ are adjacent if and only if the two corresponding vertices in the other graph are adjacent. Such a bijection is called an *isomorphism*.

A sequence $(v_0, e_1, v_1, e_2, v_2, \ldots, v_{k-1}, e_k, v_k)$ of vertices and edges such that every edge e_i has the end-vertices v_{i-1} and v_i is called a *walk*. Usually the vertices are omitted and the walk is denoted by a sequence $(e_1, e_2, \ldots, e_k)$. We say that the walk *connects* v_0 with v_k and call v_0 and v_k the start- and end-vertex of the walk, respectively. If all edges of a walk are distinct the walk is called a *path*, and if additionally all vertices are distinct the walk is called a *simple path*. The *length* of a walk or path is given by its number of edges. A path with the same vertex as start- and end-vertex is a *cycle*. A graph without cycles is called an *acyclic graph*. For example, in the graph in Fig. 2.2, the sequence $(\{1, 2\}, \{2, 3\}, \{3, 6\}, \{6, 3\}, \{3, 1\})$ is a walk and the sequence $(\{1, 2\}, \{2, 3\}, \{3, 1\})$ is a path which is furthermore a cycle.

Two vertices of a graph are called *connected* if there exists a walk between them. If any pair of different vertices of the graph is connected, the graph is *connected*. A *connected component* of a graph G is a maximal connected subgraph of G. For example, the graph in Fig. 2.5 consists of four connected components.

A *shortest path* between two vertices is a path with minimal length. The *distance* between two vertices is the length of a shortest path between them or ∞ if no such path exists. For example, in Fig. 2.2, the path $(\{1, 3\}, \{3, 6\})$ is a shortest path between vertex 1 and vertex 6 and the distance between these two vertices is 2. Note that there may be several different shortest paths between two vertices in a graph.

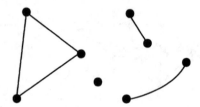

FIGURE 2.5   An unconnected graph consisting of four connected components.

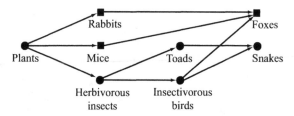

FIGURE 2.6   Some attributes connected to the vertices of a food web: textual vertex labels, different geometrical objects (mammals: squares, others: dots), and coordinates for vertices (the positions of the vertices) representing the layout of the graph.

2.2.3   Graph Attributes

Often attributes such as text, numerical values, types, colors, and coordinates are associated with the vertices and edges of a graph. Typical examples are the stoichiometry of reactions in metabolic networks represented as numerical values along the edges, protein classes for proteins represented as vertex types or textual vertex labels in protein interaction networks, and the coordinates of the vertices represented as numerical value pairs. Figure 2.6 shows a typical example.

Attributes can be represented as functions from the vertices or edges to the attribute type. For example, the mentioned stoichiometry of metabolic reactions can be represented as edge weights, that is, numerical values connected to edges. The function $\omega : E \rightarrow \mathbb{R}$ assigns each edge $e \in E$ a weight $\omega(e)$.

2.3   SPECIAL GRAPHS

There are many different biological networks with different properties. Often the graph model has to be tailored to the specific network under consideration. Typical graph models for the different networks are considered in the following section.

2.3.1   Undirected, Directed, Mixed, and Multigraphs

Graphs can be undirected, directed, or mixed, see Fig. 2.7. In an *undirected graph*, an edge between the vertices u and v is represented by the unordered vertex pair $\{u, v\}$. The graphs defined in the previous section are undirected. Typical examples from

FIGURE 2.7   An undirected, a directed, and a mixed graph (from left to right).

biology are protein interaction networks, phylogenetic networks, and correlation networks.

In a *directed graph,* an edge between the vertices u and v is represented by the ordered vertex pair (u, v). In visualizations of graphs, the direction of an edge is usually represented by an arrowhead. The two edges (u, v) and (v, u) between the vertices u, v can be represented either by two lines as shown in Fig. 2.7 or by one line with arrowheads at both ends. Typical examples of biological networks modeled by directed graphs are metabolic networks, gene regulation networks, and food webs.

In a *mixed graph* undirected and directed edges occur. This type of graph is also relevant in biology, an example is special protein interaction networks where some interactions are undirected (e.g., obtained by two-hybrid experiments) and others are directed representing activation, phosphorylation, and other directed interactions.

An undirected edge has *end-vertices,* a directed edge (u, v) has a *source vertex u* (also called origin or head) and a *target vertex v* (destination, tail). In a directed graph, a vertex has an *out-degree* that is defined as the number of edges going out of it and an *in-degree* defined as the number of edges coming into it. The degree of the vertex is the sum of its in- and out-degrees. In directed graphs the definitions for walks, paths, and cycles are similar to undirected graphs, but take the edge direction into account. For example, a walk in a directed graph is a sequence $(v_0, e_1, v_1, e_2, v_2, \ldots, v_{k-1}, e_k, v_k)$ of vertices and edges such that every edge e_i has the source vertex v_{i-1} and the target vertex v_i. We say that in a directed graph the walk *strongly connects* v_0 with v_k if the edge direction is taken into account, otherwise (i.e., if each edge is considered undirected) the walk simply *connects* v_0 with v_k. Two vertices of a graph are called *strongly connected* if there exists such a walk between them. If any pair of different vertices of the graph is strongly connected, the graph is *strongly connected.* A *strongly connected component* of a graph G is a maximally strongly connected subgraph of G.

Multigraphs are graphs containing multiple edges, that is, two or more edges that are incident to the same two vertices and in case of directed graphs have the same direction. Such edges are also called parallel edges or a multiedge, see Fig. 2.1 for an example of a multigraph. Multiple edges are, for example, useful for the modeling of metabolic pathways where the same substances can be transformed by different reactions. Undirected, loop-free graphs without multiple edges are called *simple graphs.*

2.3.2   Hypergraphs and Bipartite Graphs

There are biological networks where more than two elements are connected by an interaction. An example are metabolic networks where often several substances react with each other to build other substances, see Fig. 2.8. To model such networks, hypergraphs are used. A hypergraph $G = (V, E)$ consists of a set of *vertices V* and a set of *hyperedges E,* each hyperedge is a nonempty subsets of V. Hypergraphs can be directed or undirected.

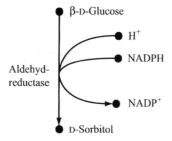

FIGURE 2.8   A metabolic network where an edge connects more than two elements. This hyperedge is labeled with the name of the enzyme catalyzing the reaction.

Hypergraphs are not commonly used in graph theory, and many algorithms developed for graphs cannot be directly applied to hypergraphs. Therefore, such graphs are seldom used to model biological networks. Instead these networks are modeled by bipartite graphs, a structure generally used to represent hypergraphs.

A graph $G = (V, E)$ is called *bipartite* if there is a partition of its vertex set $V = S \cup T$ such that each edge in E has exactly one end-vertex in S and one end-vertex in T (see Fig. 2.9).

To model a hypergraph $G = (V, E)$ by a bipartite graph $G' = (V', E')$ with $V' = S' \cup T'$, the bipartite graph is build in the following way. Each vertex $v \in V$ is represented by a vertex in S' and each hyperedge $e \in E$ by a vertex in T'. For each vertex $v \in V$ and each hyperedge $e \in E$ incident with v, an edge is inserted into E', which connects a vertex $s \in S'$ representing the vertex v of the hypergraph and a vertex $t \in T'$ representing the hyperedge e. Figure 2.10 shows a hypergraph and its representation as a bipartite graph, and Fig. 2.11 shows a typical modeling of metabolic networks by bipartite graphs.

2.3.3 Trees

The last type of special graphs we will consider in this introduction are trees. Trees play an important role in biology where they represent, for example, the evolutionary relationships between species as a phylogenetic tree (see Chapter 11).

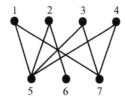

FIGURE 2.9   A bipartite graph $G = (S \cup T, E)$ with vertex set $S = \{1, 2, 3, 4\}$ and vertex set $T = \{5, 6, 7\}$.

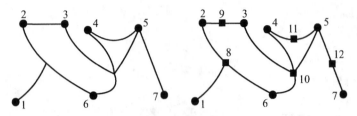

FIGURE 2.10   The hypergraph $G = (V, E)$ with vertex set $V = \{1, 2, 3, 4, 5, 6, 7\}$ and hyper-edge set $E = \{\{1, 2, 6\}, \{2, 3\}, \{3, 4, 5, 6\}, \{4, 5\}, \{5, 7\}\}$ and its corresponding bipartite graph. The two vertex sets S and T are represented by dots and squares, respectively.

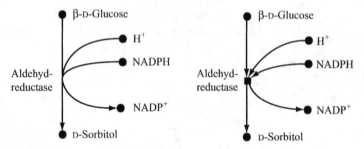

FIGURE 2.11   A metabolic network and its modeling as bipartite graph.

A *tree* is an undirected, connected, acyclic graph. The vertices of a tree with degree 1 are its *leaves*, all other vertices are *inner vertices*. A *rooted tree* consists of a tree $G = (V, E)$ and a distinguished vertex $r \in V$ called the *root*. The *depth* of a vertex is the length of the path between the root and this vertex, the *height* of a tree is the maximum depth of a vertex. A *binary tree* is a tree where each vertex has at most degree 3. See Fig. 2.12 for a tree and a binary tree.

A rooted tree is often regarded as a directed graph such that all edges are directed away from the root. For a directed tree $G = (V, E)$ and an edge $(u, v) \in E$, the vertex u is the *parent* of v and v is the *child* of u.

For a connected, undirected graph G, a special tree can be computed, the *spanning tree T of G*. The spanning tree T is composed of all the vertices of G and a minimal set of edges (some or perhaps all of the edges of G) that connect all vertices. This

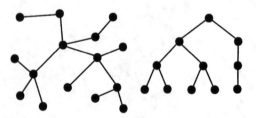

FIGURE 2.12   A tree and a binary tree. Again, there are many different graphical represen-tations of a tree.

vertex	1	2	3	4	5	6	7
1	0	1	1	0	0	0	0
2	1	0	1	0	0	0	0
3	1	1	0	0	0	1	0
4	0	0	0	0	1	0	0
5	0	0	0	1	0	0	1
6	0	0	1	0	0	0	0
7	0	0	0	0	1	0	0

FIGURE 2.13   An adjacency matrix representation for the graph shown in Fig. 2.2.

tree contains a subset of the edges of G that form a tree spanning every vertex of G, hence the name spanning tree.

2.4 GRAPH REPRESENTATION

To use graphs in a computer program, we have to represent them in the computer. There are two common representations: adjacency matrix and adjacency list. The choice of one or the other depends on the operations needed to deal with the graph and whether the graph is dense or sparse. We will discuss this aspect in Section 2.5.

2.4.1 Adjacency Matrix

A graph G with n vertices can be represented by a $(n \times n)$ *adjacency matrix A*, see Fig. 2.13 for an example. The rows and columns correspond to the vertices and a matrix-element $A_{ij} = 1$ if and only if there is an edge between the vertices v_i and v_j and $A_{ij} = 0$ otherwise. The adjacency matrix of an undirected graph is symmetric, that is, $A_{ij} = A_{ji}$.

The simplest way to implement an adjacency matrix is as an array $[1 \ldots n, 1 \ldots n]$ of numbers or boolean values. Adjacency matrices are often used to represent biological networks as their structure is very simple and matrix operations can be directly applied. However, adjacency matrices need a lot of memory, n^2 places for a network with n elements. Furthermore, several algorithms have a longer running time if they are based on this network representation. In particular for graphs with a low number of edges in relation to the number of vertices, another representation, the adjacency list, is usually more efficient.

2.4.2 Adjacency List

A graph G with n vertices can be represented by n lists, see Fig. 2.14 for an example. For each vertex $v \in V$, a list L_v contains all edges incident to this vertex (and therefore all vertices adjacent to it).

A common way to implement an adjacency list is an array $[1 \ldots n]$ of lists.

$$L_1: (\{1, 2\}, \{1, 3\})$$
$$L_2: (\{2, 1\}, \{2, 3\})$$
$$L_3: (\{3, 1\}, \{3, 2\}, \{3, 6\})$$
$$L_4: (\{4, 5\})$$
$$L_5: (\{5, 4\}, \{5, 7\})$$
$$L_6: (\{6, 3\})$$
$$L_7: (\{7, 5\})$$

FIGURE 2.14   An adjacency list representation for the graph shown in Fig. 2.2.

2.5   GRAPH ALGORITHMS

Many problems concerning biological networks can be answered using standard graph algorithms. Let us consider some of the questions raised in the introduction of this chapter. The protein with the highest number of interactions can be found by visiting all vertices of the graph and counting for each the number of its neighbors. And the shortest path between two elements in a metabolic network can be computed with the Dijkstra algorithm (e. g. see [6]). This section gives an introduction to graph algorithms and discusses how to make a good choice between different algorithms computing the same result.

2.5.1   Running Times of Algorithms

Running time and memory requirement are key properties of an algorithm. They are usually specified in the $\mathcal{O}$ notation. This is a theoretical measure of the algorithm's running time or space needed for a given size n of input data. For networks this problem size n is often the number of elements.

We will focus on the running time of algorithms. The $\mathcal{O}$ notation is used to compare the running times of algorithms for a large enough problem size and to decide whether an algorithm and a related data structure are adequate or will always be too slow for a large problem size. Formally, for two functions f, g we say that f is in $\mathcal{O}(g)$ if there are positive constants $n_0 \in \mathbb{N}_0$ and $c \in \mathbb{R}$ such that $f(n) \leq cg(n)$ for all $n \geq n_0$.

Let us consider an example. We compare two typical sorting algorithms, Quick-Sort and BubbleSort (e.g., see [7]). Both use different strategies to sort a set of n unsorted items. We will not discuss these strategies; however, QuickSort's running time is $\mathcal{O}(n \log n)$ on average, whereas BubbleSort needs $\mathcal{O}(n^2)$. For small sizes of input data, the running times of both algorithms do not differ much. But if we want to sort one million elements, QuickSort may still give us the result in reasonable time, whereas BubbleSort may take an excessively long time even on a supercomputer.

We want to sort all vertices of a graph depending on the number of neighbors a vertex has. Let n be the number of vertices and m be the number of edges of the graph. The running times above are only the times for the sorting of a set of unsorted elements. Now let us consider how the graph representation may influence the time to sort all vertices depending on the number of neighbors a vertex has. For this the number of neighbors of each vertex has to be computed first. The time needed for this counting depends on the chosen representation. First, let us consider the adjacency matrix: For each vertex we have to test all elements of the row representing it and add 1 to the number of neighbors if the matrix element is 1. To count the number of neighbors for one vertex, we need $\mathcal{O}(n)$, and to compute the number of neighbors for all vertices, we need $\mathcal{O}(n^2)$. Now consider the adjacency list: Given a vertex we have to test each element in the corresponding list and add 1 to the number of neighbors for each. The length of the list may be different for each vertex; however, there are a total of m edges and therefore a total of m list items. The running time to count the numbers of neighbors for all vertices using the adjacency list is $\mathcal{O}(n + m)$.

The adjacency list representation is therefore much better for counting the neighbors of vertices in graphs containing a low number of edges but high number of vertices. Furthermore, combining a poorer graph representation for a specific problem such as the adjacency matrix representation for counting neighbors with the QuickSort algorithm means that the overall running time is no longer in $\mathcal{O}(n \log n)$ but in $\mathcal{O}(n^2)$.

2.5.2 Traversal

Graph traversal algorithms are used to visit all vertices and subsequently perform an action with each vertex. The vertices may be visited in an arbitrary order, or a specific order may be requested. For example, we could be interested in visiting genes in a gene regulation network in an order that follows the regulatory steps. There are many different possibilities to traverse the vertices of a graph, the most important ones are *depth first search* (DFS) and *breadth first search* (BFS).

The DFS algorithm shown below works as follows. In the beginning all vertices are marked as unvisited. The algorithm starts with a given vertex, visits this vertex, and then recursively visits all neighbors of this vertex, see Fig. 2.15.

depth_first_search_component_algorithm (vertex v)
        visit(v);
        mark v as visited;
        for each edge $\{v, w\}$
          if (w is unvisited)
                depth_first_search_component_algorithm(w)

This algorithm visits all vertices within one connected component. To visit all components in a unconnected graph, an enclosing loop is needed:

depth_first_search_algorithm (vertex u)
        for each vertex u
          if (u is unvisited)
                depth_first_search_component_algorithm(u)

If the graph is represented by an adjacency list the running time is $\mathcal{O}(n + m)$. For a representation by an adjacency matrix the running time is $\mathcal{O}(n^2)$. Modifications of

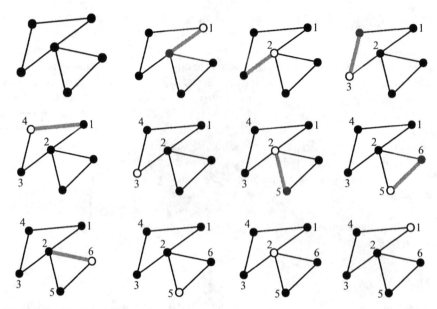

FIGURE 2.15   Depth first search of a graph. A vertex already visited is marked by its number (in the order of the visits), the currently visited vertex is represented by a circle, and the currently chosen edge to the next neighbor by a thick gray line.

the depth first search algorithm can be used to solve a number of problems such as Is the network connected or are there separate parts? Or, is the network a tree or does it contain cycles?

Another classic algorithm for the traversal of graphs is BFS. Whereas DFS follows a path into the graph as long as possible, BFS visits all neighbors of a vertex before it visits other vertices. The algorithm for BFS is as follows:

> **breadth_first_search_component_algorithm** (vertex v)
>   visit(v);
>   mark v as visited;
>   queue $Q = [v]$;
>   while Q is nonempty {
>   remove vertex w from the front of queue Q;
>   visit(w);
>   for each neighbor x of w
>   if (x is unvisited) {
>   mark x as visited;
>   add x to the end of queue Q
>   }
>   }

Again, this algorithm visits all vertices within one connected component. To visit all components in a unconnected graph, an enclosing loop similar to DFS is needed.

2.6 SUMMARY

Biological networks are commonly represented by graphs, and different graph models are used for specific networks. This chapter gives a brief yet concise introduction to most of the graph-related terminology used in the book and presents some simple algorithms for graphs.

The first part of this book discusses graph-based analysis methods in more detail. The focus is on the following topics.

Global network properties: The structure of biological networks is significantly different from random networks, and new models have been introduced [1,10]. Chapter 3 deals with the analysis of global network properties and relevant models for biological networks.

Network centralities: In biological networks some vertices are often more important or central than others. For example, highly connected vertices in protein interaction networks can be functionally important and the removal of such vertices is related to lethality [5]. *Centrality indices* are used to rank vertices, and Chapter 4 presents some of the more important centrality indices and their application to biological networks.

Network motifs: A way to understand complex biological networks is to break them down into units of commonly used network architecture. Such patterns of local interconnection are called *network motifs* [8]. They have been found in many different networks, but are particularly important for the understanding of signal transduction and gene regulation networks. Chapter 5 discusses network motif analysis and presents insights gained with this method.

Network clustering: Biological networks are hierarchically structured from network motifs and pathways at the lowest level, to functional modules, to the large-scale organization of the networks [9]. Chapter 6 studies the clustering of network elements into modules and their application to biological networks.

2.7 EXERCISES

1. Different biological networks are modeled by different graphs. Which types of graphs are typically used to model the following networks: gene regulation networks, protein interaction networks, metabolic networks?

2. Consider the undirected graph $G = (V, E)$ shown in Fig. 2.2 (right). For each vertex $v \in V$ do the following:
   (a) Compute the degree of v.
   (b) List all neighbors of v.
   (c) Find paths to all other vertices that are in the same connected component as the vertex v.

3. For the graph shown in Fig. 2.2 (right), find a different graphical representation of this graph and show how the graph is represented using an adjacency matrix and an adjacency list representation.

4. Take a graph with nine vertices, four of them of degree 2 and four of degree 1. Is this graph connected?

5. An undirected loop-free graph with n vertices has at most $n(n-1)/2$ edges. Is this statement correct? Can you prove it?

6. Draw an undirected, connected graph G with 10 vertices and 20 edges. Construct two different spanning trees of G.

7. Metabolite networks can be constructed from metabolic networks modeled as bipartite graphs by removing all vertices representing reactions and connecting substrates (vertices with an outgoing edge to the reaction) with products (vertices with an incoming edge from the reaction) directly. Construct a metabolite network from the metabolic network given in Fig. 2.11 (right).

8. Apply the algorithms DFS and BFS to traverse the graph in Fig. 2.2 (right). Start with vertex 1, then apply the algorithms again starting with vertex 5.

9. A Eulerian path is a path $(v_0, e_1, v_1, e_2, v_2, \ldots, v_{k-1}, e_k, v_k)$ in an undirected graph that contains each edge of the graph exactly once. Write an algorithm to check whether an undirected graph $G = (V, E)$ has an Eulerian path.

REFERENCES

1. A. L. Barabási and R. Albert. Emergence of scaling in random networks. *Science*, 286(5439):509–512, 1999.

2. U. Brandes and T. Erlebach, editors. *Network Analysis*, volume 3418 of *Lecture Notes in Computer Science*, Springer-Verlag, Berlin, 2005.

3. T. H. Corman, C. E. Leiserson, R. L. Rivest, and C. Stein. *Introduction to Algorithms*, MIT Press, Cambridge, MA, 2001.

4. R. Diestel. *Graph Theory*, Springer-Verlag, Berlin, 2005.

5. H. Jeong, S. P. Mason, A. L. Barabási, and Z. N. Oltvai. Lethality and centrality in protein networks. *Nature*, 411(6833):41–42, 2001.

6. D. Jungnickel. *Graphs, Networks and Algorithms*, volume 5 of *Algorithms and Computation in Mathematics*. Springer-Verlag, Berlin, 2002.

7. D. E. Knuth. *The Art of Programming: Sorting and Searching (Volume 3)*, Addison-Wesley, Reading, MA, 1998.

8. R. Milo, S. Shen-Orr, S. Itzkovitz, N. Kashtan, D. Chklovskii, and U. Alon. Network motifs: Simple building blocks of complex networks. *Science*, 298(5594):824–827, 2002.

9. Z. N. Oltvai and A.-L. Barabási. Life's complexity pyramid. *Science*, 298(5594):763–764, 2002.

10. D. J. Watts and S. H. Strogatz. Collective dynamics of small-world networks. *Nature*, 393(6684):440–442, 1998.

PART II

NETWORK ANALYSIS

3

GLOBAL NETWORK PROPERTIES

RALF STEUER AND GORKA ZAMORA LÓPEZ

3.1 INTRODUCTION

Complex dynamical systems are often characterized by a large number of nonlinearly interacting elements, giving rise to emergent properties that transcend the principle of linear superposition. In particular, within the biological sciences, one of the primary challenges is to investigate how the collective behavior of cells, tissues, or organisms can be understood in terms of the properties of their molecular constituents.

To investigate this intricate connectivity of cellular systems, the analysis of complex networks has become an important part of molecular biology. A large number of biological phenomena and processes can be translated into the abstract concept of a complex network, making biological problems mathematically tractable. Prominent examples include the representation of transcriptional regulation as a network, where vertices represent genes or proteins and edges represent regulatory interactions, as well as cellular metabolism, where vertices represent metabolites and edges represent biochemical interconversions. However, beyond these rather straightforward examples, also more abstract processes can sometimes be translated into the language of complex networks. For example, different configurational states of a protein may be represented as vertices, with edges indicating transitions between them.

Analysis of Biological Networks, Edited by Björn H. Junker and Falk Schreiber
Copyright © 2008 John Wiley & Sons, Inc.

Once a biological process or phenomenon is represented by a network, the tools of complex network theory allow for a systematic characterization of its structural properties. The analysis of network topology then seeks to uncover the functional organization, the underlying design principles, and unknown organizing principles of cellular systems. Indeed, as realized rather recently, many empirically derived complex networks, ranging from technological and sociological to biological examples, share common topological features. The organizing principles of empirical networks often reflect crucial system properties, such as robustness, redundancy, or other functional interdependencies between network elements. A quantitative analysis of the large-scale characteristics of complex networks thus contributes to a better understanding of the organization of cellular functions and has already made significant impact on our current view of molecular biology.

While not aiming at a comprehensive review, this chapter seeks to summarize and describe several basic measures and characteristics of network topology. The chapter is organized as follows: The main emphasis is placed on an overview of basic measures and indices that characterize the topology of networks, given within Section 3.2. In Section 3.3, several basic prototype models of complex networks are discussed. The subsequent Section 3.4 is devoted to a brief outline of global features of complex networks, such as hierarchies, modularity, attack tolerance, and robustness. Finally, Section 3.5 provides notes on the statistical testing of network properties and describes several known pitfalls and possible misinterpretations in the statistical analysis of network properties. The working example throughout this chapter is a reconstructed version of the *S. cerevisiae* metabolic network [23], consisting of 810 metabolites and 843 reactions. The original bipartite graph was collapsed, such that two metabolites are connected if they participate in a common reaction. A graphical representation is shown in Fig. 3.1.

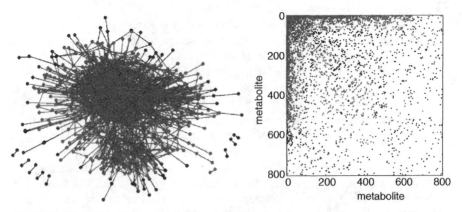

FIGURE 3.1 The substrate graph G_S of the *S. cerevisiae* metabolic network [23], consisting of $N_V = 810$ vertices (metabolites) and $N_E = 3419$ edges. Directional information is omitted. *Left:* A visualization of the substrate graph using the freely available software package Pajek [9]. *Right:* A visualization of the adjacency matrix, with vertices (metabolites) ranked according to their degree. Each dot indicates whether the corresponding vertices (metabolites) are connected by an edge. The figures are adapted from Ref. [61].

3.2 GLOBAL PROPERTIES OF COMPLEX NETWORKS

Following the nomenclature of Chapter 2, a network is formally represented by a *graph* $G = (V, E)$, consisting of a set V of N_V *vertices* and a set E of N_E *edges*. We distinguish between *undirected graphs*, whose vertices are connected by edges without any directional information, and *directed graphs* (*digraphs*), whose edges posses directional information. Additionally, in *weighted graphs*, each edge (directed or undirected) is associated with a scalar value, quantifying a possible interaction strength, a cost, or a flow on the respective edge.

In most cases, a network is represented by its *adjacency matrix* $\mathbf{A}$, with entries $A_{ij} = 1$ indicating that there exists an edge between vertex n_i and n_j, and $A_{ij} = 0$ otherwise. For undirected networks, the adjacency matrix is symmetric $A_{ij} = A_{ji}$. For weighted networks, the elements of the adjacency matrix are replaced by nonbinary scalar values.

However, in particular for *sparse networks*, that is, networks where the number of edges is much smaller than the number of possible edges $N_E \ll N_V^2$, the adjacency matrix becomes computationally inefficient in terms of memory allocation. Alternatively, the network can be specified by a set of *adjacency lists*, consisting of N_V lists that enumerate to which other vertices each vertex connects, see also Chapter 2. The adjacency matrix, as well as the adjacency lists, have their unique advantages and disadvantages in terms of computational efficiency. A schematic example of both representations is given in Fig. 3.2.

3.2.1 Distance, Average Path Length, and Diameter

In a network consisting of N_V vertices, the *distance* d_{ij} between any two vertices n_i and n_j is given by the length of the *shortest path* between the vertices, that is, the minimal number of edges that need to be traversed to travel from vertex n_i to n_j. The shortest path between two vertices does not have to be unique, often there exist several alternative paths with identical path length. For directed networks, the distance between two vertices n_i to n_j is usually not symmetric $d_{ij} \neq d_{ji}$. Likewise, for directed, as well as *disconnected networks*, that is, networks consisting of two or more isolated components, there might not always be a path that connects vertex n_i to n_j. In such a case, the distance between the respective vertices is infinite $d_{ij} = \infty$. See Fig. 3.2 for examples.

The *diameter* $d_\mathrm{m} = \max(d_{ij})$ of a network is defined as the maximal distance of any pair of vertices. The *average* or *characteristic* path length $d = \langle d_{ij} \rangle$ of a network is defined as the average distance between all pairs of vertices. In the case of infinite distances, the average inverse path length $d_\mathrm{eff} = \langle 1/d_{ij} \rangle$, also referred to as *efficiency*, can be used to specify the average path length within the network. In this case, a fully connected network $d_{ij} = 1\ \forall i, j$ has an efficiency $d_\mathrm{eff} = 1$, whereas large distances and disconnected components (using the limit $1/d_{ij} \to 0$ for $d_{ij} \to \infty$) reduce the efficiency of the network.

The situation is slightly less straightforward if weighted networks are considered. Then, we are faced with the possibility to take additional information into account.

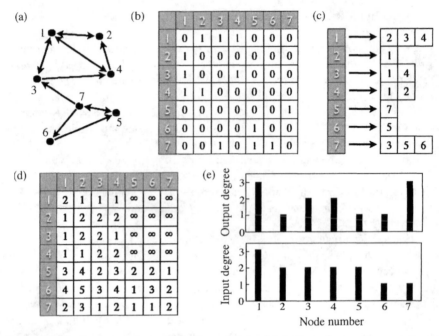

FIGURE 3.2   Representations of complex networks. (*a*) A directed network, consisting of $N_V = 7$ vertices and $N_E = 13$ directed edges. (*b*) The adjacency matrix **A** of the network. (*c*) The set of adjacency lists, specifying to which other vertices each vertex connects. (*d*) The distance matrix **D** with elements d_{ij}. Note that the distances are not symmetric and may be infinite, indicating that not all vertices can be reached from all other vertices. (*e*) The input degree k_i^{in} and output degree k_i^{out} of each vertex.

For example, within a network of train connections, the shortest path length (distance) between two stations can be defined according to physical distances, or, taking travel time into account, as the total time needed to travel from one station to another. Furthermore, the fastest connection must not always be the cheapest; thus we might wish to define the distance between two stations according to the amount of money needed to travel from one station to another. In either case, the term distance between vertices can be generalized to accommodate additional scalar information, given by a *weight factor* that is associated with each edge.

Computationally, the estimation of the distance between two vertices is not trivial. Within the extensive literature on the shortest paths problem, the most common choices are the *Dijkstra* and the *Floyd–Warshall* algorithm [6]. The Dijkstra algorithm returns the lowest cost path between a source vertex n_i and all other vertices in the network in $\mathcal{O}(N_V^2)$ time. For efficiency reasons, the algorithm return just *one* shortest path; enumerating all shortest paths between two vertices is computationally more tricky and expensive. To calculate the all-to-all distances, the Floyd–Warshall algorithm is the method of choice. The algorithm returns the distance matrix in $\mathcal{O}(N_V^3)$. Both algorithms straightforwardly allow to incorporate weighted edges. Negative weights may induce cycles that reduce the cost of a path each time the cycle is traversed. In this case, the definition of the "lowest cost" path has to be modified.

Note that distances, path length and diameter also depend on network size and density (number of vertices and links) and are therefore no genuine classifiers that straightforwardly allow to compare different networks.

3.2.2 Six Degrees of Separation: Concepts of a Small World

One of the striking properties of almost all empirical networks is that, despite their huge size of sometimes several millions of vertices, the average path length is usually surprisingly small. For example, within cellular metabolism, represented by a network of metabolites (vertices) linked by biochemical reactions (edges), the average path length between two metabolites is approximately $d \approx 3$ only, independent of the specific organism [22,33,71]. A recent study of the World Wide Web (www), represented by a network of web documents (vertices) that are connected by directed hyperlinks (URLs), estimated that the average path length between any two vertices is $d \approx 16$ only [1], extrapolated for a network of 200 million documents.

The term *small world* network itself originated in the social sciences, reflecting the assertion that within networks of social acquaintances (or friendships) all people (vertices) on the planet are separated from each other by just a small number of intermediate friends or acquaintances ("*six degrees of separation*," although the specific value six must not be taken too literally).

However, strictly speaking, the term small world is not a genuine network property, that is, there is no measure or statistical test that allows to check whether a given specific empirical network belongs to the class of small world networks. As stated above, the average distance between vertices also depends on the size of the network: The more vertices a network has, the more distant the vertices tend to be. The small world property is thus mainly understood to apply to network models whose average path length d increases slower or equal than the logarithm of the network size $d \sim \log N_V$ for $N_V \to \infty$. A further distinction includes *ultrasmall* networks [13], whose average path length scales as $d \sim \log \log N_V$.

3.2.3 The Degree Distribution

One of the most basic properties of a vertex n_i is its *degree* k_i, defined as the number of edges adjacent to the vertex. In a network without *self-loops* (edges that connect a vertex to itself) and *multiple links* (two vertices are connected by more than one edge), the degree equals the number of neighbors of the vertex. In the case of directed networks, we distinguish between the *input degree* k_i^{in} and the *output degree* k_i^{out}. Taking all vertices of a network into account, we can ask for the probability $p(k)$ that the degree of a randomly chosen vertex equals k. The *degree distribution* $p(k)$ has become one of the most prominent characteristics of network topology.

One of the key discoveries that triggered the renewed interest in complex network theory was that the distribution $p(k)$ of many empirical networks approximately follows a power law $p(k) \sim k^{-\gamma}$, where γ denotes the *degree exponent*. In contrast to

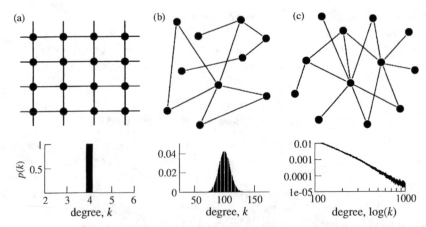

FIGURE 3.3 Degree distributions of complex networks. (*a*) A lattice-like network. Each vertex has the same degree k (for periodic boundary conditions or large networks, such that vertices at the border can be neglected). (*b*) An Erdös–Rényi random network. The degree distribution is homogeneous, the degrees of the vertices are centered around the average value. (*c*) A scale-free network. The degree distribution is highly inhomogeneous and follows a power law of the form $p(k) \sim k^{-\gamma}$, where γ denotes the *degree exponent*. While most vertices have a low number of connections only, a smaller number of vertices is highly connected.

the until then prevailing picture, where vertices are connected randomly and each vertex has approximately the same number of links, many empirical networks are strongly inhomogeneous: While the vast majority of vertices posses only a small number of links, a small number of vertices ("*hubs*") are highly connected. Examples of prototypical degree distributions are depicted in Fig. 3.3.

Though being one of the most basics characteristics of network architecture, a statistically stringent numerical estimation of the degree distribution is far from trivial [25]. In the simplest case, $p(k)$ can be straightforwardly estimated from an (usually binned) histogram of degrees. However, for many real networks with strongly inhomogeneous degree distributions, the simple histogram approach provides insufficient statistics at high degree vertices and is a notorious source of misinterpretations [25]. More reliable in terms of numerical estimation is the *cumulative degree distribution* $p_c(k)$, defined as the probability that a randomly chosen vertex has a degree larger than k. The cumulative degree distribution $p_c(k)$ is a monotonously decreasing function of k and its estimation requires no binning. For a power-law distribution $p(k) \sim k^{-\gamma}$, the cumulative degree distribution is of the form $p(k) \sim k^{-(\gamma-1)}$. An exponential distribution $p(k) \sim \exp(-k)$ corresponds to an invariant cumulative distribution $p_c(k) \sim \exp(-k)$. Computationally even more straightforward is to rank the vertices according to their degree and plot the degree versus the rank of each vertex. Examples of different representations of the degree distribution are shown in Fig. 3.4. It should be noted that all empirical networks necessarily show deviations from an strict mathematical degree distribution.

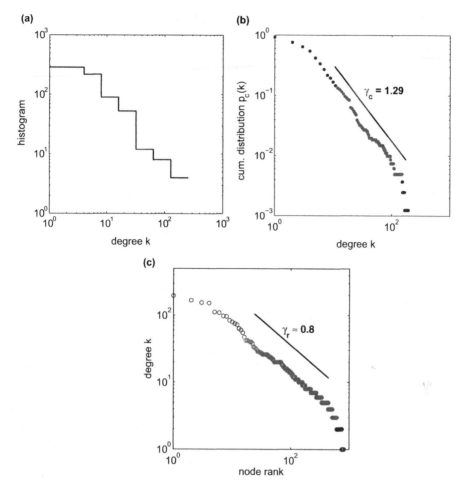

FIGURE 3.4 Different representations of the degree distribution of the metabolite substrate network described in Fig 3.1. (a) A binned histogram. Shown is the number of vertices with a degree k, using a logarithmic binning. (b) The cumulative degree distribution $p_c(k)$, that is, the probability that a vertex has a degree larger or equal k. Note that the cumulative distribution does not require binning but is obtained from the (normalized) number of vertices with degree larger or equal k. (c) The rank plot of metabolites, ranked according to their degree k. A power law of the form $p \sim k^{-\gamma_r}$ in the rank plot corresponds to a degree exponent $\gamma = 1 + 1/\gamma_r \approx 2.3$ in the original degree distribution $p(k)$ and $\gamma_c \approx 1.3$ in the cumulative distribution. The straight lines are not fitted and serve as a guide to the eye only.

In particular for power law distributions, the size (number of vertices) of the network puts constraints on the estimation of the degree exponent. Highly connected vertices are rare, and their probability is thus difficult to estimate for small networks. Likewise, the number of vertices with small degree is restricted by network size. Consequently, the formula $p \sim k^{-\gamma}$ often applies only to an intermediate region of the empirical degree distribution and has to be adjusted with an exponential cutoff at high degrees. More importantly, as shown in the recent literature, the reported degree exponent of many empirical networks correlates with network size and thus might not reflect the actual exponent of the underlying networks [17,18]. Furthermore, for small degree exponents the variance of the degree distribution is infinite, thus any empirical sample of vertex degree is no "typical" observation.

However, for many biological problems it is often more important to note that the degree distribution is highly inhomogeneous and long-tailed, as opposed to the question whether the degree distribution fits a power law in a strict statistical sense. For weighted networks, the concept of degree can also be extended to account for the weights of the edges by defining the *strength* of a vertex as the sum of the absolute values of the weights.

3.2.4   Assortative Mixing and Degree Correlations

Despite its importance in the topological characterization of complex networks, the degree distribution itself does provide only little information about the internal structure and organization of the network. More interesting is thus to look for correlations between the degrees of adjacent vertices. A network is called *disassortative* if vertices with high degree connect preferentially to vertices with low degree. Vice versa, a network is called *assortative* if vertices with high degree preferentially also connect to other vertices with high degree. As pointed out in the recent literature [49], social networks tend to be assortative, that is, persons (vertices) with many friends (connections) tend to be also connected to other persons with many friends, while most technological and biological networks are disassortative.

Formally, the degree correlation can be obtained from the joint probability distribution $p(k_i, k_j)$ that two connected vertices n_i and n_j have degrees k_i and k_j, respectively. For uncorrelated degrees, the joint probability is given by the product of the marginal degree distributions $p(k_i, k_j) = p(k_i)p(k_j)$. A measure for the deviation from statistical independence is given by the *mutual information* [64,66].

Unfortunately, a direct numerical estimation of $p(k_i, k_j)$ is computationally demanding and often not feasible due to the limited size of the (empirical) network (but see also Ref. [66] for the numerical estimation of probability distributions and a discussion of finite size effects). More straightforward is thus to consider the Pearson correlation coefficient between the degree of two adjacent vertices. The correlation coefficient or *assortativity coefficient* r lies in the range $-1 \leq r \leq 1$, with $r < 0$ corresponding to a disassortative network and $r > 0$ to an assortative network. Note that the assortativity coefficient r, similar to the usual Pearson correlation, has its limits for strongly inhomogeneous degree distributions and fails to correctly quantify nonlinear degree correlations, for example, networks that are assortative for low degree vertices and disassortative for high degree vertices.

Another popular, and closely related, measure to evaluate degree correlations is the *average neighbor degree* [53]. For each vertex n_i, the average degree $k_{i,nn} = \frac{1}{k_i} \sum_{j=1}^{N_V} A_{ij} k_j$ of its neighbors is calculated. Subsequently, these values are

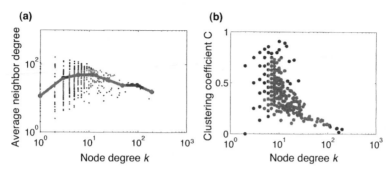

FIGURE 3.5   Vertex degree correlation in the substrate graph. *Left:* The average neighbor degree $k_{i,nn}$ of each vertex n_i, plotted versus the degree k_i. The solid line gives the (binned) average over all vertices with the same degree k. For large degrees a weak negative correlation is observed. *Right:* The clustering coefficient C_i of each vertex versus the degree k_i. Highly connected vertices exhibit a low clustering coefficient, that is, highly connected vertices preferentially connect to vertices that are not mutually connected, indicating a hierarchical structure.

averaged for all vertices having the same degree k, resulting in the average neighbor degree $k_{nn}(k)$. See Fig. 3.5 for examples of vertex degree correlations.

To evaluate the degree correlations for weighted and directed networks requires slight modifications in the respective definitions. In the case of directed networks, two distinct correlation indices are most interesting: (*i*) Do the in-degrees k_i^{in} of vertices correlate with their neighbors out-degrees k_i^{out}, and (*ii*) do the out-degrees k_i^{out} of vertices correlate with their neighbors in-degrees k_i^{in}? In the case of weighted networks, the degrees can again be replaced by their weighted counterparts.

3.2.5   The Clustering Coefficient

Another basic measure that accounts for the internal structure of a network is the *clustering coefficient C*. The clustering coefficient relates to the local cohesiveness of a network and measures the probability that two vertices with a common neighbor are connected. In the case of undirected networks, given a vertex n_i with k_i neighbors, there exist $E_{max} = k_i(k_i - 1)/2$ possible edges between the neighbors. The clustering coefficient C_i of the vertex n_i is then given as the ratio of the actual number of edges E_i between the neighbors to the maximal number E_{max},

$$C_i = \frac{2E_i}{k_i(k_i - 1)} \quad . \tag{3.1}$$

See Fig. 3.6 for a schematic example. Note that, strictly speaking, the clustering coefficient C_i is not a property of the vertex n_i itself, but rather a property of its neighbors. The global or mean clustering coefficient $C = \langle C_i \rangle$ of the network is the average cluster coefficient of all vertices.

Many empirical networks exhibit a rather high clustering coefficient, indicating a local cohesiveness and a tendency of vertices to form clusters or groups. Indeed,

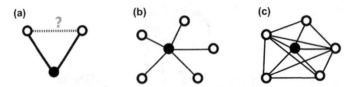

FIGURE 3.6   The clustering coefficient relates to the local cohesiveness of a network. (a) The clustering coefficient is defined as the probability that two vertices with a common neighbor are connected. (b) A highly connected vertex with a low clustering coefficient, indicating a (at least locally) hierarchical structure. (c) A a vertex with high clustering coefficient $C_{vertex} = 0.8$.

for example, in social networks, it seems intuitive that two persons (vertices) who have a common friend are much more likely to be also friends, as compared with two randomly chosen persons. Interestingly, this also directly relates to the notion of degree correlations and dynamics on networks. As persons that share a common friend are likely to become acquainted themselves, they will acquire new friends over time. In particular, a highly connected person will induce new connections among his friends (neighboring vertices). In this sense, within social networks, a situation with disassortative degree correlations and low clustering coefficients is dynamically unstable and must be expected to evolve gradually toward more clustering and thus assortative degree correlations.

However, despite its conceptual simplicity, the interpretation and statistical testing of the clustering coefficient holds some pitfalls, which are discussed in more detail in Section 3.5. Furthermore, the clustering coefficient depends on the number of edges within the network.

To claim a nontrivial local clustering within the network, an estimated value of C thus has to be compared with an appropriate null model to validate whether the value is indeed statistically significant, that is, whether the respective network indeed exhibits a higher degree of clustering than a corresponding random network. Difficulties also arise for specific types of graphs, such as bipartite graphs, that exhibit a nontrivial clustering coefficient inherent to the bipartite structure [1,52], see Section 3.5 for a detailed discussion.

An alternative, but equivalent, definition of C can be given with respect to the number of triads (triples of vertices where each vertex is connected to both others) within a network. Note that the number of edges between the neighbors of a vertex is equal to the number of triads that vertex is part of. The global clustering coefficient is then defined as the proportion of triads in a network with respect to the total number of connected triples (triples where at least one vertex is connected to both others).

$$C = \frac{3 \times \text{number of triads}}{\text{number of connected triples}} \qquad (3.2)$$

The factor 3 accounts for the fact that each triad contributes to three connected triples [1]. A characterization of the clustering coefficient with respect to the number of triads holds some advantages with respect to numerical estimation and can be generalized to other structures, such as the number of squares [31].

Of particular interest is also the correlation of the clustering coefficient C_i with other properties of a vertex n_i. For example, as described by Newman [49], many empirical networks exhibit a negative correlation between the degrees k_i and the clustering coefficients C_i, indicating a modular structure of the network. See Fig. 3.5 for an example.

3.2.6 The Matching Index

Within many empirical networks, two vertices that are functionally similar do not necessarily have to be connected. For example, within a network of protein interactions, two proteins that are involved in the regulation of similar processes and should be considered as closely related, must not necessarily bind to each other. Correspondingly, the normalized *matching index* M_{ij} quantifies the "similarity" between two vertices based on the number of common neighbors shared by two vertices n_i and n_j.

$$M_{ij} = \frac{\sum \text{common neighbors}}{\sum \text{total number of neighbors}} = \frac{\sum_{k,l}^{N} A_{ik} A_{jl}}{k_i + k_j - \sum_{k,l}^{N} A_{ik} A_{jl}} \qquad (3.3)$$

Note that for the measure to be properly normalized, the denominator counts only the number of distinct neighbors, that is, neighbors that are shared by both vertices are counted only once. One of the virtues of the matching index is that it can be straightforwardly applied to networks consisting of different types of vertices, such as bipartite graphs. For example, two transcription factors may regulate the expression of similar genes, without necessarily regulating (or binding to) each other. A schematic illustration of the matching index is given in Fig. 3.7.

The matching index can be generalized beyond the immediate neighbors of a vertex or extended to multiple vertices [40]. Furthermore, at the most general level, two vertices can be regarded (or defined) as "similar" if their distance to all other vertices within the network is approximately the same, irrespective of whether they are directly connected or not [75]. An advantage of this definition lies in the fact that the actual pair-wise similarity of two vertices must not be specified. The definition draws upon the notion only that two entities (vertices) must be considered "similar" if they perceive the rest of the

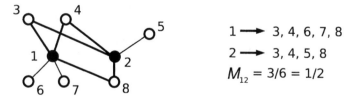

$$1 \longrightarrow 3, 4, 6, 7, 8$$
$$2 \longrightarrow 3, 4, 5, 8$$
$$M_{12} = 3/6 = 1/2$$

FIGURE 3.7 Vertices that are functionally related do not necessarily have to be connected. The matching index counts number of common neighbors shared by two vertices, normalized by the total number of distinct neighbors. The right panel shows the adjacency list of the vertices n_1 and n_2, along with the corresponding matching index M_{12}.

world (here the distance to all other vertices within the network) in a similar way.

3.2.7   Network Centralities

Closely related to distance measures, network centrality indices seek to characterize each vertex or edge with respect to their position within the network. Centrality measures will be discussed in more detail in Chapter 4, here we will briefly outline some basic features.

Intuitively, a basic measure of the importance of a vertex n_i is its degree k_i (degree centrality). And indeed, several studies on biological network report a significant relationship between vertex degree and functional importance of vertices [2]. For example, within protein interaction networks, the removal of highly connected proteins is more likely to have lethal effects than removal of proteins with only a small number of links [32]. However, the degree is clearly not the only determinant of the functional importance of a vertex. Often more relevant, is the contextual location of the vertex within the network. For example, we can ask from which vertex a signal should be sent to reach all other vertices in minimal time. Or, vice versa, which vertices can be reached fastest from any other vertex within the network? In this respect, the *closeness centrality* specifies which vertices have the shortest paths to all others, measured, for example, by the (inverse of the) average distance from a vertex to all other vertices. For detailed definitions, see Chapter 4.

Probably the most well-known centrality measure is the *betweenness centrality* (*BC*). The betweenness centrality can be defined with respect to vertices and edges, and measures how often a vertex or edge is present in the set of all shortest paths. As can be seen in Fig. 3.8, low degree vertices can be crucial to establish communication or mass flow within a network. Thus, with respect to robustness properties of a network, a selective attack on vertices with high *BC* was often found to be more relevant than a removal of vertices with high degree. Computationally, the estimation of the betweenness centrality is rather demanding and described in Chapter 4.

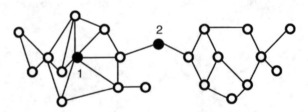

FIGURE 3.8   The degree of a vertex does not necessarily reflect importance with respect to function of a network. While vertex n_1 has a high degree, its removal does not necessarily affect communication within the network. However, removal of vertices with low degree may have significant effects on communication or mass flow within the network, as seen for vertex n_2.

3.2.8 Eigenvalues and Spectral Properties of Networks

An important property of network topology are the spectral properties of the adjacency matrix **A**. Though as yet only hardly used in biological research, the spectra of random graphs are among the oldest characteristics of network topology with a plethora of applications in many branches of physics [1].

For an undirected graph, the symmetric adjacency matrix **A** has N_V real eigenvalues λ_i. The spectral density $\rho(\lambda)$,

$$\rho(\lambda) = \frac{1}{N_V} \sum_{i=1}^{N_V} \delta(\lambda - \lambda_i), \tag{3.4}$$

approaches a continuous function for increasing network size $N_V \to \infty$. An extensive amount of work about the mathematical properties of the spectral density is available, including the famous Wigner semicircle law [1,21].

Of more relevance to the biological sciences, the eigenvalues of network matrices are becoming increasingly important with respect to two different fields of research: First, in networks of coupled oscillators, that is, in networks where each vertex corresponds to an oscillator coupled to other oscillators via an adjacency matrix, the global dynamics of the system are determined by the structure of the adjacency matrix. In particular, the stability of the synchronized state, that is, the state of the network where almost all vertices oscillate synchronously, can be related to the eigenvalues of the *Laplacian* matrix of the network [54], defined in close analogy to the adjacency matrix. Recent studies also take into account the effect of weighted edges [74].

Second, along similar lines, the eigenvalues of network matrices determine the stability and local dynamics of networks composed of interacting elements. For example, the vertices of a metabolic network denote metabolites, whose concentrations change according to the adjacent edges (metabolic reactions). Formally this system is represented by a differential equation for all metabolite concentrations. However, at least locally, this (usually unknown) system of differential equations can be approximated by a weighted interaction matrix, denoted as the Jacobian **J** of the system. The Jacobian matrix already governs essential aspects of the dynamics and predicts specific dynamic behavior even if detailed knowledge about the underlying reactions and interactions is not available [63,65,68].

3.3 MODELS OF COMPLEX NETWORKS

The various network indices discussed until now characterize and quantify the topological structure of a given network. However, to understand and elucidate whether an estimated value indeed corresponds to nontrivial structure within the network requires to consider basic prototype models of complex networks. We emphasize that none of the models described below aims to mimic the detailed features of any real network. Rather they represent minimal models, each invented to exhibit distinct generic features of complex networks. The purpose of prototype models is twofold: First, they provide null models to understand whether an observed feature is a generic feature of certain network classes or whether it deviates from what could be expected for a simplistic model. Second, prototype models often provide insight on how certain features of complex networks arise from the construction rules of the prototype models, allowing to probe to what extent (for example evolutionary) mechanisms can account for the observed features of empirical networks. Again, more detailed mathematical treatises on random network models are given elsewhere [1,17,50], here we outline the basic ideas only.

3.3.1 The Erdös–Rényi Model

Probably the most basic model of a random network is given by the Erdös–Rényi (ER) network [20]. The ER network consists of N_V vertices, connected by N_E (undirected) edges that are chosen randomly from the set of $N_V(N_V - 1)/2$ possible edges (excluding multiple connections and links from a vertex to itself). The probability p that two randomly chosen vertices are connected is thus $p = 2N_E/N_V(N_V - 1)$.

Alternatively, the ER model can be defined as a set of N_V vertices, with each pair of vertices connected with an equal probability $p \leq 1$. The number of edges N_E is then a random variable, with the expectation value $\langle N_E \rangle = pN_V(N_V - 1)/2$ [1].

The ER model has been the primary subject of random graph theory, resulting in extensive knowledge about its mathematical properties and typical features. Here we summarize some basic properties only.

The degree distribution of the ER model is given by a *binomial distribution* that becomes approximately *Poissonian* in the limit of large networks ($N_V \to \infty$). The probability of a vertex to have degree k is

$$p(k) \simeq e^{-\langle k \rangle} \frac{\langle k \rangle^k}{k!} \tag{3.5}$$

with $\langle k \rangle = pN_V$ denoting the *average degree*. A typical realization of the ER model is rather homogeneous, most vertices have a similar degree, distributed approximately symmetrically around the average degree $\langle k \rangle$, as shown in Fig. 3.3b. Most analytical work on the ER model has concentrated on questions related to percolation theory, that is, the connectedness of the network and the emergence of paths that enable a traversal of the whole network. For small p the network is disconnected and consists of a large number of isolated components [50]. At $p \approx 1/N_V$ (thus for average degree $\langle k \rangle \approx 1$) a phase transition occurs, giving rise to a *giant component* that encompassed most of the vertices of the network. For $p \geq \log(N_V)/N_V$, all vertices are connected for almost all realizations of the random network. The ER model exhibits the *small-world* property. Above the percolation threshold, the average path length is very small and scales as the logarithm of the number of vertices $l \sim \log N_V$ (with $\langle k \rangle$ kept constant for increasing number of vertices).

By construction, the clustering coefficient of the ER network $C = p = \langle k \rangle/N_V$, that is, the probability that two vertices with a common neighbor are connected equals the probability that any pair of randomly chosen vertices are connected. The ER model does not show any local cohesiveness. Likewise, the degree of connected vertices is uncorrelated, the ER model does not display degree correlations.

The Erdös–Rényi model remains one of the most important prototype models in graph theory. However, the main limitations for a direct comparison of network properties with empirical networks are its homogeneous degree distribution, the absence of local structure, and the lack of degree correlations. A close variant of the ER model, the *configuration model*, will be discussed in Section 3.5.

3.3.2   The Watts–Strogatz Model

While the Erdös–Rényi model correctly reproduces the *small world* property, it fails to account for the local clustering that characterizes many empirical networks. In particular for social networks, that is, networks of mutual friendships or acquaintances, most studies indicate a clustering coefficient that is orders of magnitude higher than the value obtained for a corresponding ER network.

In one of the seminal papers of complex network theory, Watts and Strogatz proposed a model for coexistence of local structure on the one hand, and a small average path length on the other hand [72]. The starting point of the model is the limiting case of a regular lattice-like network: Each vertex (arranged on an one-dimensional ring in the original model) is connected to its $n/2$ nearest neighbors. In social terms, this would resemble a strictly local medieval-like world, where each person knows people in his or her immediate vicinity only, such as neighbors and people in nearby villages. Consequently, the model exhibits strong local cohesiveness (a high clustering coefficient), but the spread of information is slow, that is, the average path length scales linearly with system size.

Extending the regular lattice-like network, shortcuts between distant vertices are introduced, that is, with a probability p_{rew} a link is rewired, such that one end is detached from its original vertex and connected to a randomly chosen vertex. In social terms, this would correspond to a merchant or traveler, who is also acquainted to a small number of more distant people within the country. As the probability p_{rew} increases and more links are rewired, the model approaches a random network of the ER type. In the limit $p_{\text{rew}} \to 1$ the ER model is recovered. The network thus again exhibits no local structure (small clustering coefficient) and the average path length scales as the logarithm of network size.

One of the intriguing result of the WS model is that already a very small number of shortcuts ($p_{\text{rew}} \ll 1$) is sufficient to rapidly decrease the average path length [28]. However, for small p_{rew}, the local clustering remains almost unaffected and the clustering coefficient decreased significantly only for $p_{\text{rew}} \to 1$. Thus, for an intermediate region of p_{rew}, the WS model exhibits a coexistence of high local clustering and short average path length (small world property), as also observed in many empirical networks. A schematic representation of the WS model is given in Fig. 3.9.

The main significance of the WS model results from the fact that it emphasizes a difference between local and global properties of networks. The clustering coefficient, a local property, is determined by the immediate neighborhood of a vertex and is almost unaffected by the introduction of additional "shortcuts" within the network.

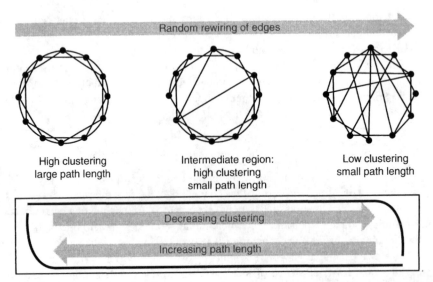

FIGURE 3.9   The Watts–Strogatz model: Starting point is a regular network, constructed such that each vertex is connected to its two nearest neighbors, resulting in a maximal clustering coefficient $C = 1$. With probability p_{rew} links are randomly rewired. In the limit $p_{rew} \to 1$, the ER model is recovered.

On the contrary, the average path length, a global property, rapidly decreases upon the introduction of just a few shortcuts. This has, for example, profound implications on the spread of infectious diseases across continents. A change in average path length to distant vertices is not detectable at the local level, that is, your social neighborhood might remain almost unaltered, while the "distance" (in network terms) to infected persons can rapidly decrease with only a small number of transcontinental travelers.

However, apart from the coexistence of high local clustering and short average path length, the WS models captures almost no other feature found in empirical networks. Its importance as a null model for biological networks thus remains limited.

3.3.3   The Barabási–Albert Model

Among the most important limitations of the models discussed above is that neither captures nor accounts for the inhomogeneous degree distribution found in many empirical networks. To this end, Barabási and Albert [7] proposed a simple network model that gives rise to a scale-free degree distribution and still provides the conceptual basis for most current network models described in the literature.

Closely related to (and actually a simplification of) an earlier model by Price [4,16,45,50], the BA model is based on two essential ingredients: (i) Growth: In contrast to the models discussed above, the BA model does not assume that the number of vertices within the network is fixed. Mimicking the dynamics of many real networks, vertices are continuously added and the network grows as a function of time. (ii) Preferential attachment: New edges are not introduced randomly, but the

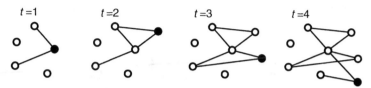

FIGURE 3.10   The Barabási–Albert model [7]: Starting with an initial small network, consisting of N_0 unconnected vertices, a new vertex is introduced at each time step and connected with $m < N_0$ edges (here shown with $m = 2$).

probability that a vertex receives a new edge depends on its present degree k_i, again reflecting dynamic properties of real networks.

The growth process is organized as follows: starting with an initial small network, consisting of N_0 unconnected vertices, a new vertex is introduced at each time step. The new vertex is connected with $m \leq N_0$ edges to the already present vertices. The probability $\rho(n_i)$ that already present vertex n_i receives a new edge is proportional to its degree k_i:

$$\rho(n_i) = \frac{k_i}{\sum_j k_j} \ . \tag{3.6}$$

After t time steps, the network consists of $N_V(t) = N_0 + t$ vertices, connected by $N_E(t) = mt$ edges. Due to the preferential attachment mechanism, older vertices tend to have accumulated more links and thus have an even higher probability to receive yet more links (a *"rich-get-richer"* dynamics). Likewise, new vertices only have a small number of links, and thus a low probability of receiving additional links. A schematic illustration of the growth process is given in Fig. 3.10.

In the long time limit $t \gg 1$, the BA model exhibits a scale-free degree distribution $p(k) \sim k^{-\gamma_{BA}}$ with a degree exponent $\gamma_{BA} = 3$ that is invariant with time. The degree exponent is independent of the free parameters m and N_0. The BA model captures the small world property; BA networks are found to have shorter average path length than ER and WS models of the same size and density. The degrees are uncorrelated, and analytical estimates of the clustering coefficient are available [36].

One of the merits of the BA model is that it provides a possible mechanism to explain the observed scale-free distribution of many empirical networks. Indeed, the time evolution of many empirical networks is governed by preferential attachment-like processes. For example, within a social network, people (vertices) with already many friends (edges) are more likely to acquire new friends, as compared to people with few edges. Likewise, already famous actors will obtain more offers to act in a new movie than young unknown actors. Scientific papers that are already frequently cited are more likely to be read and cited again than less frequently cited papers.

Importantly, the preferential attachment rule also provides several testable predictions for complex networks. For example, if metabolic networks are reported as scale-free, then, according to this growth rule, highly connected metabolites should have an early evolutionary origin. Indeed, as emphasized by Wagner and Fell [71], many of the highly connected metabolites, mainly intermediates of the TCA cycle and glycolysis, as well as some ubiquitous co-factors, are among the evolutionary oldest.

However, explanations in terms of evolutionary mechanisms also hold some pitfalls that are unfortunately rarely—if ever—discussed in the literature. Most importantly, not only the formation of a network itself, but often also the acquisition of data about the network is governed by similar mechanisms. For example, minor movies with famous actors are more likely to be included in the respective databases than local movies starring only unknown actors. Likewise, putative biochemical regulations or reactions adjacent to the TCA cycle are more likely to be investigated, and thus reported in publications, than putative regulations within the outskirts of metabolism. In this sense, an observed feature of an empirical network might also always reflect properties of the data acquisition process, rather than genuine properties of the network itself.

3.3.4 Extensions of the BA Model

The BA model constitutes the conceptual basis for a large variety of extensions and modifications and has triggered an exceptional amount of further work in the complex network models. Most extensions can roughly be subdivided into two (though often overlapping) categories: (i) Modifications that aim to generate networks with specific tunable features, such as different degree exponents, tunable cluster coefficients, or degree correlations. (ii) Modifications that aim to mimic the evolutionary growth processes of specific networks in more detail, such as aging in social networks or capacity restrictions in transportation networks.

For example, the exponential cutoff at high degrees observed in many real networks can be accounted for by aging of vertices, that is, vertices that have been present for a given time T stop acquiring new edges or are removed from the network [3]—as could be expected in social networks. Similarly, an airport within a transportation network will not acquire new connections beyond a certain capacity, again resulting in an exponential cutoff at high degrees. Other processes to modify the properties of the network include rewiring of edges according to defined rules. For example, within a social network people that have a common friend are more likely to become acquainted themselves, resulting in an increased local clustering of the network [15]. Further extensions and modifications include memory effects and high clustering [36], degree correlations [56,73], tunable degree exponents [38], information accessibility [48], among many more. An overview of early modifications and extensions of the original BA model can also be found in Table III of Ref. [1].

3.4 ADDITIONAL PROPERTIES OF COMPLEX NETWORKS

Within the first section, most emphasis was placed on quantitative measures that describe the properties of individual vertices and edges. However, complex networks are also characterized by emergent features that transcend the properties of individual vertices and relate to the organization of the network as a whole. In the following, the basic emergent global properties of complex networks, such as robustness or modularity, are outlined.

3.4.1 Structural Robustness and Attack Tolerance

Most biological systems share a common feature: robustness [8,35,60,69]. Constituting one of the fundamental organizing principles of biology, cellular networks must be able to maintain their function in the face of constant perturbations and fluctuations that affect the internal or external parameters of the system. In the context of complex network analysis, robustness is mainly understood as the persistence of topological network properties, such as average path length or connectedness, upon removal of vertices or links [1,2,10,33].

Indeed, most empirical networks show a surprising tolerance against removal of vertices. Focusing on topological aspects of robustness only, a number of studies revealed significant differences between distinct network topologies upon removal of vertices or edges [2,29]. In general, we have to distinguish between random and intentional attacks on network topology. While for ER networks, due to the homogeneity of vertex properties, the response to random and intentional attacks is roughly similar, the situation for scale-free networks is markedly different. Most properties of scale-free networks were found to be exceptionally robust against random removal of vertices. However, at the same time, scale-free networks are vulnerable with respect to intentional attacks. This difference is due to the heterogeneous degree distribution. Low degree vertices are far more frequent than high degree vertices, but play only a minor role in overall network topology. While random attacks will most likely affect low degree vertices, a selective attack on high degree vertices has far more dramatic consequence on global network indices [1,2,29]. A schematic illustration is given in Fig. 3.11.

In general, the difference between random and intentional attacks is at the core of most current research on network robustness. The "robust, yet fragile" nature of complex systems refers to the fact that many complex systems are robust against random attacks, whereas they remain fragile against selective attacks. In particular, highly optimized systems are extremely robust against anticipated attacks, whereas

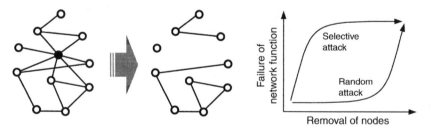

FIGURE 3.11 The "robust, yet fragile" nature of scale-free networks. Properties of scale-free networks are highly robust against random removal of vertices, but vulnerable against selective intentional removal of vertices.

optimization concomitantly leads to vulnerability against unanticipated perturbations, related to the principle of "highly optimized tolerance" (HOT) [12].

It should be noted though that the restriction to topological aspects of robustness allows only for a rather restricted view on network robustness. Dynamic aspects of functional robustness thus receive increasing interest recently [35,47,60,63,65].

3.4.2   Modularity, Community Structures and Hierarchies

Related to the idea of functional robustness is the notion of modules and community structures within complex networks. In general, it is assumed that many complex networks are built up from (interacting and possibly overlapping) *modules* or *communities*. The detection of such community structures has attracted substantial interest recently and defines an important aspect of complex network analysis [57,75,76].

Unfortunately, there is no generally accepted definition of what constitutes a module or community structure within a complex network. As a working definition, modules or communities are often understood as subsets of vertices that are densely connected among each other, but are only sparsely connected to other vertices outside the community. In this respect, probably the most influential definition of network modularity comes from the work of Newman and Girvan, who define a score function for network modularity based on this principle [24,51].

Computationally, the detection of community structures is closely related to the clustering of elements, that is, the assignment of "entities" into distinct categories based on a suitably defined notion of similarity. In the context of complex network analysis, similarity of vertices can be defined in different manners, for example, with respect to the shortest path between two vertices, the total number of paths between vertices (weighted according to path length), among several other possibilities. Likewise, and again resembling the situation in clustering algorithms, modules can be detected using agglomerative or divisive methods, that is, by grouping similar vertices into larger units or by iteratively breaking down a complex network into smaller and smaller subsets. Although there is a plethora of proposed novel algorithms specifically designed to detect modules within complex networks, it should be noted that a recent study [26,27] demonstrates that standard ways of clustering data, for example, based on k-means or hierarchical clustering, often outperform newer methods.

The relevance of modularity in complex networks can be seen from two different perspectives: First, modularity represents an inherent design principle of many complex networks. Indeed, in particular for biological networks shaped by evolution, it seems plausible that modular networks have advantages in terms of their ability to adjust to new evolutionary constraints by incorporating new functionality within the network [26,34]. Second, a decomposition into modules if often helpful to understand the functional organization of complex networks. For example, many technological networks, such as integrated electronic circuits, cannot be rationalized on the basis of individual resistors or transistors. Only by the construction of functional modules, such as shift registers, counters, gates, inverters, each consisting of many lower level elements, an effective manipulation and design can be achieved.

As already implicit in the example of an integrated electronic circuit, modules are not restricted to one specific level of representation. Rather, modules may overlap and consist of smaller modules, giving rise to a hierarchical organization of modularity within complex networks [14,57,58,70]

3.4.3 Subgraphs and Motifs in Networks

Closely related to the detection of community structures or functional modules is the notion of *motifs* as building blocks of complex networks [31,39,43,44,59]. Providing a bridge between local vertex-related properties and global functional properties of networks, the basic idea of network motifs is that large complex networks are essentially composed of small interlinked subgraphs. Similar to the notion of sequence motifs, we can thus look for reoccurring patterns within the network, that is, small sets of vertices that exhibit an identical local topological structure. Indeed, as shown in previous studies [44], the transcriptional interaction network of *Escherichia coli* is essentially composed of repeated appearances of three highly overrepresented network motifs. Furthermore, similar studies reveal that the distribution of motifs is characteristic for certain classes of networks, that is, networks with similar overall functionality (such as communication networks or food-webs) also exhibit a highly similar motif distribution [43]. Figure 3.12 shows all possible directed subgraphs composed of three vertices. Unfortunately, the systematic enumeration of network motifs is computationally demanding.

In general, the concept of network motifs is not restricted to subgraphs with a fixed number of vertices. Rather, it allows to look for any reoccurring patterns, including more complicated topological structures, such as multi-input motifs, regulator chains, or dense overlapping regulons (DORs) [39,59]. The most intriguing aspect of network motifs, however, results from the asserted connection of local topological structure to

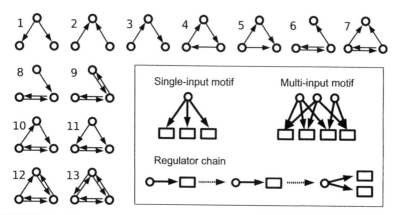

FIGURE 3.12   All 13 possible three-vertex motifs, according to Refs. [44,59]. *Inset:* The concept of motifs can be generalized to account for specific repeated subgraphs within a complex network. Shown are possible motifs of the yeast transcription regulatory network, with circles denoting regulators and rectangles denoting gene promoters, according to Ref. [39].

dynamic function of a network motif. While such a correspondence between topology and function, that is, each motif has also a specific "hard-wired" function, was claimed in several early studies on network motifs [44,59], later results indicate that this is not the case. Though there is increasing interest in the analysis of dynamic aspects of complex networks, the precise relationship between structure and function of complex networks is still largely unclear [30,37,55,63,68]. Additional aspects in the interpretation of motif distributions, including possible pitfalls in their statistical estimation, are discussed in the next section. More details on network motifs are given in Chapter 5.

3.5   STATISTICAL TESTING OF NETWORK PROPERTIES

The most crucial and probably most widely underestimated aspect of complex network analysis is the statistical testing of network properties. As yet, all network indices were described as numerical quantities that can usually be straightforwardly estimated from any given network topology. However, in most applications, network indices are also associated with a biological meaning or interpretation—we seek to uncover those features of the network that are characteristic of the underlying system. Thus, for example, in the case of a metabolic network, the question is not whether the clustering coefficient takes a specific numerical value, but rather whether this value distinguishes the network from other networks of similar size, that is, whether the metabolic network can be regarded as "highly clustered." Only in the latter case, that is, if the clustering coefficient indeed deviates from what could be expected for networks of similar size, it represents a distinctive feature of the respective network. But then, how should such a deviation be detected or quantified? What values of clustering coefficient should be considered "usual" or "typical" for a network of given size?

One answer to these questions, in addition to the prototype models described in the last section, lies in the formulation of appropriately randomized null models of complex networks. We create an ensemble of surrogate networks, usually of identical size and density, and compare the values of network indices obtained from the empirical network with those obtained from the ensemble of randomized surrogate networks.

More general, statistical testing always means to set up a null hypothesis, that is, a process or mechanism that is assumed to account for an observed feature, and a subsequent test whether the observed feature is actually consistent with the null hypothesis. In the context of complex network analysis, a possible null hypothesis is, for example, that an observed clustering coefficient is consistent with values arising from Erdös–Rényi networks of the similar size. Given a certain probability threshold, the null hypothesis is rejected if the probability to actually find the observed clustering coefficient for Erdös–Rényi networks is below the defined threshold. In this case, the deviation of the clustering coefficient with respect to the null hypothesis is *significant*.

However, apart from some straightforward cases, the statistical testing of network properties holds several potential pitfalls and possible sources of misinterpretations.

In the following, we briefly outline some of the most widely used null models for complex network analysis and point out possible ambiguities in their interpretation.

3.5.1 Generating Networks and Null Models

The most basic null model is a network of identical size (number of vertices and edges) but lacking any other internal structure. Conceptually equivalent to an Erdös–Rényi network of the same size, such an ensemble can be constructed by randomly rewiring links within the network—as already done in the construction of the Watts–Strogatz model.

Usually more appropriate, however, is to preserve the degree distribution of the empirical network. An ensemble of randomized surrogate networks with preserved degree distribution is obtained by iteratively swapping randomly selected edges, as schematically depicted in Fig. 3.13: For a directed network, at each iteration two edges $(a \rightarrow b)$ and $(c \rightarrow d)$ are selected at random and rewired as $(a \rightarrow d)$ and $(c \rightarrow b)$, provided that the respective edges do not already exist. Repeating this sufficient times, that is, such that most edges have a statistical chance to be selected, the resulting network has a preserved degree distribution, but lacks any other internal structure of the initial empirical network. The approach can be generalized to account for other features of complex networks. For example, the analysis of network motifs [44,59] makes use of a similar approach to generate networks with a preserved three-vertex motif distribution, swapping two edges if and only if the resulting motif distribution is preserved. Closely related to network randomization is the *configuration model* of complex networks: To construct a network with a specified degree distribution, each vertex is assigned a number k_i of adjacent edges, such that the total number of assigned edges is even. Subsequently, pairs of the, as yet unconnected, "stubs" are randomly chosen and connected [52,59].

In any case, a network index Q of interest is subsequently compared against the values found for the ensemble of surrogate networks and can be evaluated according to a significance score

$$S = \left| \frac{Q^{\text{network}} - \langle Q^{\text{surrogate}} \rangle}{\sigma_{\text{surrogate}}} \right| , \qquad (3.7)$$

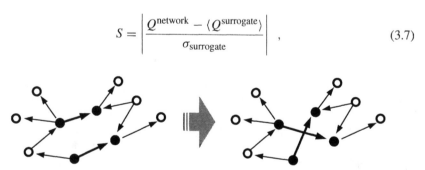

FIGURE 3.13 Generating random networks with preserved degree distribution: At each iteration, two edges $(a \rightarrow b)$ and $(c \rightarrow d)$ are selected at random and rewired as $(a \rightarrow d)$ and $(c \rightarrow b)$, provided that the respective edges do not already exist. Note that for undirected networks, there are two possible ways to rewire the links.

where $\langle Q^{\text{surrogate}} \rangle$ denotes the average found within the ensemble of surrogate networks and $\sigma_{\text{surrogate}}$ the standard deviation. Unfortunately, the construction of randomized networks that preserve features other than the degree or motif distribution is far from straightforward.

3.5.2 The Conceptualization of Cellular Networks

The most difficult problem of complex network analysis is often the choice of an appropriate null model or null hypothesis. By definition, all statistical tests rely on the definition of a null model, and any notion of "significance" is defined with respect to this null model only. Thus, if the null model is erroneous or trivial, so will be any result obtained from an evaluation of the significance of network properties against this null model. In particular, the choice of the null model is additionally complicated by the fact that networks are usually abstract representations of more complex biological processes. While this "reduction in complexity" is often necessary to make biological questions mathematically tractable, it also holds the temptation to neglect properties of the underlying system—resulting in an erroneous or misleading interpretation of network properties.

An illustrative example of such a case was discussed in the context of a recent study of motif distributions in complex networks [43]. Therein, the significance profile of small subgraphs (motifs) within a neuronal network was evaluated and compared with simple degree-preserving randomized networks (see also Chapter 5). The study concluded that the neural information processing networks exhibits a highly characteristic significance profile for its motif distribution, suggesting evolutionary mechanisms that result in key circuit elements to perform specific tasks. However, as pointed out later [5], a neural network, that is, a network of neurons connected by synapses, is not just a network of vertices connected by edges. Rather, neurons have a spatial position and a tendency to form local clusters, hence neighboring neurons have greater chance of forming connections than distant neurons. As the spatial properties are not reflected in the null hypothesis, the statistical test misclassifies a completely random but spatially clustered network as one that is nonrandom and exhibits significant network motifs. Indeed, a simple toy model that preserves the spatial position of neurons and connects neurons preferentially to nearby neurons is able to reproduce an almost identical significance profile for the motif distribution, without the need to invoke any evolutionary mechanisms to select for specific functional tasks [5].

The example serves to illustrates two important aspects of statistical testing of network properties: (i) The observed motif distribution within the neural information processing network was indeed highly significant, as compared with degree-preserving random networks of the same size—there was no computational error involved. The significance profile of motifs is thus indeed a true characteristic of the system. (ii) However, the significance profile of motifs tests against a trivial null hypothesis: the assumption of a completely random network. Strictly speaking, rejecting this null hypothesis only trivially proves that neural networks are not random. However, only little can be learned about the possible biological function of network motifs from the sole fact that their distribution is not random. As exemplified here for the neural network, the significant deviation from random networks is most likely a simple and straightforward consequence of the (neglected) spatial structure of the system.

Thus, as a general rule, network properties that are found to be significant with respect to simple randomized networks must not necessarily be important with respect to function. Even though the statistical estimation of significance might be technically correct, significance here only implies deviations from randomness—which is often a trivial consequence of the underlying process or system itself. Two important classes of complex networks where the construction of the networks itself implies deviations from randomness, and thus implies highly significant network properties, are discussed below.

3.5.3 Bipartite Graphs

A large number of complex networks in biology and other fields are derived from bipartite graphs. A prominent example is the metabolic network, consisting of two distinct types of vertices: substrates and reactions. Likewise, one of the benchmark examples of complex network analysis, the movie–actor network, originates from a bipartite graph: Based on a public movie database (http://www.imdb.org), two actors are connected by an edge if they have performed in a movie together. The movie–actor network often serves as a prototypical (and computationally much better accessible) example of a social acquaintance network. In particular, the coexistence of high local clustering and small average path length, typical for social networks, has been demonstrated using the movie–actor network as a paradigmatic example [72]. Comparing the observed clustering coefficient of the movie–actor network to Erdös–Rényi networks of the same size, indeed reveals a highly significant clustering that is orders of magnitudes higher than for corresponding random networks.

However, by construction, the movie–actor networks is composed of fully connected subnetworks only: All actors performing in one movie are mutually connected and form a fully connected subgraph. Since an actor participates in several movies, the cliques for each movie are then joined together and organized into the final movie–actor network. Thus, by construction, the underlying bipartite structure clearly distinguishes the network from actual social acquaintance networks and already implies a "significant" local structure.

A similar reasoning holds for metabolic networks: Using the same scheme as previous studies [71], the substrate graph of a metabolic network is constructed by connecting two metabolites if they share a common reaction. The resulting network, analyzed in a number of recent studies [22,33,41,71], shows a significantly higher clustering coefficient than could be expected for random networks with preserved degree distribution. Repeating the analysis for the *S. cerevisiae* substrate graph, already used in the previous sections and depicted in Fig. 3.1, confirms these results. The observed clustering coefficient of the metabolic network significantly deviates from the values obtained for the ensemble of random networks with preserved degree distributions, as shown in Fig. 3.14 (left plot). However, does this result really imply a significant clustering of the network beyond the fact that it was derived from a bipartite graph? To test this assertion, a scheme that generates random instances of the original

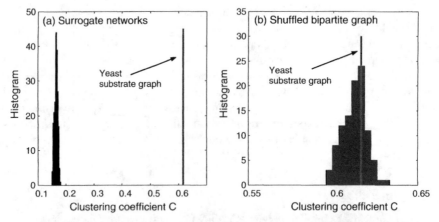

FIGURE 3.14 The clustering coefficient of the *S. cerevisiae* substrate graph, depicted in Fig. 3.1. *Left plot:* The clustering coefficient is compared against randomized networks with preserved degree distribution. The yeast substrate graph shows a highly significant clustering, as compared with randomized networks generated according to the scheme shown in Fig. 3.13. *Left plot:* The clustering coefficient is compared against randomized networks that preserve the original bipartite structure. In this case, no significant deviation between the clustering coefficient of the yeast substrate graph and its shuffled counterparts are detected.

network, but always retains the underlying bipartite structure, was proposed [61]. The randomization scheme is shown in Fig. 3.15. Note that the randomization of the bipartite graph leads to nontrivial modifications when only the projection onto the substrate graph is considered. Again comparing the resulting clustering coefficients of the randomized networks to the clustering coefficient obtained from the yeast metabolic network, reveals no significant deviation. Thus, we have to conclude that the metabolic network of *S. cerevisiae* considered here shows no significant clustering

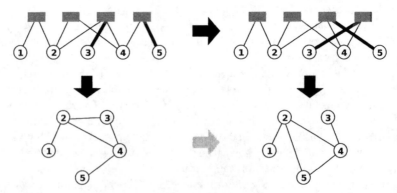

FIGURE 3.15 To preserve the underlying bipartite structure of a network, a modified randomization scheme is proposed. Using the original bipartite graph, two edges are randomly selected and rewired. Note that this leads to nontrivial modifications of the substrate graph projection.

beyond the trivial clustering imposed by the construction rule of the network—namely a projection that was constructed from a bipartite graph.

3.5.4 Correlation Networks

A similar reasoning also holds for *correlation networks*, representing an increasingly popular way to investigate and visualize large-scale empirical data [11,19,42,46,62,67]. A correlation network is constructed by connecting two vertices, such as metabolites or proteins, if their respective concentration is correlated over a suitable set of experiments. That is, the correlation matrix $\mathbf{C}$ is converted into a binary (adjacency) matrix $\mathbf{A}$, with elements $A_{ij} = 1$ whenever the correlation between two elements n_i and n_j exceeds a given threshold, usually defined with respect to a given significance level.

However, networks that are constructed from correlation matrices posses, by construction, a nontrivial structure and are highly clustered. In particular, correlations are transitive, that is, if an entity A correlates with two entities B and C, then B and C are also (likely to be) correlated themselves. Formally, this transitivity can be expressed as an inequality that holds between any triplet of correlation coefficients [61]. To test for the (nontrivial) structure of correlation networks thus implies to generate networks that preserve all features inherent to correlation matrices, but are random otherwise. Specifically, randomized networks with preserved degree distributions, generated according the scheme shown in Fig. 3.13, violate the inherent structure of correlation networks. If recent studies [19] thus report an "unsuspected assortative feature" for biological correlation networks, this is most probably a result of an inappropriate null hypothesis, rather than a distinctive feature of the process in question. However, no satisfying solution to the problem of constructing appropriate null models for correlation networks is known. The interpretation of correlation networks is discussed in more detail in Ref. [62].

These examples serve to illustrate that a proper interpretation of network properties, even when tested for significance against randomized networks, is sometimes far from straightforward. Unfortunately, to what extend the underlying structure of the original network trivially determines its topological features, is often not as obvious as in the case of bipartite graphs or correlation networks.

3.6 SUMMARY

The analysis of complex networks will continue to play a major role in many scientific disciplines. In particular the analysis of large systems, consisting of many interacting elements, often necessitates an abstract representation of the system and its interactions in terms of a complex network. In respect thereof, the theory of complex networks, emanating from the mathematical field of graph theory, has attracted renewed interest over the last decade. While many concepts of complex network theory have their roots in graph theory, novel problems have also triggered the development of a variety of novel concepts, shedding light on the topological organization and design principles of complex networks. In this chapter, we have provided a brief overview over some typical properties and concepts used in the analysis of large complex networks. Though certainly no comprehensive review, we have focused mainly on the best known indices to characterize the topological structure of complex networks, such as distance and average path length, the degree distribution, the clustering coefficient, assortative mixing, and network centralities. Several of the indices and concepts described in this chapter will play an important role in the remainder of this book and are discussed in more detail in later chapters.

However, complex network analysis does not stop at the description and numerical estimation of individual indices characterizing the topological structure. An important aspect of complex network theory is also the development of paradigmatic prototype models that elucidate and explain how topological features arise from the construction rules and design principles of complex network. Of further interest are also global features of networks that result from the local properties of vertices and often relate to the function of a network as a whole. In this respect, we have briefly discussed concepts of network functionality and organization, such as robustness, attack tolerance, modularity, and hierarchies within complex networks. One of the most important aspects of complex network analysis, however, is the statistical testing of network properties. Any interpretation of the functional relevance of topological structures is valid only as far as it has been properly tested and validated against appropriate null models. Not always receiving the attention it deserves, a sound validation of network properties is far from trivial. As demonstrated in this chapter, an inappropriate choice of null models will lead to "significant" results, even when the network property in question is actually a simple and straightforward consequence of the way the network is constructed. In this sense, the great simplification achieved by the representation of a complex system as a network, concomitantly gives rise to a large number of potential pitfalls and misinterpretations in the analysis. One should always keep in mind that complex biological processes, such as protein binding or cellular metabolism, are not actually networks. Rather, they are physical or biological systems that can be represented as complex networks—making biological problems mathematically tractable, but only at the cost of also omitting some aspects of their entireness.

3.7 EXERCISES

1. Name several typical characteristics of vertices and describe how they can be estimated. In particular, give examples for vertex properties that can be estimated using only information about the vertex and its neighbors and properties that relate to the topology of the network as a whole.

2. Give examples of typical degree distributions. How is the degree exponent defined and how can it be estimated from a given empirical network.

3. Why does a negative correlation between vertex degree and clustering coefficient indicate a hierarchical structure?

4. Are functionally related vertices always connected? If not, what are alternative approaches to define "relatedness" of vertices within a complex network?

5. What are "small-world" networks? Is the Erdös–Rényi model a "small-world" network?

6. What is the average clustering coefficient of Erdös–Rényi random networks?

7. Outline how the Barabási–Albert network model is constructed. What are the main differences to Erdös–Rényi and Watts–Strogatz network models?

8. Outline how an ensemble of random networks can be constructed that preserves the degree distribution of a given empirical network.

9. Outline a scheme to randomize a network, such that the three-vertex motif distribution is preserved.

10. What does the term "significant" mean in the context of statistical validation of network properties?

REFERENCES

1. R. Albert and A.-L. Barabási. Statistical mechanics of complex networks. *Reviews of Modern Physics*, 74:47–97, 2002.

2. R. Albert, H. Jeong, and A.-L. Barabási. Error and attack tolerance of complex networks. *Nature*, 406:378–382, 2000.

3. L. A. N. Amaral, A. Scala, M. Barthelemy, and H. E. Stanley. Classes of small world networks. *Proceedings of the National Academy of Science (USA)*, Vol. 97(21), 11149–11152, 2000.

4. M. Arita. Scale-freeness and biological networks. *Journal of Biochemistry*, 138(1):1–4, 2005.

5. Y. Artzy-Randrup, S. J. Fleishman, N. Ben-Tal, and L. Stone. Comment on "network motifs: Simple building blocks of complex networks" and "superfamilies of evolved and designed networks." *Science*, 305(5687):1107c, 2004.

6. J. Bang-Jensen and G. Gutin. *Digraphs: Theory, Algorithms and Applications*. Springer Monographies in Mathematics. Springer-Verlag, London, 2000.

7. A.-L. Barabási and R. Albert. Emergence of scaling in random networks. *Science*, 286:509–512, 1999.

8. A.-L. Barabási and Z. N. Oltvai. Network biology: Understanding the cell's functional organization. *Nature Reviews Genetics*, 5:101–113, 2004.

9. V. Batagelj and A. Mrvar. Pajek—program for large network analysis. *Connections*, 21(2):47–57, 1998.

10. D. S. Callaway, M. E. J. Newman, S. Strogatz, and D. J. Watts. Network robustness and fragility: Percolation on random graphs. *Physics Review Letters*, 85(25):5468–5471, 2000.

11. D. Camacho, A. de la Fuente, and P. Mendes. The origin of correlations in metabolomics data. *Metabolomics*, 1:53–63, 2005.

12. J. M. Carlson and J. Doyle. Highly optimized tolerance: Robustness and design in complex systems. *Phys. Rev. Lett.*, 84(11):2529–2532, 2000.

13. R. Cohen and S. Havlin. Scale-free networks are ultrasmall. *Physics Review Letters*, 90:058701, 2003.

14. L. da F. Costa. The hierarchical backbone of complex networks. *Physics Review Letters*, 93:098702, 2004.

15. J. Davidsen, H. Ebel, and S. Bornholdt. Emergence of a small-world from local interactions: Modeling acquaintance networks. *Physics Review Letters*, 88(12):128701, 2002.

16. D. J. de Solla Price. Network of scientific papers: The pattern of bibliographic references indicates the nature of the scientific research front. *Science*, 149:510–515, 1965.

17. S. N. Dorogovtsev and J. F. F. Mendes. Evolution of networks. *Advances in Physics*, 51(4):1079–1187, 2002.

18. S. N. Dorogovtsev, J. F. F. Mendes, and A. N. Samukhin. Size-dependent degree distribution of a scale-free growing network. *Physical Review E*, 63(6):062101, 2001.

19. V. M. Eguiluz, D. R. Chialvo, G. Cecchi, M. Baliki, and A. V. Apkarian. Scalefree brain functional networks. *Physical Review Letters*, 94:018102, 2005.

20. P. Erdös and A. Rényi. On random graphs. *Publicationes Mathematicae*, 6:290–297, 1959.

21. I. J. Farkas, I. Derényi, A.-L. Barabási, and T. Vicsek. Spectra of 'real-world' graphs: Beyond the semicircle law. *Physical Review E*, 64(2):026704, 2001.

22. D. A. Fell and A. Wagner. The small-world of metabolism. *Nature Biotechnology*, 18:1121–1122, 2000.

23. J. Förster, I. Famili, P. Fu, B. O. Palsson, and J. Nielsen. Genome-scale reconstruction of the *Saccharomyces cerevisiae* metabolic network. *Genome Research*, 13:244–253, 2003.

24. M. Girvan and M. E. J. Newman. Community structure in social and biological networks. *Proceedings of the National Academy of Science USA*, 99(12):7821–7826, 2002.

25. M. L. Goldstein, S. A. Morris, and G. G. Yen. Problems with fitting to the power-law distribution. *The European Physical Journal B*, 41(2):255–258, 2004.

26. M. Gustafsson. *Large-scale topology, stability and biology of gene networks*, volume 1256 of *Linköping Studies in Science and Technology Thesis*. Linköpings university, Norrköpping, 2006.

27. M. Gustafsson, M. Hörnquist, and A. Lombardi. Comparison and validation of community structures in complex networks. *Physica A*, 367:559–576, 2006.

28. H. Herzel. How to quantify 'small-world networks'? *Fractals*, 6(4):301–303, 1998.

29. P. Holme, B. J. Kim, C. N. Yoon, and S. K. Han. Attack vulnerability of complex networks. *Physical Review E*, 65:056109, 2002.

30. P. J. Ingram, M. P. H. Stumpf, and J. Stark. Network motifs: structure does not determine function. *BMC Genomics*, 7:108, 2006.

31. S. Itzkovitz, R. Milo, N. Kashtan, G. Ziv, and U. Alon. Subgraphs in random networks. *Physical Review E*, 68(2):026127, 2003.

32. H. Jeong, S. P. Mason, A.-L. Barabási, and Z. N. Oltvai. Lethality and centrality in protein networks. *Nature (Brief Communications)*, 411:41–42, 2001.

33. H. Jeong, B. Tombor, R. Albert, Z. N. Oltvai, and A.-L. Barabási. The large-scale organization of metabolic network. *Nature*, 407:651–654, 2000.

34. N. Kashtan and U. Alon. Spontaneous evolution of modularity and network motifs. *Proceedings of the National Academy of Science USA*, 102(39):13773–13778, 2005.

35. H. Kitano. Biological robustness. *Nature Review Genetics*, 5:826–837, 2004.

36. K. Klemm and V. M. Eguíluz. Highly clustered scale-free networks. *Physical Review E*, 65:036123, 2002.

37. E. Klipp, W. Liebermeister, and C. Wierling. Inferring dynamic properties of biochemical reaction networks from structural knowledge. *Genome Informatics Series*, 15(1):125–137, 2004.

38. P. L. Krapivsky, S. Redner, and F. Leyvraz. Connectivity of growing random networks. *Physics Review Letters*, 85(21):4629–4632, 2000.

39. T. I. Lee, N. J. Rinaldi, F. Robert, D. T. Odom, Z. Bar-Joseph, G. K. Gerber, N. M. Hannett, C. T. Harbison, C. M. Thompson, I. Simon, J. Zeitlinger, E. G. Jennings, H. L. Murray, D. B. Gordon, B. Ren, J. J. Wyrick, J.-B. Tagne, T. L. Volkert, E. Fraenkel, D. K. Gifford, and R. A. Young. Transcriptional regulatory networks in *Saccharomyces cerevisiae*. *Science*, 298:799–804, 2002.

40. A. Li and S. Horvath. Network neighborhood analysis with the multi-node topological overlap measure. *Bioinformatics*, 23(2):222–231, 2007.

41. H. Ma and A.-P. Zeng. Reconstruction of metabolic networks from genome data and analysis of their global structure for various organisms. *Bioinformatics*, 19(2):270–277, 2003.

42. P. Mendes, D. Camacho, and A. de la Fuente. Modelling and simulations for metabolomics data analysis. *Biochemical Society Transactions*, 33(6):1427–1429, 2005.

43. R. Milo, S. Itzkovitz, N. Kashtan, R. Levitt, S. Shen-Orr, I. Ayzenshtat, M. Sheffer, and U. Alon. Superfamilies of evolved and designed networks. *Science*, 303(5663): 1538–1542, 2004.

44. R. Milo, S. Shen-Orr, S. Itzkovitz, N. Kashtan, D. Chklovskii, and U. Alon. Network motifs: Simple building blocks of complex networks. *Science*, 298:824–827, 2002.

45. M. Mitzenmacher. A brief history of generative models for power law and lognormal distributions. *Internet Mathematics*, 1(2):226–251, 2004.

46. K. Morgenthal, W. Weckwerth, and R. Steuer. Metabolomic networks in plants: Transitions from pattern recognition to biological interpretation. *BioSystems*, 83(2-3):108–117, 2006.

47. M. Morohashi, A. E. Winn, M. T. Borisuk, H. Bolouri, J. Doyle, and H. Kitano. Robustness as a measure of plausibility in models of biochemical networks. *Journals of Theoretical Biology*, 216:19–30, 2002.

48. S. Mossa, M. Barthélémy, H. E. Stanley, and L. A. N. Amaral. Truncation of power law behavior in scale-free network models due to information filtering. *Physics Review Letters*, 88:138701, 2002.

49. M. E. J. Newman. Assortative mixing in networks. *Physics Review Letters*, 89:208701, 2002.

50. M. E. J. Newman. The structure and function of complex networks. *SIAM REVIEW*, 45(2):167–256, 2003.

51. M. E. J. Newman and M. Girvan. Finding and evaluating community structure in networks. *Physical Review E*, 69:026113, 2004.

52. M. E. J. Newman, S. H. Strogatz, and D. J. Watts. Random graphs with arbitrary degree distributions and their applications. *Physical Review E*, 64:026118 [17 pages], 2001.

53. R. Pastor-Satorras, A. Vázquez, and A. Vespignani. Dynamical and correlation properties of the internet. *Physics Review Letters*, 87(25):258701, 2001.

54. L. M. Pecora and T. L. Carroll. Master stability function for synchronized coupled systems. *Physics Review Letters*, 80(10):2109–2112, 1998.

55. R. J. Prill, P. A. Iglesias, and A. Levchenko. Dynamic properties of network motifs contribute to biological network organization. *PLoS Biology*, 3(11):1881–1892, 2005.

56. A. Ramezanpour, V. Karimipour, and A. Mashaghi. Generating correlated networks from uncorrelated ones. *Physical Review E*, 67:046107, 2003.

57. E. Ravasz, A. L. Somera, D. A. Mongru, Z. N. Oltvai, and A.-L. Barabási. Hierachical organization of modularity in metabolic networks. *Science*, 297:1551–1555, 2002.

58. E. Ravasz and A.-L. Barabási. Hierarchical organization in complex networks. *Physical Review E*, 67(2):026112, 2003.

59. S. S. Shen-Orr, R. Milo, S. Mangan, and U. Alon. Network motifs in the transcriptional regulation network of *Escherichia coli*. *Nature Genetics*, 31:64–68, 2002.

60. J. Stelling, U. Sauer, Z. Szallasi, F. J. Doyle III, and J. Doyle. Robustness of cellular function. *Cell*, 118:675–685, 2004.

61. R. Steuer. Nonlinear dynamics and molecular biology: From gene expression to metabolic networks. Ph.D. thesis, University Potsdam, 2005.

62. R. Steuer. On the analysis and interpretation of correlations in metabolomic data. *Briefings in Bioinformatics*, 7(2):151–158, 2006.

63. R. Steuer. Computational approaches to the topology, stability and dynamics of metabolic networks. *Phytochemistry*, 68(16–18), pp. 2139–2151, 2007.

64. R. Steuer, C. O. Daub, J. Selbig, and J. Kurths. Measuring distances between variables by mutual information. In D. Baier and K.-D. Wernecke, editors, *Innovations in Classification, Data Science, and Information Systems*, volume Proc. 27th Annual GfKl Conference, pages 81–90. Springer-Verlag, Berlin, 2004.

65. R. Steuer, T. Gross, J. Selbig, and B. Blasius. Structural kinetic modeling of metabolic networks. *Proceedings of the National Academy of Science USA*, 103(32):11868–11873, 2006.

66. R. Steuer, J. Kurths, C. O. Daub, J. Weise, and J. Selbig. The mutual information: Detecting and evaluating dependencies between variables. *Bioinformatics*, 18(Suppl. 2):S231–S240, 2002.

67. R. Steuer, J. Kurths, O. Fiehn, and W. Weckwerth. Observing and interpreting correlations in metabolomic networks. *Bioinformatics*, 19(8):1019–1026, 2003.

68. R. Steuer, A. Nunes-Nesi, A. R. Fernie, T. Gross, B. Blasius, and J. Selbig. From structure to dynamics of metabolic pathways: Application to the plant mitochondrial TCA cycle. *Bioinformatics*, 23(1), pp. 1378–1385, 2007.

69. L. J. Sweetlove and A. R. Fernie. Regulation of metabolic networks: Understanding metabolic complexity in the systems biology era. *New Phytologist*, 168:9–24, 2005.

70. A. Trusina, S. Maslov, P. Minnhagen, and K. Sneppen. Hierarchy measures in complex networks. *Physics Review Letters*, 92(17):178702, 2004.

71. A. Wagner and D. A. Fell. The small-world inside large metabolic networks. *Proceedings of the Royal Society of London B.*, 268:1803–1810, 2001.

72. D. J. Watts and S. H. Strogatz. Collective dynamics of 'small-world' networks. *Nature*, 393:440–442, 1998.

73. R. Xulvi-Brunet and I. M. Sokolov. Reshuffling scale-free networks: from random to assortative. *Physical Review E*, 70:066102, 2004.

74. C. Zhou, A. E. Motter, and J. Kurths. Universality in the synchronization of weighted random networks. *Physics Review Letters*, 96:034101, 2006.

75. H. Zhou. Network landscape from a Brownian particle's perspective. *Physical Review E*, 67:041908, 2003.

76. E. Ziv, M. Middendorf, and C. H. Wiggins. Information-theoretic approach to network modularity. *Physical Review E*, 71:04611, 2005.

4

NETWORK CENTRALITIES

Dirk Koschützki

4.1 INTRODUCTION

In the social sciences it is a common task to determine which individuals of a group of people are more influential than others. To perform such an analysis the communication between these people is modeled as a network consisting of vertices, one for each person, and edges, modeling the communication between them. Based on the structure of such a network, it is possible to compute which individuals are more influential than others. This information is used, for example, for marketing purposes or to find a potential bottleneck within the communication structure of a team working together on a large project [42,46].

Similar methods can be applied to analyze biological networks. Take, for example, a protein interaction network for an organism under analysis. The task given is to order the proteins such that the most important proteins can be used first in an experiment. For gene regulation networks, a similar question can be stated: which gene regulates many other genes and can therefore be considered as the global regulator for the organism under analysis? The creation of a list of persons that have to be vaccinated first is also a good example of a combination of medicine and sociology where this method, centrality analysis, is applied [11].

The general question of which network elements are the most important ones cannot be answered unambiguously. Take, for example, the network in Fig. 4.1a. In this network three groups of vertices exist: the vertex marked A, which connects the two halves of the network, the two vertices marked B1 and B2, which have

Analysis of Biological Networks, Edited by Björn H. Junker and Falk Schreiber
Copyright © 2008 John Wiley & Sons, Inc.

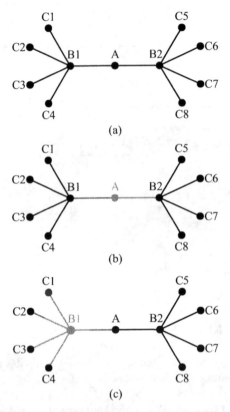

FIGURE 4.1   (*a*) A small network to explain the motivation for different centrality measures. Which of the vertices is more important, the one in the middle marked A or the vertices marked B1/B2 that connect many of the vertices marked C1–8? (*b*) The same network as above with the vertex marked A and the corresponding edges hidden. Clearly, the network without A is separated into two components. Is therefore A the most important vertex? (*c*) The network with the vertex marked B1 hidden. In this case the network breaks into several components. Is one of the B vertices more important than the A vertex?

many connections to other vertices, and the vertices marked C1 to C8, which can be considered as the border of the network. Depending on the context, one of the vertices is more important than the others. If the vertex marked A is removed, then the network is separated into two components, see Fig. 4.1b. On the contrary, if the vertex marked B1 is removed, then the network is separated into many smaller components, and communication between many of the remaining vertices is not possible, see Fig. 4.1c. Therefore the importance of one vertex over others is dependant on the specific question.

In this chapter different concepts and algorithms for the determination of the importance of vertices are described. In total seven different methods for the *centrality analysis* of a network are introduced. At first a definition for the general concept of a centrality is given, then the trivial centrality degree and centralities based on shortest paths are explained, afterwards centralities based on the concept of feedback are introduced and finally a short overview of available tools for the centrality analysis of (biological) networks is provided.

4.2 CENTRALITY DEFINITION AND FUNDAMENTAL PROPERTIES

Ranking of objects is usually based on numerical values. Take, for example, the result of the annual season of your favorite ball game, for example, water polo. By the end of the season a table showing the points, awarded for winning a game, is available. From these points a clear ranking, in the sense which team played best is established.

The same concept can be applied to rank vertices of (biological) networks. A numerical value assigned to every vertex allows their ranking. Table 4.1 shows some values assigned to every vertex of the network in Fig. 4.1a. According to these values the vertices are ranked in the order B1 and B2 on the first position, then A and finally all vertices marked C1–8.

A function that assigns a numerical value to each vertex of a network is called a *centrality*.[1] There exist many different concepts for computing such a centrality [31] and some of them will be introduced in this chapter. All centrality measures introduced will fulfill the following definition:

Definition 4.1 (centrality)   Let $G = (V, E)$ be a directed or undirected graph. A function $C: V \mapsto \mathbb{R}$ is called a centrality.

Centralities assign every vertex a real number. They allow a pairwise comparison of the vertices, and a vertex v_1 is said to be more central or more important than a vertex v_2 if $C(v_1) > C(v_2)$. A more formalized definition for centrality exists (see Ref. [31]) but is not required for the understanding of the general concepts presented in this chapter.

Two things must be considered during centrality analysis of a network: (a) centrality values are comparable inside a specific network only and (b) some centrality measures can be applied to networks that are connected only.

TABLE 4.1   The Degree Centrality Values for the Vertices in Fig. 4.1a

Vertices	Value
A	2
B1–2	5
C1–8	1

---

[1]Centrality, centrality measure, and centrality index are used synonymously in the following.

FIGURE 4.2   A network to explain the comparison of centrality values.

4.2.1   Comparison of Centrality Values

For the same network, centrality values of two different centrality measures are incomparable. Take, for example, the network in Fig. 4.2 and the corresponding centrality values for three different centralities in Table 4.2. The vertices A and B have a centrality value of 1.5 for the centrality shortest path (SP) betweenness. According to this value both vertices are ranked top under this centrality. In contrast, the vertices C, D, and E all have a centrality value of 2 for the degree centrality. Even if this value is higher than all values of the shortest path betweenness values, these vertices are ranked into the second group by the degree centrality. Under the eccentricity centrality the five vertices all receive the same centrality value of 0.5 and are therefore indistinguishable. Therefore centrality values of different centralities are (in general) not comparable even when applied to the same network.

In the same way centrality values for the same centrality measures are incomparable between different networks. Take, for example, the network in Fig. 4.1a and the corresponding values in Table 4.1, and the network in Fig. 4.2 and the values in the degree column in Table 4.2. The vertices B1/B2 in the first figure have a (degree) value of 5 and are the top ranked ones. In the second figure the top ranked vertices have a degree value of 3. Therefore it is impossible to infer the rank of a vertex from its centrality value alone.

Comparable values of centralities can be derived via normalization, which is not covered in this chapter, but described in Ref. [32].

4.2.2   Disconnected Networks

Some of the centrality measures described in the following require that the network to be analyzed is (strongly) connected (see Section 2.2). This applies, for example, to the closeness centrality as the length of a shortest path between two vertices that

TABLE 4.2   Centrality Values for Three Different Centralities for the Network Shown in Fig. 4.2

Vertex	Degree	Eccentricity	SP betweenness
A	3	0.5	1.5
B	3	0.5	1.5
C	2	0.5	0.5
D	2	0.5	0.5
E	2	0.5	0.0

are unconnected is defined as infinity. One method to overcome this restriction is the reduction of the network to (strongly) connected components (see Section 2.2). Each component is then analyzed as a separate network, and the resulting centrality values for the vertices of the components must be interpreted in the context of the component and not in the context of the whole network.

4.3   DEGREE AND SHORTEST PATH-BASED CENTRALITIES

In this section, four centralities that have already been applied to the analysis of biological networks are described. The first centrality, degree, is almost trivial: It counts the number of edges attached to a vertex. The other three centralities use information about shortest paths between vertices of the network. All centralities are defined for undirected and unweighted networks in this section. The necessary extensions for directed and/or weighted networks are mentioned in the algorithm section (Section 4.3.5) near the end of this section.

4.3.1   Degree Centrality

The result of an election can be modeled as a directed network: Every person partic-ipating in the election is modeled as a vertex and votes are shown as arrows between the voter and the elected person, pointing toward the elected person. Clearly the per-son with the most votes wins the election and ranks that follow the top rank are interpreted as substitutes. Figure 4.3 and Table 4.3 show the result of a hypothetical election. Clearly Sepp wins the election and Jan and Klaus are two substitutes. The vote is modeled as a directed network as, for example, the vote from Pit to Klaus is not returned by Klaus.

Counting the number of edges connected to a vertex leads to the first centrality:

Definition 4.2   (degree centrality)   Let $G = (V, E)$ be an undirected graph. The degree centrality is defined as:

$$\mathcal{C}_{\deg}(v) := |\{e \mid e \in E \wedge v \in e\}|$$

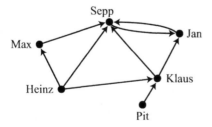

FIGURE 4.3   A network modeling an election.

TABLE 4.3   Result of the Election in Fig. 4.3

Person	Number of votes received
Sepp	4
Jan	2
Klaus	2
Max	1
Heinz	0
Pit	0

For directed networks, two degree centralities, the in-degree centrality and the out-degree centrality, exist. Degree centrality is a local centrality measure: Only the immediate neighborhood of the vertex of interest is considered. Degree can be computed for all kinds of networks. The network to be analyzed does not have to fulfill stricter requirements like connectedness.

For the analysis of biological networks degree centrality has been applied in numerous situations.

The correspondence of a high centrality value for a protein in the protein interaction network and the effect of the knockout of this protein for the organism have been discussed several times. For *Saccharomyces cerevisiae*, it has been shown that proteins with a high degree centrality value are more likely to be essential for the organism than proteins with a lower degree value [23].

A larger study compared three centralities (degree, closeness, and betweenness) for the identification of essential proteins in three different organisms: *Saccharomyces cerevisiae*, *Caenorhabditis elegans*, and *Drosophila melanogaster*. In all three networks and for all three centralities it was shown that the mean centrality value for essential proteins is significantly higher than the centrality value of nonessential proteins [20].

An even larger study was performed for the protein interaction network of *S. cerevisiae*. In this study six centralities (degree, closeness, betweenness, eigenvector centrality, information centrality, and subgraph centrality) were applied. Again all centralities performed better in the identification of essential proteins than a random selection strategy [12].

Similar results were received for gene coexpression networks for *S. cerevisiae*, *E. coli*, and *C. elegans*. Genes with a high degree in the coexpression network of these organisms are more likely to be essential [3].

The ranking of the metabolites of the intermediate metabolism for energy generation and small building block synthesis of *E. coli* based on the degree and the closeness centrality (called importance number in the publication) leads to comparable results. For both centralities metabolites from the tricarboxylic acid cycle are highly overrepresented in the list of top ranked metabolites. NAD, ATP, and their derivatives were removed from the ranking, otherwise these metabolites ranked the highest [45].

In a comparison with two types of networks, a random small-world network and an Erdös–Rényi random network, centralities were benchmarked to identify high-risk individuals for an infection that is transferred by contact between individuals. The hypothesis that a higher centrality value indicates a higher risk and an earlier time of infection was confirmed. All compared centralities (degree, closeness, shortest path betweenness, and random walk betweenness) correlated and degree performed as well as the other centralities [11].

4.3.2 Eccentricity Centrality

The following three centralities use information about the length of the shortest paths within a network. They assume that the network models communication between objects, which are represented by vertices. It is hypothesized that only shortest paths are used for the transmission of a message or that shortest paths communication is at least a plausible model for communication in the network.

The first of the three centralities can be easily explained by an example. A map of a city is given, roads are modeled as edges, and vertices represent potential places for a hospital to be constructed within this city. The position for the hospital should be chosen such that it is reachable from all other places with the least moves (measured by the shortest path distance) possible. In Fig. 4.4, the vertex marked 4 is therefore the best position for the hospital, as a maximum of three moves are necessary to reach the hospital from any other place.

This idea defines a measure of centrality. For every vertex compute the maximum distance, which is the length of the longest shortest path to all other vertices (see Section 2.2). This gives a value for every vertex and the vertices that are reachable in less moves receive a low value. As the Definition 1 requires a high centrality value for central vertices, the reciprocal of the computed value is used as the centrality value.

Definition 4.3 (eccentricity centrality)   Let $G = (V, E)$ be an undirected and connected graph. The eccentricity centrality [19] is defined as

$$C_{ecc}(s) := \frac{1}{\max\{dist(s, t): t \in V\}}$$

where $dist\,(s,t)$ denotes the distance between the vertices s and t, that is, the length of a shortest path between s and t.

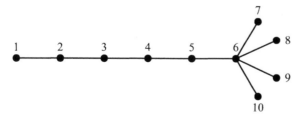

FIGURE 4.4   A network to explain the three shortest path-based centralities.

TABLE 4.4   Centrality Values for the Three Shortest Path-Based Centralities for the Network in Fig. 4.4

Vertex	C_{ecc}	C_{clo}	C_{spb}
1	0.167	0.026	0
2	0.200	0.032	8
3	0.250	0.040	14
4	0.333	0.048	18
5	0.250	0.053	20
6	0.200	0.053	26
7	0.167	0.037	0
8	0.167	0.037	0
9	0.167	0.037	0
10	0.167	0.037	0

Eccentricity uses information about the length of shortest paths between any two vertices of a network. In networks that are not connected, the distance *dist* between vertices that are not connected is defined as infinity, see Section 2.2. In this case, the value for the eccentricity centrality is the same for every vertex ($C_{ecc}(v) = 1/\infty$ for all $v \in V$) as the maximum distance in an unconnected network is infinity for every vertex. Therefore eccentricity centrality can be computed for connected networks only. In the case of directed networks even strongly connected is required.

Table 4.4 shows the eccentricity centrality values for the network in Fig. 4.4. The vertex named 4 is the most central one, all other vertices receive a lower centrality value.

Eccentricity centrality together with the centralities closeness and centroid value (not covered here) were used for the analysis of three networks of *E. coli* and *S. cerevisiae*. For the metabolic network of *E. coli*, all three centralities produce a very similar ranking for the top ranked metabolites and these rankings coincide well with a ranking based on the degree centrality. In contrast to other studies, the eccentricity centrality applied to the protein interaction network of *S. cerevisiae* was not able to distinguish essential from nonessential proteins. For the network of cooccurring protein domains in *S. cerevisiae*, the centrality centroid value ranked protein domains that take part in cell–cell contacts and signal transduction highest [47].

4.3.3   Closeness Centrality

The closeness centrality can be explained in the same context as the eccentricity centrality. Instead of a hospital a shopping mall has to be placed onto the map. For a shopping mall the constraint is that most customers can reach it comfortably. Therefore it is placed at a point where the shortest path distances for all vertices to the position of the mall is minimized. In Fig. 4.4, the vertices marked 5 and 6 are both the best positions for the shopping mall as at both positions the sum of all distances for all customers is minimized.

Following the same argument as for the eccentricity centrality (Definition 4.3), the closeness centrality is defined as the reciprocal of the sum of all pairwise distances within the network.

Definition 4.4   (closeness centrality)   Let $G = (V, E)$ be an undirected and connected network. The closeness centrality [41] is defined as

$$C_{clo}(s) := \frac{1}{\sum_{t \in V} dist(s, t)}$$

where $dist(s, t)$ denotes the distance between the vertices s and t, that is, the length of a shortest path between s and t.

The closeness centrality, as the eccentricity centrality, uses the length of the shortest paths between all pairwise vertices. Therefore, closeness centrality has to follow the same constraints as the eccentricity centrality: The network to be analyzed has to be (strongly) connected.

Table 4.4 shows the closeness centrality values for the network in Fig. 4.4. According to this centrality, the two vertices 5 and 6 are equally important in this network.

According to a slight modification of the closeness centrality, the top 8 of the top 10 metabolites of the metabolic network of E. coli are part of the glycolysis and citrate acid cycle pathways [35].

4.3.4   Shortest Path Betweenness Centrality

The previous two centralities measure how good a message originating from a single vertex can reach other vertices via shortest paths. The first, eccentricity centrality, focused on a single communication and the second, closeness centrality, considers communication to all other vertices. In contrast to these the centrality now introduced measures the ability of a vertex to monitor communication between other vertices.

Every vertex that is part of a shortest path between two other vertices can monitor communication between them. Counting how many communications a vertex may monitor leads to an intuitive definition of a centrality: A vertex is central if it can monitor many communications between other vertices.

In the following, let σ_{st} denote the number of shortest paths[2] between two vertices s and t and let $\sigma_{st}(v)$ denote the number of shortest paths between s and t that use v as an interior vertex.[3] The rate of communication between s and t that can be monitored by an interior vertex v is denoted by $\delta_{st}(v) := \sigma_{st}(v)/\sigma_{st}$. If no shortest path between s and t exists ($\sigma_{st} = 0$), then we set $\delta_{st}(v) := 0$.

With these definitions the shortest path betweenness centrality[4] can be defined.

---

[2]Between two vertices more than one shortest path may exist.
[3]A vertex that is not the start or the end vertex of the path.
[4]Previously called betweenness centrality, without the prefix shortest path. This name is not precise enough as other betweenness measures exist [7,16,37].

Definition 4.5 (shortest path betweenness centrality)   Let $G = (V, E)$ be an undirected network. The shortest path betweenness centrality [1,15] is defined as

$$C_{spb}(v) := \sum_{s \in V \wedge s \neq v} \sum_{t \in V \wedge t \neq v} \delta_{st}(v)$$

In contrast to the other two centralities that use information about shortest paths, the shortest path betweenness centrality does not require the network to be connected. If two vertices s and t are not connected, then the corresponding value $\delta_{st}(v)$ is 0 by definition. Therefore unconnected pairs of vertices do not change the centrality values of other vertices.

Table 4.4 shows the shortest path betweenness centrality values for the network in Fig. 4.4. Vertex 6 is the most important vertex according to this centrality. Vertices at the border of the network receive a centrality value of zero as they do not participate as interior vertices in any shortest path communication between other vertices.

In the *S. cerevisiae* protein interaction network, it was reported that proteins with a high betweenness centrality value cover a broad range of degree centrality values. In particular, proteins with a high betweenness and low degree value (HBLC, high betweenness low connectivity proteins) are prominent as they are supposed to support modularization of the network [25].

Betweenness centrality was applied to mammalian transcriptional regulatory networks and it was noted that betweenness appears to be the most representative topological characteristic in regard to the biological significance of distinct elements [39].

4.3.5 Algorithms

The level of difficulty of the algorithms for computing the four presented centralities ranges from nearly trivial, in the case of the degree centrality, to more advanced, in the case of the shortest path betweenness. As for the definition of the centralities above, algorithms for only undirected and unweighted networks are given in this section. The extension toward weighted networks is either simple, for example, for the degree centrality, or requires the use of the Dijkstra algorithm [27] instead of the breadth first search algorithm (BFS, see Section 2.5) to compute the shortest paths between two vertices. Directed networks require a modification of the algorithms such that the direction of the edge is taken into account.

Degree centrality is computed for an arbitrary network with the following (trivial) algorithm:

```
degree_centrality_algorithm (network G = (V, E))
    for each vertex v ∈ V
    C_deg(v) := 0;
    for each edge {v, w} ∈ E {
    C_deg(v) := C_deg(v) + 1;
    C_deg(w) := C_deg(w) + 1;
    }
    return (C_deg);
```

This algorithm computes the degree centrality in time $\mathcal{O}(n + m)$ as we have to initialize the resulting centrality vector to 0 for every vertex ($\mathcal{O}(n)$) and then iterate over all edges ($\mathcal{O}(m)$) to increase the centrality value of each of the two connected vertices of the edge.

Eccentricity and closeness centrality are defined very similarly. They differ only in the operation performed with the pairwise distances of the vertices. Therefore in the following the algorithm for only eccentricity centrality is discussed, the modifications for closeness centrality have been left as an exercise.

Eccentricity centrality needs the pairwise distances between all vertices. They are usually computed with the breadth first search (BFS) algorithm (see Section 2.5). Using an existing implementation of this algorithm to compute the distance matrix, a naïve algorithm for the eccentricity centrality is

naïve_eccentricity_centrality_algorithm (network $G = (V, E)$)
    for each vertex $v \in V$
    $C_{ecc}(v) := 0$;
    dist := compute_distance_matrix (G);
    for each vertex $s \in V$
      for each vertex $t \in V$
        if $C_{ecc}(s) < dist(s, t)$ then
        $C_{ecc}(s) := dist(s, t)$;
    $C_{ecc}(s) := \frac{1}{C_{ecc}(s)}$;
    return (C_{ecc});

This algorithm computes the eccentricity centrality in time $\mathcal{O}(n)$ for the initialization, $\mathcal{O}(nm)$ for the computation of the distance matrix via BFS in the unweighted case, and $\mathcal{O}(n^2)$ for the computation of the maximal and reciprocal value of each vertex. This results in a time complexity of $\mathcal{O}(n^2)$ for this naïve eccentricity centrality algorithm.

A different algorithm, based on a modification of the breadth first search algorithm, has complexity $\mathcal{O}(n * (n + m))$ for unweighted networks:

bfs_eccentricity_centrality_algorithm (network $G = (V, E)$)
    for each vertex $s \in V$
    $C_{ecc}(s) := $ single_vertex_bfs_eccentricity (s, G);

single_vertex_bfs_eccentricity_algorithm (vertex s, network $G = (V, E)$)
    result := 0;
    for each vertex $v \in V$
      distance $(v) := -1$;
    distance $(s) := 0$;
    list $L = [s]$;
    while L is nonempty {

```
remove vertex w from the front of list L;
for each neighbor x of w {
  if (distance (x) == -1) {
    distance (x) := distance (w) + 1;
    if (result < distance (x))
      result := distance (x);
    add x to the end of list L
  }
}
}
return 1/result;
```

The complexity of this algorithm is $\mathcal{O}(n * (n + m))$ as the algorithm for computing a centrality value for a single vertex is called for each vertex (n times); the BFS has complexity $\mathcal{O}(m)$ and the initialization of the distance vector needs time proportional to $\mathcal{O}(n)$.

A naïve algorithm for the shortest path betweenness centrality has complexity $\mathcal{O}(n^3)$ as the summation over both s and t runs over the set of vertices twice and this operation has to be performed for every vertex v. Therefore the complexity is $\mathcal{O}(n^3)$ even without considering the complexity of computing σ_{st} and $\sigma_{st}(v)$.

A much more efficient algorithm than the naïve one exists. The algorithm by Ulrik Brandes [5] has a running time of $\mathcal{O}(nm)$ for unweighted networks and $\mathcal{O}(nm + n^2 \log n)$ for weighted networks. Several implementations in different programming languages are available for this algorithm [26,44]. Additionally, this algorithm can be augmented to compute eccentricity and closeness centrality in the same run; therefore all three shortest path based centralities can be computed in a single run of the extended algorithm. In this chapter, neither the naïve nor the Brandes algorithm are given as the first one is trivial and the second one is extensively discussed in several places [5,22].

4.3.6   Example

Based on the distance matrix in Table 4.5, the shortest path based centralities for the network in Fig. 4.4 can be computed.

The maximum distance from vertex 4 to all the other vertices is 3, therefore the eccentricity centrality value of vertex 4 is 1/3. The closeness centrality value for the same vertex is the reciprocal of $3 + 2 + 1 + 0 + 1 + 2 + 3 + 3 + 3 + 3 = 21$, therefore $1/21 \approx 0.048$.

Computing the shortest path betweenness requires the values of σ_{st} and $\sigma_{st}(v)$ for all s, t, v combination of vertices. In this example, the value of σ_{st} is 1 for every combination of s and t as there exists only one shortest path from one vertex to any other vertex. The vertex 2 is an interior vertex for all shortest paths that start at vertex 1 and end at the vertices 3, 4, 5, 6, 7, 8, 9, and 10. As there are eight different shortest paths that use vertex 2 as the interior vertex, the shortest path betweenness value for vertex 2 is 8/1.

TABLE 4.5   Shortest Path Distances Between All Vertices for the Network Shown in Fig. 4.4

	1	2	3	4	5	6	7	8	9	10
1	0	1	2	3	4	5	6	6	6	6
2	1	0	1	2	3	4	5	5	5	5
3	2	1	0	1	2	3	4	4	4	4
4	3	2	1	0	1	2	3	3	3	3
5	4	3	2	1	0	1	2	2	2	2
6	5	4	3	2	1	0	1	1	1	1
7	6	5	4	3	2	1	0	2	2	2
8	6	5	4	3	2	1	2	0	2	2
9	6	5	4	3	2	1	2	2	0	2
10	6	5	4	3	2	1	2	2	2	0

4.4   FEEDBACK-BASED CENTRALITIES

The centralities described in the previous section were based on degree and shortest path information. In this section, three centralities are described that use feedback as the underlying concept. Two of them have already been applied to the analysis of biological networks.

In the basic definition (Definition 4.1), a centrality is a function from the set of vertices to the set of the reals. In this section, we will often give a centrality as a vector. Therefore we have to enumerate the set of vertices accordingly.

4.4.1   Katz's Status Index

The status index introduced by Leo Katz can be interpreted as a generalization of the degree centrality (Definition 4.2): A vertex is of high importance if many other vertices choose this vertex. For the status index, instead of only counting direct votes, even indirect ones are considered. It is assumed that both direct and indirect votes raise the importance of a vertex and that the effect of indirect votes decreases with the length of the voting chain.

The status index is defined via a power series:

Definition 4.6   (Katz's status index)   Let $G = (V, E)$ be a directed or undirected and loop-free network and let A be the adjacency matrix (see Section 2.4) of G. The Katz's status index [29] is defined as

$$C_{\text{Katz}}(i) := \sum_{k=1}^{\infty} \sum_{j=1}^{n} \alpha^k (A^k)_{ji}$$

Here A^k denotes the k-times multiplication of the adjacency matrix A of the network G with itself (the kth power of A). α is a positive scaling factor that has to be chosen such that α is smaller than the reciprocal of the absolute value of the largest

TABLE 4.6 Katz Status Index for the Election Network in Fig. 4.3 for Two Different Damping Factors α

Person	$\alpha = 0.1$	$\alpha = 0.9$
Sepp	0.457	47.937
Jan	0.266	46.563
Klaus	0.2	1.8
Max	0.1	0.9
Heinz	0.0	0.0
Pit	0.0	0.0

eigenvalue of the adjacency matrix A ($0 \le \alpha < 1/|\lambda_{max}|$). In this case the following formula using matrix inversion can be used to compute the status index:

$$\vec{C}_{Katz} = \left((I - \alpha A^T)^{-1} - I \right) \vec{1}$$

Here I is the matrix of size $n \times n$ consisting of ones on the diagonal and zeros otherwise, $\vec{1}$ is a vector of size n consisting of ones, A^T denotes the transposed matrix of A and $^{-1}$ denotes matrix inversion.

The damping factor α can be chosen in the range 0 to $1/|\lambda_{max}|$. A damping factor near zero reduces the influence of longer chains of votes, therefore a damping factor of zero results in a centrality that is equivalent to the degree centrality.

To apply the status index to the network modeling an election in Fig. 4.3 the damping factor has to be chosen accordingly. The largest eigenvector in absolute value of the adjacency matrix for this network is 1, therefore α might be selected in the range [0, 1[. Table 4.6 shows the results for two different values of α: 0.1 and 0.9. Compared with the degree-based ranking (see Table 4.3), the status index is able to distinguish between Jan and Klaus: Both received two direct votes, but Jan receives a vote from Sepp who received many direct and indirect votes himself.

Currently no application of the status index to the analysis of biological networks is known.

4.4.2 Bonacich's Eigenvector Centrality

Another idea for centrality was presented by Phillip Bonacich [4]. He proposed that the centrality value of a vertex is directly dependent on the centrality values of its connected neighbors: A high centrality value of the neighbors should result in a high centrality for the vertex under consideration.

This idea can be formalized into a linear equation system:

$$C(v_1) = a_{11}C(v_1) + \cdots + a_{1n}C(v_n)$$

$$\vdots$$

$$C(v_n) = a_{n1}C(v_1) + \cdots + a_{nn}C(v_n)$$

Here the a_{ij} denote the corresponding entry in the adjacency matrix A (see Section 2.4) for the network under analysis. If no edge between the vertices i and j exists, this entry is 0 and therefore the corresponding term is eliminated by the multiplication with $\mathcal{C}(v)$.

In matrix notation this equation system is written as $\vec{\mathcal{C}} = A\vec{\mathcal{C}}$. For mathematical reasons only a specific form of this equation system using eigenvectors and eigenvalues can be solved in general and for that the network even has to be connected [14,18]. As there exists n eigenvectors for a matrix A, we select the eigenvector corresponding to the largest eigenvalue in absolute value.

Definition 4.7 (eigenvector centrality)   Let $G = (V, E)$ be a strongly connected directed or a connected undirected network and A the adjacency matrix for G. The eigenvector centrality [4] is the eigenvector $\mathcal{C}$eiv of the largest eigenvalue $\lambda_{\max}$ in absolute value of the following equation system:

$$\lambda\vec{\mathcal{C}}_{\text{eiv}} = A\vec{\mathcal{C}}_{\text{eiv}}$$

Eigenvector centrality was evaluated as one of the six centralities in the comparison of centralities for determining essential proteins in the protein interaction network of *S. cerevisiae* [12]. Eigenvector centrality was slightly outperformed by the subgraph centrality, a centrality that uses similar concepts from spectral graph theory. Subgraph centrality [13] is not covered in this chapter.

4.4.3 PageRank

The third centrality explained in this section is related to ranking web pages. Web pages can be interpreted as vertices of a network (sometimes called the web graph). Links connecting these pages are (directed) edges within this network. Several algorithms for the ranking of search results do exist and some of them are clearly centralities [30,34,38]. In the following, we will describe the most popular representative of the centralities related to the web graph: PageRank, the centrality underlying the Google search engine.

PageRank models the behavior of a visitor of web pages (surfer) and tries to establish a ranking of the pages based on his behavior. As the true behavior of a surfer is unknown, an idealized model of a random surfer is used. A random surfer starts at a random page and switches from page to page via the outgoing links. At each page he selects a link uniformly at random, jumps to the selected page, and starts his selection again. Additionally, he might jump to any other page (by entering the address of the web page). After many iterations a probability for hitting each page is computed. The higher this probability is, the more often the random surfer visited this page and the more interesting this page is.

The formal definition of the PageRank centrality is not presented here. Langville and Meyer review PageRank and two other algorithms (HITS, SALSA) to rank web pages in detail [33].

A modification of the original PageRank algorithm, called GeneRank, was recently described [36]. This modification integrates preexisting expression information about genes into the ranking process.

4.5 TOOLS

For the analysis of biological networks many tools with different focuses are available. Cytoscape [43], Osprey [9], and VisANT [21] are the three that are often cited. Currently, none of these three support the centrality analysis of biological networks. As all of them have an active development community and Cytoscape and VisANT even provide a plug-in mechanism for the easy integration of extensions, this might change in the future.

Besides specific tools for the analysis of biological networks, a large number of tools for the analysis of networks exist. One of them, Pajek [2], is already heavily used for the analysis and visualization of biological networks.

The libraries JUNG [26] and Boost Graph Library [44] provide implementations of some algorithms for the computation of centralities in Java and C++, respectively. Within the Bioconductor project [17], several packages implementing graphs and graph based algorithms do exist. The package RBGL [10] integrates centrality algorithms from the Boost Graph Library into the programming language R [40].

Two tools that allow the computation of a large number of different centralities are Visone [8] and CentiBiN [28]. Visone is a system for the visual exploration of social networks. Version 1.1 of Visone supports the computation of 10 different centrality measures and allows the direct visualization of the network together with the centrality values. CentiBiN is a tool focusing on the computation and exploration of centralities in biological networks. It computes more than 15 different centrality measures for directed and undirected networks. CentiBiN supports the visualization of the centrality distributions for all computed centralities and the visualization of the network with different layout methods.

4.6 SUMMARY

In this chapter, different concepts for the centrality analysis of networks have been described. Besides the trivial degree centrality three shortest path based centralities were introduced; some applications of these centralities for the analysis of biological networks were summarized and the algorithmic foundations of these centralities were presented. Three centralities based on the concept of feedback were briefly introduced. Finally, a short overview of available tools for the computation of centralities for biological and nonbiological networks has been given.

Besides the centrality measures discussed in this chapter numerous others do exist. A more complete coverage of these including random walk-based centralities, centralities based on maximum flow, and several centralities used for the web graph are summarized in a recent review [22,31].

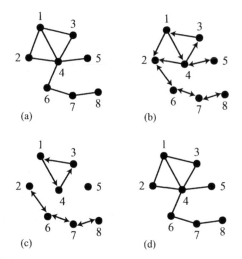

FIGURE 4.5 Four small networks for the exercises.

4.7 EXERCISES

1. Please *name* the centralities that can be computed for the four networks in Fig. 4.5.

2. Please *compute* the possible centralities for the four networks in Fig. 4.5. Choose sensible values for required parameters.

3. Explain why the centralities eccentricity and closeness cannot be computed for some of the networks in Fig. 4.5. Is it possible to compute shortest path betweenness for these networks?

4. Modify both algorithms for eccentricity such that closeness centrality is computed.

5. Compute the centralities Katz status index, eigenvector centrality, and Page-Rank with CentiBiN or Visone for the networks in Fig. 4.5.

6. Search the literature for at least three more definitions of centralities.

REFERENCES

1. J. M. Anthonisse. The rush in a directed graph. Technical Report BN 9/71, Stichting Mathematisch Centrum, 1971.

2. V. Batagelj and A. Mrvar. Pajek - Analysis and visualization of large networks. In Jünger and Mutzel [24], pages 77–103.

3. S. Bergmann, J. Ihmels, and N. Barkai. Similarities and differences in genome-wide expression data of six organisms. *PLoS Biology*, 2(1):85–93, 2004.

4. P. Bonacich. Factoring and weighting approaches to status scores and clique identification. *Journal of Mathematical Sociology*, 2:113–120, 1972.

5. U. Brandes. A faster algorithm for betweenness centrality. *Journal of Mathematical Sociology*, 25(2):163–177, 2001.

6. U. Brandes and T. Erlebach, editors. *Network Analysis: Methodological Foundations*, volume 3418 of *Lecture Notes in Computer Science (LNCS) Tutorial*. Springer-Verlag, 2005.

7. U. Brandes and D. Fleischer. Centrality measures based on current flow. In *Proceedings of the 22nd Symposium of Theoretical Aspects of Computer Science (STACS '05)*, volume 3404 of *Lecture Notes in Computer Science (LNCS)*, pp. 533–544. Springer-Verlag, 2005.

8. U. Brandes and D. Wagner. Visone—analysis and visualization of social networks. In Jünger and Mutzel [24], pp. 321–340.

9. B. J. Breitkreutz, C. Stark, and M. Tyers. Osprey: A network visualization system. *Genome Biology*, 4(3), 2003.

10. V. J. Carey, J. Gentry, E. Whalen, and R. Gentleman. Network structures and algorithms in bioconductor. *Bioinformatics*, 21(1):135–136, 2005.

11. R. M. Christley, G. L. Pinchbeck, R. G. Bowers, D. Clancy, N. P. French, R. Bennett, and J. Turner. Infection in social networks: Using network analysis to identify high-risk individuals. *American Journal of Epidemiology*, 162(10):1024–1031, 2005.

12. E. Estrada. Virtual identification of essential proteins within the protein interaction network of yeast. *Proteomics*, 6(1):35–40, 2006.

13. E. Estrada and J. A. Rodriguez-Velazquez. Subgraph centrality in complex networks. *Physical Review E*, 71(056103), 2005.

14. W. L. Ferrar. *Finite Matrices*. Clarendon Press, Oxford, 1951.

15. L. C. Freeman. A set of measures of centrality based upon betweenness. *Sociometry*, 40:35–41, 1977.

16. L. C. Freeman, S. P. Borgatti, and D. R. White. Centrality in valued graphs: A measure of betweenness based on network flow. *Social Networks*, 13(2):141–154, 1991.

17. R. C. Gentleman, V. J. Carey, D. M. Bates, B. Bolstad, M. Dettling, S. Dudoit, B. Ellis, L. Gautier, Y. Ge, J. Gentry, K. Hornik, T. Hothorn, W. Huber, S. Iacus, R. Irizarry, F. Leisch, C. Li, M. Maechler, A. J. Rossini, G. Sawitzki, C. Smith, G. Smyth, L. Tierney, J. Y. H. Yang, and J. Zhang. Bioconductor: Open software development for computational biology and bioinformatics. *Genome Biology*, 5:R80, 2004.

18. C. Godsil and G. Royle. *Algebraic Graph Theory*, volume 207 of *Graduate Texts in Mathematics*. Springer, 2001.

19. P. Hage and F. Harary. Eccentricity and centrality in networks. *Social Networks*, 17:57–63, 1995.

20. M. W. Hahn and A. D. Kern. Comparative genomics of centrality and essentiality in three eukaryotic protein-interaction networks. *Molecular Biology and Evolution*, 22(4):803–806, 2005.

21. Z. Hu, J. Mellor, J. Wu, T. Yamada, D. Holloway, and C. DeLisi. VisANT: data-integrating visual framework for biological networks and modules. *Nucleic Acids Research*, 33(Web Server issue):W352–W3577, 2005.

22. R. Jacob, D. Koschützki, K. A. Lehmann, L. Peeters, and D. Tenfelde-Podehl. Algorithms for centrality indices. In Brandes and Erlebach [6], pp. 62–82.

23. H. Jeong, S. P. Mason, A.-L. Barabási, and Z. N. Oltvai. Lethality and centrality in protein networks. *Nature*, 411:41–42, 2001.

24. M. Jünger and P. Mutzel, editors. *Graph Drawing Software*. Mathematics and Visualization. Springer-Verlag, New York, 2004.

25. M. P. Joy, A. Brock, D. E. Ingber, and S. Huang. High-betweenness proteins in the yeast protein interaction network. *Journal of Biomedicine and Biotechnology*, 2:96–103, 2005.

26. JUNG – the java universal network/graph framework. http://jung.sourceforge.net/, Accessed 2006-11-14.

27. D. Jungnickel. *Graphs, Networks and Algorithms*, volume 5 of *Algorithms and Computation in Mathematics*. Springer, New York, 2002.

28. B. H. Junker, D. Koschützki, and F. Schreiber. Exploration of biological network centralities with CentiBiN. *BMC Bioinformatics*, 7(219), 2006.

29. L. Katz. A new status index derived from sociometric analysis. *Psychometrika*, 18(1):39–43, 1953.

30. J. M. Kleinberg. Autoritative sources in a hyperlinked environment. *Journal of the ACM*, 46(5):604–632, 1999.

31. D. Koschützki, K. A. Lehmann, L. Peeters, S. Richter, D. Tenfelde-Podehl, and O. Zlotowski. Centrality indices. In Brandes and Erlebach [6], pp. 16–61.

32. D. Koschützki, K. A. Lehmann, D. Tenfelde-Podehl, and O. Zlotowski. Advanced centrality concepts. In Brandes and Erlebach [6], pp. 83–111.

33. A. N. Langville and C. D. Meyer. A survey of eigenvector methods for web information retrieval. *SIAM Review*, 47(1):135–161, 2005.

34. R. Lempel and S. Moran. The stochastic approach for link-structure analysis (SALSA) and the TKC effect. *Computer Networks: The International Journal of Computer and Telecommunications Networking*, 33:387–401, 2000.

35. H.-W. Ma and A.-P. Zeng. The connectivity structure, giant strong component and centrality of metabolic networks. *Bioinformatics*, 19(11):1423–1430, 2003.

36. J. L. Morrison, R. Breitling, D. J. Higham, and D. R. Gilbert. GeneRank: Using search engine technology for the analysis of microarray experiments. *BMC Bioinformatics*, 6:233, 2005.

37. M. E. J. Newman. A measure of betweenness centrality based on random walks. *Social Networks*, 27:39–54, 2005.

38. L. Page, S. Brin, R. Motwani, and T. Winograd. The pagerank citation ranking: Bringing order to the web. Technical report, Stanford Digital Library Technologies Project, 1998.

39. A. P. Potapov, N. Voss, N. Sasse, and E. Wingender. Topology of mammalian transcription networks. *Genome Informatics*, 16(2):270–278, 2005.

40. R Development Core Team. *R: A Language and Environment for Statistical Computing*. R Foundation for Statistical Computing, Vienna, 2005.

41. G. Sabidussi. The centrality index of a graph. *Psychometrika*, 31:581–603, 1966.

42. J. Scott. *Social Network Analysis: A Handbook*. Sage Publications, Thousand Oaks, CA, 2000.

43. P. Shannon, A. Markiel, O. Ozier, N. S. Baliga, J. T. Wang, D. Ramage, N. Amin, B. Schwikowski, and T. Ideker. Cytoscape: A software environment for integrated models of biomolecular interaction networks. *Genome Research*, 13(11):2498–2504, 2003.

44. J. G. Siek, L.-Q. Lee, and A. Lumsdaine. *The Boost Graph Library: User Guide and Reference Manual*. Addison-Wesley, Reading, MA, 2002.

45. A. Wagner and D. A. Fell. The small world inside large metabolic networks. *Proceedings of the Royal Society London B*, 268:1803–1810, 2001.

46. S. Wasserman and K. Faust. *Social Network Analysis: Methods and Applications*. Cambridge University Press, Cambridge, UK, 1994.

47. S. Wuchty and P. F. Stadler. Centers of complex networks. *Journal of Theoretical Biology*, 223:45–53, 2003.

5

NETWORK MOTIFS

Henning Schwöbbermeyer

5.1 INTRODUCTION

Recent progress in molecular biology and advances in experimental methodology, particularly in genome sequencing and high-throughput technologies, have led to an unprecedented growth in data. The availability of detailed molecular data allows the reconstruction of the structure and dynamics of biological processes and systems. For an understanding of the function and regulation of these complex biological systems, a transition from the molecular level to the systems level is necessary [14,18]. In this regard, the application of mathematical and computational techniques for the analysis of biological data at the systems level is of great importance due to the complexity of the systems and the wealth of data.

An area of mathematics that can be applied to modeling complex biological systems is graph theory. The elements of a system are represented as vertices of a graph and the interactions between them are represented as edges. Graph algorithms can then be used to analyze, simulate, and visualize the system. Graphs have been used to represent, for example, metabolic, protein interaction, and gene-regulatory networks. In these networks entities such as metabolites, proteins, or genes are represented by vertices and relationships between entities such as reactions, protein interactions, or regulatory interactions are represented by edges.

Processes of life are highly regulated. A cell as the smallest entity of life has the ability to respond to various signals and can adapt to changing conditions of its environment while keeping its internal environment homoeostatic. Different

Analysis of Biological Networks, Edited by Björn H. Junker and Falk Schreiber
Copyright © 2008 John Wiley & Sons, Inc.

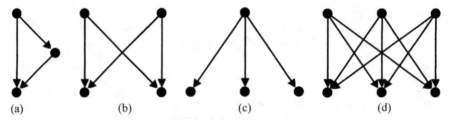

FIGURE 5.1 Network motifs that have been shown to be functionally relevant in biological networks: (*a*) feed-forward loop motif [6,7,20,34,45], (*b*) bifan motif [6,7,34], (*c*) single-input motif [20,45], and (*d*) multi-input motif [6,20].

mechanisms are recruited for regulation, either short term regulation by changing the activity of enzymes or long term regulation by changing the expression level of genes. A major goal of systems biology is in understanding the complex regulatory mechanisms of biological systems in detail that constitute the function and behavior of these systems. The analysis of regulatory circuits of biological networks can give important insights for the understanding of biological systems.

The term *network motif* has been introduced for particular subgraphs representing patterns of local interconnections between network elements. These motifs have been described as basic building blocks and design patterns of complex networks [34], and several motifs have been shown to be functionally relevant in biological networks. Some of these motifs are illustrated in Fig. 5.1. For example, the feed-forward loop motif was shown to have particular information filtering capabilities within the process of gene regulation. In Fig. 5.2, some occurrences of a network motif within a gene-regulatory network of yeast (*Saccharomyces cerevisiae*) are shown.

Network motifs have been originally introduced as statistically significant over-represented patterns of local interconnections in complex networks. In the case of biological networks, the structure has been shaped during evolution. The overabundance of particular motifs in these networks has been supposed to be a consequence of a positive selection for these interaction patterns due to their functional or structural properties. Various aspects of network motifs have been analyzed so far and different applications of motifs have been studied, for example, the comparison of networks on the basis of their local structure represented by network motifs.

5.2 DEFINITIONS AND BASIC CONCEPTS

5.2.1 Definitions

A *motif* is a small connected graph G'. Usually, the size of a motif is given in the number of vertices. A *match* of a motif within a target graph G is a graph G'', which is isomorphic to the motif G' and a subgraph of G, see Fig. 5.3 for an example. In general, a match does not have to be an induced subgraph of G. The frequency of a motif is the number of its matches in the target graph. Different frequency concepts are discussed in Section 5.2.3.

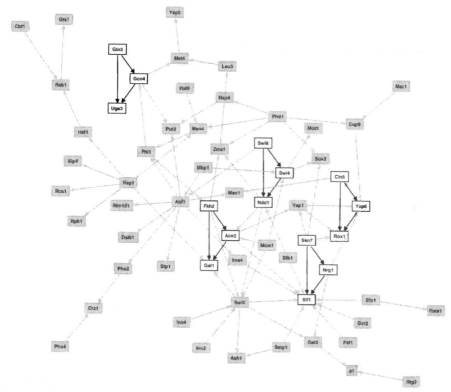

FIGURE 5.2   Some occurrences of the feed-forward loop motif (see Fig. 5.1a) within a part of the gene-regulatory network of yeast (*Saccharomyces cerevisiae*, data taken from Ref. [47]). This figure was created with MAVisto (Section 5.5.3) using a motif-preserving layout for the highlighted motif matches (see Section 5.2.3).

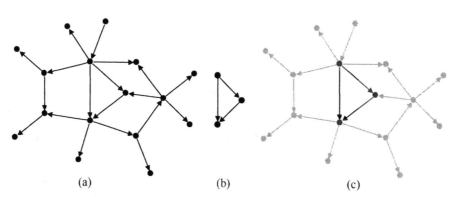

FIGURE 5.3   Illustration of (*a*) a target graph G, (*b*) a motif G', and (*c*) a highlighted match G'' of the motif G' in the target graph G.

The networks typically considered in network motif analysis are directed or undirected, connected, simple, and loop-free as defined in Chapter 2. In some cases also mixed networks containing directed and undirected edges have been studied for network motifs.

5.2.2   Modeling of Biological Networks

Biological data can often be represented as graphs or networks, as described in Chapter 1. Interaction between proteins can be modeled as graphs with proteins represented by vertices and interactions between proteins modeled as edges. Usually, only the presence or absence of an interaction between two proteins is detected, without any direction. Therefore, the edges in protein interaction networks are undirected.

In gene-regulatory networks genes correspond to vertices and interactions between genes are represented by edges. These interactions model the regulatory control of genes via their products on their target genes. As control is directed from a gene toward the regulated gene, gene-regulatory networks are usually directed.

Network motif analysis is applied to both types of networks, directed and undirected, and in addition also mixed networks containing directed and undirected edges have been studied for motifs [52,53]. However, the emphasis in network motif analysis lies on the study of directed networks.

5.2.3   Concepts of Motif Frequency

The frequency of a motif in a particular network is the number of different matches of this motif. There are three reasonable concepts for the determination of the frequency of a motif based on different restrictions on the sharing of network elements (vertices or edges) for the matches [43]. These concepts have different properties and are used to analyze different aspects of the motifs. Concept F_1 has no restrictions and considers all matches, therefore showing the full potential of a particular motif even if elements of the target graph have to be used several times. Usually, the frequency of network motifs is given by concept F_1 if no further information is supplied. Concept F_2 allows the sharing of vertices but not of edges and therefore calculates the number of instances of a motif that have disjoint edges. F_2 shows, for example, in networks where edges represent information flow how many motif instances can be "active" at a time. For concept F_3 matches have to be vertex and edge disjoint and can be seen as nonoverlapping clusters. This clustering of the target graph allows specific analysis and navigation methods such as motif-preserving layout of the network, where each selected motif match is uniformly laid out within the network corresponding to a given layout of the motif [19]. Figure 5.2 shows an application of the motif-preserving layout. The frequency calculated by concept F_3 is also known as the uniqueness value of a motif [15]. The application of the three frequency concepts is illustrated by an example shown in Fig. 5.4.

The restrictions on the reuse of graph elements for concepts F_2 and F_3 have consequences for the determination of motif frequency in case of overlapping matches, as

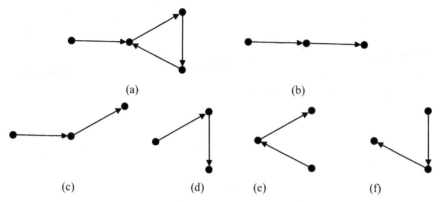

FIGURE 5.4 Illustration of the application of the different concepts of motif frequency on the basis of (*a*) a target graph *G*, (*b*) a motif *G′*, and (*c*)–(*f*) all four matches *G″* of the motif *G′* in the target graph *G*. The application of frequency concept $\mathcal{F}_1$ results in a frequency of four by counting all different matches (*c*)–(*f*), for $\mathcal{F}_2$ the frequency is two by counting the matches at (*c*) and (*f*), and for concept $\mathcal{F}_3$ the frequency is one as only one match out of the four matches (*c*)–(*f*) can be selected.

not all matches can be counted for the frequency. To determine the maximum number of different matches of a motif, the maximum set of nonoverlapping matches has to be calculated. This is known as the maximum independent set problem. Since this problem is $\mathcal{N}\mathcal{P}$-complete [10], usually a heuristic is used to compute a lower bound for the frequency. Note that there is not necessarily one unique set of nonoverlapping matches with maximum size, but rather different sets with maximum size can exist containing different collections of motif matches.

5.3   MOTIF STATISTICS AND MOTIF-BASED NETWORK DISTANCE

5.3.1   Determination of Statistical Significance of Network Motifs

Network motifs are originally defined as patterns of interconnections occurring in networks at numbers that are significantly higher than those in randomized networks [34]. Even though various different aspects of network motifs have been considered [20,45,46,51], the statistical significance is still an important property. Generally, in statistics a result is considered to be significant if the probability that this result has occurred by chance is low. In order to test for statistical significance, a *null hypothesis* is formulated based on the distribution of such results. A significance level is defined as a probability threshold on that distribution to either accept or reject the null hypothesis. If the probability of a result is below the defined probability threshold, the null hypothesis is rejected and the result is considered as statistically significant.

In network motif analysis, a commonly used null hypothesis is based on the frequency distributions of motifs in a sufficiently large ensemble of random networks of appropriate structure (Section 5.3.2), which are used as null model networks. To calculate the statistical significance of a motif of a network of interest, the frequency of this motif in the considered network is tested against the frequency distribution of this motif in the null model networks. Based on these frequency values, measures for the statistical significance of the motif can be calculated, see Section 5.3.5.

Random networks are considered as null model networks because their structure is generated by a process free of any type of selection acting on their motifs. Therefore, a significant overrepresentation of a motif in a real-world network compared with null model networks has been taken to represent evidence of functional constraints and design principles that have shaped the network structure at the level of the motifs through selection [1,34].

5.3.2   Randomization Algorithm for Generation of Null Model Networks

In network motif analysis, a commonly used algorithm for the generation of randomized versions of a given network arbitrarily rewires the connections of the network locally [24,25,45]. The algorithm reconnects two edges (A,B) and (C,D) in such a way that A becomes connected to D and C to B, provided that none of the newly created edges already exist in the network, see Fig. 5.5. Additional restrictions can be applied to preserve the number of bidirectional edges in the randomized network or to preserve the number of motifs of size $n - 1$ when searching for motifs of size n. This rewiring step is repeated a great number of times to generate a properly randomized network. The essential feature of this algorithm is the conservation of the degree of each vertex. The degree distribution of a network is a characteristic network property and has been used to characterize the large-scale topological structure of biological networks [3]. The applied randomization algorithm changes the network topology at the local level and preserves the degree distribution at the global level. Therefore, it is assumed that this algorithm provides an appropriate null model for the calculation of statistical significance of network motifs [25].

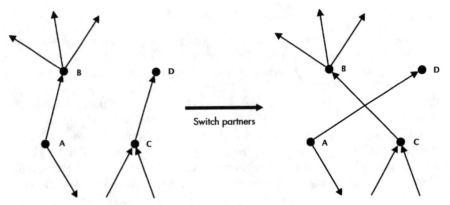

FIGURE 5.5   Example of a rewiring step of the randomization algorithm. Two edges (A,B) and (C,D) are reconnected in such a way that A becomes connected to D and C to B.

5.3.3   Influence of the Null Model on Motif Significance

Different randomization methods for the generation of the null model networks have been applied in a study on neuronal networks from macaque and cat [42], in which significance profiles (see Section 5.3.5) of motifs of size three were investigated. The three applied randomization methods for the generation of the null model networks preserved (1) the number of vertices and edges of the real-world networks, (2) the number of vertices and edges and the degrees of the vertices, and (3) the number of vertices and edges, the degrees of the vertices and the number of motifs of size two of the real-world networks. The statistical significance computed on the basis of the different null models showed clear differences for some motifs. However, if the same null model was applied for calculation of the significance profiles of the two neuronal networks, these profiles were highly correlated. This correlation holds for each of the three null models. These results indicate that the choice of the network randomization method for the generation of the null model networks is very important for the computation of motif significance and that only results obtained by application of the same null model can be reasonably compared.

5.3.4   Limitations of the Null Model on Motif Detection

The appropriateness of the randomization algorithm to represent a random null model has been questioned [1]. In this work the authors provide an example where the same motifs have been found in a real network which has been created through the process of evolution and in a network which has been constructed randomly using a network generation model that produces a "similar" structure. The statistical significance of a motif depends on the null model used for the test. It is argued that a reformulation of the test for motif significance is demanded, which is able to discriminate functional constraints and design principles from other origins that are a consequence of the network's construction mechanisms, for example, spatial clustering [1]. In this context it was noted that the great majority of motif occurrences overlap and are embedded in larger structures (see Section 5.6.3). This is not taken into account by current null models but might be important for the calculation of the statistical importance of network motifs [28]. Furthermore, it was mentioned that biological networks represent a static view of all possible interactions and that motifs that are not active *in vivo* could emerge as a consequence of the network structure [1].

5.3.5   Measures of Motif Significance and for Network Comparison

Statistical significance of motifs for a particular network can be measured by calculating the Z-score and P-value using frequency concept $\mathcal{F}_1$. Significance profiles on the basis of the Z-scores of particular motif sets can be used to compare different networks. However, the results of the statistical analysis depend on the appropriateness of the applied null model, as described in Section 5.3.2. Furthermore, the frequency of motifs can directly be used for a comparison of networks, for example, by application of the relative graphlet frequency distance or by methods described in Section 5.6.4.

Z-Score   The *Z-score* $Z(m)$ is defined as the difference of the frequency $\mathcal{F}_1$ of a motif m in the target network and its mean frequency $\overline{\mathcal{F}_{1,r}}$ in a sufficiently large set of randomized networks, divided by the standard deviation σ_r of the frequency values of the randomized networks [25,34], see Equation 5.1. Motifs are considered as statistically overrepresented if they have a Z-score greater than 2.0 [15].

$$Z(m) = \frac{\mathcal{F}_1(m) - \overline{\mathcal{F}_{1,r}(m)}}{\sigma_r(m)} \qquad (5.1)$$

P-Value   The *P-value* represents the probability $P(m)$ of the appearance of a motif m in a randomized network, an equal or greater number of times than in the target network, see Equation 5.2. $\mathcal{F}_1$ is the frequency of a motif in the target network, $\mathcal{F}_{1,r}$, is the frequency of a motif in a randomized network and N denotes the number of randomized networks. The Kronecker delta $\delta_{c(n)}$ is one if the condition $c(n)$ holds, otherwise it is zero. Motifs with a P-value less than 0.01 are regarded as statistically significant. For a reasonable calculation of a P-value at least a thousand randomized networks have to be considered [15]. If less randomized networks are considered only the Z-score should be used.

$$P(m) = \frac{1}{N} \sum_{n=1}^{N} \delta_{c(n)} \qquad c(n): \; \mathcal{F}_{1,r_n}(m) \geq \mathcal{F}_1(m) \qquad (5.2)$$

Motif Significance Profile   The *motif significance profile* SP is defined as a vector of Z-scores $Z(m)$ of a particular set of motifs ($\{m_1, ..., m_n\}$), which is normalized to a length of one [33], see Equation 5.3. Motif significance profiles allow for a comparison of networks of different size or origin on the basis of their motifs.

$$SP(m_i) = \frac{Z(m_i)}{\sqrt{\sum_{i=1}^{n} Z(m_i)^2}} \qquad (5.3)$$

Typically, motif sets comprise all motifs of a particular size, for example, the 13 motifs of size three shown in Fig. 5.6 have been used for the classification of different directed networks into distinct groups, see Section 5.6.4. For two gene-regulatory networks of *S. cerevisiae* and *E. coli*, the significance profiles of the motifs of size three are depicted in Fig. 5.7. The profiles of these two networks show similar characteristics as expected due to their related origin. The number of supported motifs in the *S. cerevisiae* network is higher than in the *E. coli* network, which may be a consequence of

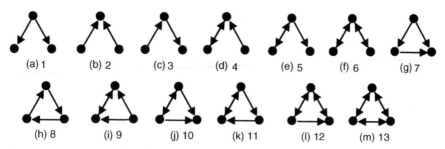

FIGURE 5.6   Structure of the 13 directed motifs with three vertices. Motif numbers are used in accordance to Figure 5.7.

a higher complexity of eucaryotic gene-regulatory networks represented by *S. cerevisiae* compared with procaryotic gene-regulatory networks represented by *E. coli*.

Relative Graphlet Frequency Distance   The *relative graphlet frequency distance* was introduced as a distance measure for undirected networks [38]. The authors termed small connected induced subgraphs as graphlets to avoid a confusion of

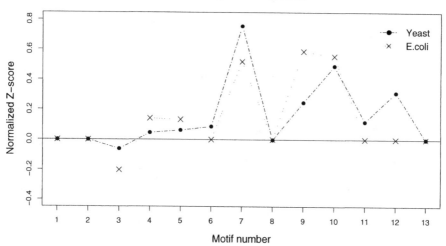

FIGURE 5.7   Significance profiles of motifs of size three of two gene-regulatory networks. The yeast (*S. cerevisiae*) network (4441 vertices, 12873 edges) was taken from Ref. [37] and the *E. coli* network (1250 vertices, 2431 edges) was obtained from the RegulonDB (Version 5.0) [39]. The Z-scores were calculated using the MAVisto program (see Section 5.5.3). The structure of the motifs is illustrated in Fig. 5.6. Note that motifs 1 and 2 have a Z-score of 0.0 since the applied randomization algorithm conserves the in- and out-degree of the vertices. Therefore, the frequency of these two motifs does not change in the randomized networks, which leads to a Z-score of 0.0. For the other motifs a Z-Score of 0.0 represents the absence of a motif within a network (in particular these are in the yeast network motif 13 but not motif 8, which has a value slightly less than 0.0 and in the *E. coli* network motifs 6, 8, 11, 12, and 13).

terminology with network motifs. The network distance calculation is based on the relative frequency F_r of 29 undirected graphlets g of size three to five, where $F(g_i)$ is the frequency of the graphlet of type i ($i \in \{1, ..., 29\}$), see Equation 5.4. Of these 29 graphlets, there are two graphlets of size three, six graphlets of size four, and 21 graphlets of size five (see Table 5.1).

$$F_r(g_i) = \frac{F(g_i)}{\sum_{i=1}^{29} F(g_i)} \tag{5.4}$$

The relative graphlet frequency distance D of two networks G_1 and G_2 is calculated by summing up the differences of the logarithm of the relative graphlet frequencies, see Equation 5.5. The logarithm is used in order to give equal weight to each relative graphlet frequency difference. This is done because the absolute frequencies of different graphlets can vary by orders of magnitude and the graphlets with highest frequency would have otherwise a dominating influence on the result.

$$D(G_1, G_2) = \sum_{i=1}^{29} |\log(F_{r,G_1}(g_i)) - \log(F_{r,G_2}(g_i))| \tag{5.5}$$

5.4   COMPLEXITY OF NETWORK MOTIF DETECTION

5.4.1   Aspects Affecting the Complexity of Network Motif Detection

Generally, for the detection of network motifs, the different motifs supported by the analyzed network have to be determined, their matches within the network have to be identified, and their statistical significance has to be calculated. The process of network

TABLE 5.1   Number of Nonisomorphic Connected Loop-Free Undirected and Directed Simple Graphs with up to 10 Vertices [11]

Vertices	Undirected	Directed[a]
1	1	1
2	1	2
3	2	13
4	6	199
5	21	9364
6	112	1530843
7	853	880471142
8	11117	1792473955306
9	261080	13026161682466252
10	11716571	341247400399400765678

[a] Including mutual (bidirectional) edges.

motif detection includes several aspects that affect the computational complexity of this task:

1. *Graph isomorphism testing*: During the search, the discovered motifs have to be compared concerning their structure in order to group isomorphic motifs together. This is known as the graph isomorphism problem. The computational complexity of this problem is not exactly classified [10], but is considered to be a hard problem with a complexity that lies between $\mathcal{P}$ and $\mathcal{NP}$. In practice, testing whether graphs are isomorphic can be done by computing a label for each graph based on the graph structure. This label is unique for a particular graph structure and is called a *canonical label*. Therefore, if two graphs have identical canonical labels, these graphs are isomorphic. Canonical labels for graph isomorphism testing are used by the motif detection programs FANMOD (Section 5.5.4), which applies the *nauty*-algorithm [29] for this task and MAVisto (Section 5.5.3) that identifies the detected motifs by canonical label.

2. *Number of motifs*: The number of nonisomorphic motifs grows exponentially with the size of the motif (number of vertices) and rapidly reaches particularly for directed motifs an enormous quantity, as shown in Table 5.1. The number of different undirected motifs is far less than the number of directed motifs, but normally the number of matches of undirected motifs is much higher than for directed motifs in networks of comparable size. In practice, only a fraction of all possible motifs is supported by typical real-world networks. To date well-known network motifs are small and usually comprise three to five vertices. For these motif sizes the number of nonisomorphic motifs is relatively small.

3. *Number of motif matches*: The maximum number of matches of a motif $G_m = (V_m, E_m)$ in a target graph, $G_t = (V_t, E_t)$ is bounded by $|E_t|^{|E_m|}$. $|E_t|$ is the number of edges in the target graph, and $|E_m|$ is the number of edges in the motif. In typical real-world networks only some motifs have a high frequency and the majority are less frequent.

4. *Calculation of statistical significance*: A commonly used method for calculation of statistical significance of network motifs compares the frequency in a target network to the frequency in an ensemble of appropriately randomized networks, see Section 5.3. For a reasonable calculation of motif statistics several hundreds or thousand randomized networks have to be considered. The computational effort multiplies with the number of randomized networks, as the motif distribution has to be detected for each of these networks.

5. *Size of analyzed networks*: The size of the target network affects the three previously described points 2–4. In general, the number of different motifs supported by the target network as well as the number of motif matches increases with the size of the target network. Typical biological networks comprise several hundreds or thousands of vertices and are often sparse, that is, with a ratio of the number of edges to the number of vertices between 1 and 3.

Despite the high complexity involved in the detection of network motifs, in practice the search can be executed in reasonable time for typical real-world networks.

Common algorithms and tools for the analysis of network motifs are described in Section 5.5.

5.4.2 Frequency Estimation by Motif Sampling

Since the computational complexity of an exhaustive enumeration of all motifs increases strongly with the network size, heuristics for the detection of network motifs have been developed to accelerate the search in large networks. A heuristic random sampling approach has been proposed for the calculation of network motif frequency whose run time is asymptotically independent of the network size [16]. However, this algorithm suffers from sampling bias and efficiently estimates motif frequencies only in networks with hubs. Another sampling algorithm has been presented, which overcomes these drawbacks and allows for an unbiased subgraph sampling [49]. Furthermore, the motif analysis tools Mfinder (Section 5.5.2) and FANMOD (Section 5.5.4) both offer a sampling method for a fast approximation of the motif number.

5.5 METHODS AND TOOLS FOR NETWORK MOTIF ANALYSIS

Different methods and tools have been developed for the analysis of network motifs. There are three focused tools that deal with the task of motif detection and analysis, see Sections 5.5.2 – 5.5.4. Furthermore, various specialized methods have been developed and applied to investigate specific questions [7,28,35,40,51,53]. These methods are usually not described in detail and the source code is not available. Publicly available are, for example, MATLAB scripts for motif detection [26,46]. An algorithm for the alignment of motifs was developed to identify motifs derived from families of mutually similar but not necessarily identical patterns [5].

5.5.1 Pajek

Pajek is a program for the analysis and visualization of large networks [4]. It offers the possibility of calculating the frequencies of certain subgraphs like triads and particular tetrads, which are subgraphs with three respectively four vertices. Triads can be connected and unconnected and their study originates in social network analysis. Pajek calculates the number of triads of a network and reports values for the expected frequencies.

5.5.2 Mfinder

The Mfinder is a software tool for network motif detection in directed and undirected networks [15,30]. It computes the number of occurrences of a motif of restricted size in the target network (concept $\mathcal{F}_1$) and a uniqueness value, which is a lower bound for the frequency of concept $\mathcal{F}_3$. A value for the frequency of concept $\mathcal{F}_2$ is not calculated. The statistical significance is determined on the basis of the number of occurrences of the motif in randomized networks and is given by a P-value and Z-score. The applied randomization method preserves the degree of each vertex. Furthermore, a sampling method is available for a fast approximation of the motif number. The results are presented in a text file and the structure of discovered motifs can be looked up in a motif dictionary.

5.5.3 MAVisto

MAVisto is a tool for the exploration of motifs in biological networks combining a flexible motif search algorithm and different views for the analysis and visualization of network motifs [27,44]. MAVisto supports the Pajek-.net- [4] and the GML-format [12] and offers graph editor functionality for network creation and manipulation. Furthermore, an advanced force-directed layout algorithm [9] is included to generate readable drawings of the network automatically while preserving the layout of motifs where possible. MAVisto's motif search algorithm discovers all motifs of a particular size, which is given either by the number of vertices or by the number of edges. All motifs of this size are analyzed and the frequencies for the three different frequency concepts are determined. Furthermore, P-values and Z-scores are computed based on a randomization method that preserves the degrees of the vertices. MAVisto supports the analysis of vertex-labeled and / or edge-labeled networks. The motif search algorithm of MAVisto is described in detail in Ref. [43].

5.5.4 FANMOD

FANMOD is a relatively recent tool for network motif detection in directed and undirected networks [8,50] and is based on an improved algorithm that outperforms the existing tools Mfinder and MAVisto in the task of network motif detection. In contrast to Mfinder and MAVisto, the FANMOD tool detects only motifs that are induced subgraphs whereas the other two discover all supported motifs. FANMOD supports the analysis of vertex-labeled and / or edge-labeled networks. It offers a graphical user interface and the results are presented as text- or HTML-files similar to the Mfinder. The HTML-pages show additionally the structure of each detected motif. FANMOD calculates besides the statistics of the motifs given by P-values and Z-scores the relative frequency for concept $\mathcal{F}_1$, based on all discovered motifs of a particular size. Further values for motif frequency are not computed. FANMOD also offers a sampling method for a fast approximation of the motif number.

5.6   ANALYSES AND APPLICATIONS OF NETWORK MOTIFS

5.6.1   Network Motifs in Complex Networks

Network motifs have been studied initially in gene-regulatory networks of *E. coli* [45] and *S. cerevisiae* [20] and were introduced by Alon and coworkers as a general property of complex networks [34]. In the work of Alon and colleagues, biological and technological networks from the fields of biochemistry, neurobiology, ecology, and engineering were analyzed. Motif detection resulted in each network in a small set of statistically significant motifs and some of the motifs found are shared by different networks, that are the feed-forward loop motif, the bifan motif and the biparallel motif, see Table 5.2. The World Wide Web network is the only one that exhibits a unique set of motifs, all other networks share at least one motif with another network and only the food webs and the gene-regulatory networks do not share any motifs with each other.

The gene-regulatory networks, the neuronal network, and the electronic circuits of forward logic chips share two, respectively, three motifs. These networks all perform information processing, therefore Alon and coworkers hypothesized that these motifs may have specific functions as elementary computational circuits within these networks. Furthermore, based on the motifs found they assumed that the World Wide Web network may reflect a design that tends to provide short paths between related pages, and that food webs evolve to allow for an energy flow from the bottom to the top of food chains. Moreover, the authors supposed that network motifs may be structures that arise because of the special constraints under which a network has evolved and that each class of network has specific types of elementary structures. Based on these findings, Alon and coworkers proposed that motifs could be used for the definition of classes of networks and network homologies [34].

5.6.2   Dynamic Properties of Network Motifs

Dynamic Properties of the Feed-Forward Loop Motif   The *feed-forward loop* motif occurs in many biological and technological networks (see Table 5.2) and is the best studied motif to date. The functional properties of the feed-forward loop motif have been analyzed in detail theoretically and experimentally in gene-regulatory networks [21,23,45,48]. In these networks, a feed-forward loop motif is built by two regulators, a general and a specific regulator, with the general regulator regulating the specific one and both jointly regulating a target gene, see Fig. 5.8a. As in gene-regulatory networks the interactions of regulators on genes can be activating $(+)$ or repressing $(-)$, there are 2^3 (two interaction types for each of the three edges) possibilities for a feed-forward loop motif with a different combination of interactions. A feed-forward loop motif is called *coherent* if the direct effect of the general regulator on the regulated gene is the same (positive or negative) as its net indirect effect through the specific regulator (e.g., if the general regulator positively regulates the specific regulator and both positively regulate the regulated gene, see Fig. 5.8b). If the net effects of the two regulators are different, the feed-forward loop motif is called *incoherent* (Fig. 5.8c).

Mathematical modeling of the dynamics of the coherent feed-forward loop motif shown in Fig. 5.8b indicated that it responds only to persistent activations of the general regulator but not to transient activations, therefore filtering out fluctuating signals. The reason for this behavior is a delayed activation of the regulated gene because at first the specific regulator has to be activated by the general regulator. If the activation period of the general regulator is too short, the specific regulator does not reach the level sufficient for activation of the regulated gene, which is done in

TABLE 5.2   The Network Motifs Discovered in a Study on Biological and Technological Networks [34]. The Analyzed Networks were Gene-Regulatory Networks (GRN) of *E. coli* and *S. cerevisiae*, a Neuronal Network (NN) of *Caenorhabditis Elegans*, Seven Food Webs (FW), Five Electronic Circuits of Forward Logic Chips (EC1), Three Electronic Circuits of Digital Fractional Multipliers (EC2), and a Network of the World Wide Web (WWW)

Structure	Motif description	GRN	NN	FW	EC1	EC2	WWW
	Feed-forward loop	✓	✓		✓		
	Bifan	✓	✓		✓	✓	
	Biparallel		✓	✓	✓		
	Three chain		✓[a]				
	Three-node feedback loop					✓	
	Four-node feedback loop					✓	
	Feedback with two mutual dyads[b]						✓

TABLE 5.2   (*Continued*)

Structure	Motif description	GRN	NN	FW	EC1	EC2	WWW
	Uplinked mutual dyad[b]						✓
	Fully connected triad[b]						✓

Note: All Motifs found in these networks are given by their structure and the originally used description. A tick indicates a statistically significant overrepresentation of a Motif in a particular network.
[a] The three chain motif was statistically significant in five out of the seven food webs.
[b] dyad and triad are used in social science for interaction patterns of two respectively three individuals.

conjunction with the general regulator, and therefore this gene is not activated. Furthermore, the feed-forward loop motif allows for a rapid shutdown once the activation of the general regulator stops, since this regulator is directly required for the activation of the regulated gene. In summary, this coherent feed-forward loop motif acts as an asymmetric delay circuit: It shows a delayed response on activation and a rapid response in the opposite direction on deactivation [45]. Such a characteristic can be advantageous for gene regulation in noisy environments with fluctuating signals. This

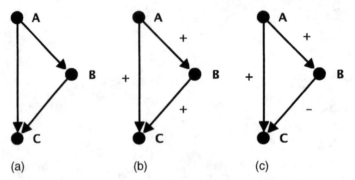

FIGURE 5.8   In gene-regulatory networks, a feed-forward loop motif as in (*a*) consists of two regulators, a general regulator *A* and a specific regulator *B*, with *A* regulating *B* and both jointly regulating a target gene *C*. In (*b*) an example of a coherent feed-forward loop motif is shown, where the direct and net indirect effect of the general regulator *A* on the regulated gene *C* are equal. The other three coherent feed-forward loop motifs have for the edges ($A{\rightarrow}C$, $A{\rightarrow}B$, $B{\rightarrow}C$), the edge types $(-,-,+)$, $(-,+,-)$, and $(+,-,-)$. In (*c*) an example of an incoherent feed-forward loop motif is given as here the direct and the net indirect effect are different. The other three incoherent feed-forward loop motifs have the edge types $(-,-,-)$, $(-,+,+)$, and $(+,-,+)$.

functional behavior of the coherent feed-forward loop motif shown in Fig. 5.8b on gene regulation was experimentally confirmed in a study of the L-arabinose utilization system of *E. coli* [23].

The kinetic behavior of all eight structural types of the feed-forward loop motif in the process of gene regulation was analyzed theoretically by the use of mathematical modeling [21]. In this study it was found that all types of feed-forward loop motifs showed an asymmetric kinetic behavior for the two directions. These directions are the transitions from the nonactivated to the activated state, respectively, from the activated to the nonactivated state. The coherent configurations of the feed-forward loop motif showed delayed responses in one direction and a direct response in the opposite direction, whereas for the incoherent configurations accelerated responses in one direction and direct responses in the opposite direction were found. This accelerated response on turn-on of the incoherent feed-forward loop motif shown in Fig. 5.8c was demonstrated experimentally for the galactose gene-regulatory system of *E. coli* [22].

A more detailed analysis of mathematical models of the kinetic behavior of the feed-forward loop motif identified numerous different kinetic patterns that depend on the parameters used for modeling the activity of the genes [48]. These results support the perspective that for an understanding of gene-regulatory networks, the signal processing properties of the underlying regulatory circuits have to be considered in detail.

Dynamic Properties of the Bifan Motif   Although different aspects of the feed-forward loop motif have been analyzed in a number of studies, as described in the previous paragraph, other motifs have received much less attention. Recently, the dynamic properties of the *bifan motif* have been examined in detail by mathematical modeling based on differential equations [13]. A bifan motif is built by two regulators and two regulated genes, with the two regulators jointly regulating each target gene, see Fig. 5.1b. Five variants of the bifan motifs have been considered, which are likely to be present in *S. cerevisiae*. These variants had different combinations of activating and repressing interactions of the regulators on the target genes. The modeling showed a broad range of dynamic behavior for the variants of the bifan motif, but no characteristic response was observed. These results indicate that an understanding of the function of gene-regulatory networks on the basis of network structure or structural motifs is unlikely, and additional information on dynamic properties is necessary [13].

Dynamic Stability of Network Motifs   Biological networks are usually static representations of large-scale dynamic systems, with only a certain fraction of the elements being active at a particular time. Whereas many structural properties of these networks have been identified, dynamic features are far less studied. To analyze network motifs in the context of system dynamics, the dynamic properties of network motifs of size three and four have been systematically determined and related to their distribution in directed gene-regulatory, signal transduction, and neuronal networks [36]. The dynamic behavior was characterized by a structural stability that represents the probability of a motif to return to a steady state after small-scale perturbation. Three stability classes have been identified based on the capability of interactions between the vertices of a motif. These classes are stable motifs without feedback interactions, moderately stable motifs with one feedback interaction between two vertices and unstable motifs with feedback interactions between three or more vertices. Examples of motifs from the three classes of structural stability are the feed-forward loop motif that represents a structurally stable motif due to the absence of feedback interaction (Fig. 5.7g), a moderately stable motif that comprises one mutual edge (Fig. 5.7j) and a feedback loop motif as an example for an unstable motif (Fig. 5.7h). To exclude impacts of the edge number on motif frequency for this comparison, the motifs were grouped into density classes with equal edge number. The comparison of the frequency of motifs with three and four vertices to random networks of different null models revealed a significant overrepresentation of motifs with higher structural stability. It was supposed that robust dynamical stability of network motifs contributes to biological network organization and that there is a deep interplay between network structure and system dynamics [36].

5.6.3 Higher Order Structures Formed by Network Motifs

Motif Cluster   The distribution of motif matches of the feed-forward loop motif and of the bifan motif has been studied in the gene-regulatory network of *E. coli* [7]. For each motif the majority of matches overlap and aggregate into regions of interconnected matches, which were termed *homologous motif clusters* by the authors. They discovered that many of these motif clusters largely overlap with modules of known biological functions within the gene-regulatory network. Furthermore, the combined motif clusters of the feed-forward loop motif and of the bifan motif build a superstructure that covers large parts of the network that is assumed to represent the core or backbone of the network and to play a central role in defining the global topological organization. This study shows that on the basis of motifs and motif clusters distinct topological hierarchies within the *E. coli* gene-regulatory network are formed.

Motif Generalization   The combination of network motifs into larger structures was analyzed in a systematic approach that defined *motif generalizations*, families of motifs of different size that share a common architectural theme [17]. Roles of motif vertices based on structural equivalence were introduced for the definition of motif generalizations. For example, the feed-forward loop motif has three roles: an input vertex A, an internal vertex B, and an output vertex C, see Fig. 5.9a. These vertex roles were termed general regulator, specific regulator, and target gene, respectively, in gene-regulatory networks (Section 5.6.2). Motif generalizations are based on the duplication (or multiplication) of one (or more) vertex role(s). Therefore, the feed-forward loop can have three simple generalizations, based on duplicating each of the three roles and their connections, which is illustrated in Fig. 5.9b–d. In this study it was discovered that networks that share a common motif can have very different generalizations of that motif. Furthermore, the genes of functionally corresponding multioutput feed-forward-loop motifs (an example is given in Fig. 5.10) of *E. coli* and *S. cerevisiae* gene-regulatory networks are not evolutionarily related, which suggests convergent evolution to the same regulation pattern [17].

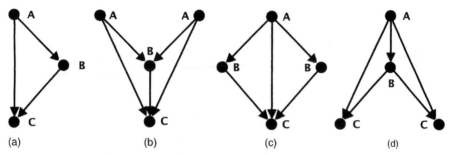

FIGURE 5.9 Illustration of the concept of motif generalizations. (*a*) The feed-forward loop motif is shown with labels indicating the roles of the vertices: input role *A*, internal role *B*, and output role *C*. Subsequently, the three simple generalizations of the feed-forward loop motif are shown duplicating (*b*) the input vertex *A*, (*c*) the internal vertex *B*, and (*d*) the output vertex *C*.

Network Theme Higher order interconnection patterns that comprise multiple occurrences of network motifs have been introduced as *network themes* in a study of an integrated network of *S. cerevisiae* containing five different types of interactions between genes and proteins [53]. One example is the *feed-forward loop theme* — a pair of transcription factors, one regulating the other, and both regulating a common set of target genes that are often involved in the same biological process, see Fig. 5.10. This network theme is also a motif generalization of the feed-forward loop motif, which multiplies the output vertex *C* as introduced in the previous paragraph (see Fig. 5.9d).

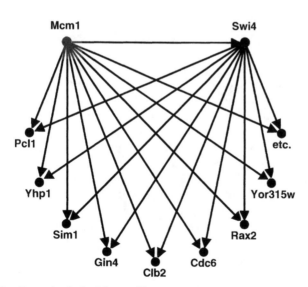

FIGURE 5.10 Example of a *feed-forward loop network theme* of the gene-regulatory network of *S. cerevisiae* [53]. *Mcm1* regulates *Swi4* and in conjunction they regulate a set of target genes. This network theme also represents a motif generalization of the feed-forward loop motif that multiplies the output vertex *C* (see Fig. 5.9d).

The authors argued that network motifs have been defined with artificial restrictions on the size and that motif occurrences often overlap within a network. Furthermore, it has been shown that network themes can be tied to specific biological phenomena and the authors suggest that network themes may represent more fundamental network design principles [53].

5.6.4 Network Comparison Based on Network Motifs

Distance Measure for Networks Based on Motif Frequencies A distance measure for undirected networks has been introduced on the basis of the frequency of undirected induced subgraphs of size three to five as the *relative graphlet frequency distance* [38], see Section 5.3.5. This network distance measure has been applied to compare protein interaction networks of yeast (*S. cerevisiae*) and fruit fly (*Drosophila melanogaster*) to different artificial network generation models. The authors discovered that these protein interaction networks were better modeled by geometric random networks than by the commonly accepted scale-free random network model with respect to the relative graphlet frequency distance. Geometric random networks were constructed by randomly distributing the vertices in a two–, three–, or four–dimensional space and connecting vertices if their distance is below a particular threshold for the applied vertex distance measure. These results were confirmed by other global network properties, network diameter (average shortest path length), and clustering coefficients. Only the degree distribution of the protein interaction networks is closer to the distribution of the scale-free random network model than to the distribution of the geometric random networks. The authors emphasized that the selection of an appropriate model of protein interaction networks is important, for example, to efficiently guide experiments on such networks [38].

Motif Frequencies as Classifiers for Network Model Selection Motif frequencies have been used as classifiers for the selection of an artificial network generation model that most suitably resembles the structure of the protein interaction network of fruit fly (*D. melanogaster*) [31]. The model has been selected out of seven network generation models that resemble different mechanisms of network evolution. For this selection, discriminative classification techniques adapted from machine learning were used. An alternating decision tree was constructed on the basis of the frequencies of the motifs of a defined set for 1000 randomly constructed networks for each model. One set contained all 148 motifs that could be constructed by walks of length eight, the other set contained all 130 motifs with up to seven edges. Then the frequencies of the motifs of the protein interaction network were used to traverse the tree, and at each branch of the tree it is checked if the frequency of a particular motif associated with this branch is above a certain threshold. Based on this decision, the path in the tree and the prediction scores for each model at this branch are determined. The model with the maximum value for the sum of all obtained prediction scores is selected as the most suitable model. Furthermore, the authors showed that although the networks of the different generation models had similar global properties (e.g., degree distribution, clustering coefficients, and network diameter), they could be distinguished by the applied discriminative classification techniques.

The presented method for identification of adequate network generation models as well as the models that resemble different mechanisms of network evolution have some limitations [41]. Real-world networks can face varying pressures within their history, and the adaptation to these particular conditions causes different changes of the structure, therefore a single network generation mechanism may not be sufficient to resemble the structure of these networks. Furthermore, the applied discriminative classification technique used an artificially selected set of motifs for which a search is computationally tractable; however, other important motifs may be missed. By considering only small-scale features represented by the distribution of a selected set of motifs, some large-scale features are not recapitulated, for example, the size of the giant component [41].

Network Superfamilies Identified by Motif Significance Profiles   Networks of different biological and technological domains have been classified into different superfamilies on the basis of motif significance profiles, respectively, motif frequency profiles [33]. For the classification of directed networks, significance profiles (see Section 5.3.5) of motifs of size three have been used. For undirected networks, the statistical significance of motifs showed a dependence on network size. Therefore, for these networks the frequency of motifs of size four relative to random networks was used without considering the statistical significance to compute a motif frequency profile. The correlation of significance profiles, respectively, frequency profiles was used to cluster the networks into distinct superfamilies. Several of these superfamilies contained networks of different domains of vastly different size, for example, one family contained a network of signal-transduction interactions, a developmental gene-regulatory network, and a neuronal network. Currently, it is not verified whether similarity in the profiles is accidental or whether the networks have similar key circuit elements because they evolved to perform similar tasks [33].

Limitations of Network Classification by Motif Significance Profiles   The classification of networks on the basis of motif significance profiles depends on the method used to generate the null model networks for calculation of the statistical significance of motifs (see Section 5.3.1). As depicted in Section 5.3.4, the same overrepresented motifs were found in real networks and networks generated using a particular network model. Therefore the appropriateness of a classification of networks into superfamilies based on motif significance profiles as described in the previous paragraph has been questioned [1]. However, by considering the full motif significance profile of motifs of size three, there were some motifs that are equally over respectively underrepresented in both the real and the random networks, but some motifs showed clear differences and allowed a distinction between artificial and real-world networks [32]. Furthermore, it was suggested that the resolution for a discrimination between networks by the use of motif significance profiles may be enhanced by the use of motifs of increased size and by the use of refined null model networks for calculation of statistical significance. An enhanced resolution of higher order motifs was confirmed by a comparison of significance profiles of motifs of size four [36]. This study puts the assignment of networks of developmental gene-regulation, signal transduction, and neuronal connections into one superfamily on the basis of three vertex motif significance profiles as described in the previous paragraph into question.

5.6.5   Evolutionary Origin of Network Motifs

Convergent Evolution of Motifs in Gene-Regulatory Networks   The evolutionary origin of network motifs has been investigated in gene-regulatory networks of *E. coli* and *S. cerevisiae* [6]. For the analyzed motifs (feed-forward loop motif, bifan motif, and others), it was discovered that the genes of different motif occurrences are in most instances not evolutionarily related. This indicates that the occurrences of these motifs have not been created by duplication of ancestral circuits. These findings suggest a convergent evolution of motif occurrences in these gene-regulatory networks as a consequence of their optimal design. Convergent evolution was also discovered for multioutput feed-forward-loop motifs [17] (Section 5.6.3).

The gene-regulatory networks of *S. cerevisiae* and four related species belonging to the class of *hemiascomycetes* have been analyzed for the cooccurrence of evolutionarily related genes in matches of different motifs. In this study it was shown that occurrences of statistically significant overrepresented motifs are to the same degree evolutionarily conserved as all other interaction patterns and that genes are not subject to evolutionary pressure to preserve corresponding interaction patterns [28]. Similar results were obtained in a study on the evolution of gene-regulatory networks, which showed that regulatory interactions in motifs are lost and retained at the same rate as the other interactions in the network [2].

Evolutionary Conservation of Motifs in a Protein Interaction Network   The proteins of the protein interaction network of *S. cerevisiae* were studied for a correlation between their evolutionary conservation and the structure of the motifs to which they belong [51]. A subset of evolutionarily conserved proteins was identified, which have known orthologs in five higher eukaryotes. The fraction of motif occurrences of motifs up to size five, which are fully assembled by proteins of the evolutionarily conserved protein subset, was determined. This fraction was significantly higher than expected at random, suggesting that motifs represent evolutionarily conserved topological units in protein interaction networks.

5.7   SUMMARY

Network motifs have been introduced in recent years as small statistically significant overrepresented patterns of local interconnections in biological and technological networks. The structure of these complex networks differs strongly from random graphs, and global structural properties such as degree distribution, network diameter, and clustering coefficients have already been extensively studied in contrast to local structural properties. In the case of biological networks, which are more and more available these days, their structure has been shaped during evolution. The overabundance of particular motifs here has been supposed to be a consequence of a positive selection for these interaction patterns due to their functional or structural properties. A common hypothesis on network motifs considers them as basic building blocks and as information processing units of complex networks.

The detection of network motifs is computationally challenging as it involves different computational and combinatorial problems. Despite the high computational complexity associated with the search for network motifs, the analysis of typical real-world networks is feasible, and tools exist for this task. The calculation of statistical significance of network motifs is usually based on a comparison of motif frequency in the analyzed real-world network to the frequency distribution in properly randomized networks. These randomized networks are considered as null model networks since their structure is generated by a process free of any type of selection acting on motifs. Generally, the detection of network motifs depends highly on the applied method for the generation of the null model networks and, therefore, the choice of an appropriate null model is very important. By this approach, network motifs are solely identified on the basis of their statistical properties and further analyses are necessary to identify motifs that represent functional properties and design principles.

Numerous studies have been carried out since their introduction addressing various aspects of network motifs. Several motifs have been found in biological networks, for example, the feed-forward loop motif and the bifan motif, for which the dynamic properties have been studied in some detail. In the regard, it was shown that the feed-forward loop motif can act as a persistence detector by filtering out noise within the process of gene regulation. Dynamic properties of network motifs have also been found and it has been shown that network motifs can act as information processing units. It was indicated that biological networks usually represent a static view of all possible interactions and that some motifs could emerge as a consequence of the network structure, but are not active *in vivo*. For gene-regulatory networks, it was shown that motif occurrences are not evolutionarily conserved nor emerge through duplication of ancestral circuits. In contrast, in protein interaction networks the tendency for motif occurrences to be completely assembled by evolutionarily conserved proteins exists.

The distribution of motifs characterizes the local structure of networks. Different methods have been applied to compare networks on the basis of their local structure using either the statistical significance or the frequency of motifs. The statistical significance of motifs has been used for the calculation of significance profiles for different networks. Networks with similar profiles have been classified into superfamilies. This method grouped networks of unrelated fields together to the same superfamilies on the basis of the statistical significance of motifs. For protein interaction networks, methods based on the frequency of motifs allow the selection of suitable network generation models, which best reflect the structure of real-world networks. For this task also a distance measure based on frequency differences of a particular set of motifs has been introduced for undirected networks and has been applied to model selection.

The occurrences of motifs usually overlap and only a small fraction exists in isolation. Different concepts have been introduced to characterize these higher order structures built by overlapping motif matches, such as motif generalizations and network themes, and different levels of network organization are assumed. Furthermore, different frequency concepts have been introduced that have particular restrictions on counting overlapping matches and characterize different aspects of network motifs. Currently, it is not clear whether an overlapping of motif occurrences affects the findings obtained when motifs were considered in isolation.

5.8   EXERCISES

1. How were network motifs originally defined in the research literature. What networks are used as null model networks for the detection of statistically significant overrepresented motifs in a real-world network?

2. How do the existing concepts of motif frequency calculation address overlapping matches? Which concept is usually used to specify the frequency of a motif? What is the uniqueness value of a motif?

3. How does a commonly used randomization algorithm work for the generation of the randomized versions of a target network? Which global network property is preserved in these randomized networks?

4. What are the roles of the vertices of the feed-forward loop motif? How do the three simple duplications of each vertex role look like?

5. Given is the following directed network by a sequence of edges, where each edge is represented by a pair of vertex ID's (<ID_source>, <ID_target>): {(1,2), (1,3), (2,3), (2,4), (2,7), (4,5), (4,6), (4,7), (6,5), (8,4), (8,5), (8,10), (9,1), (9,8), (9,10), (10,2), (10,4), (10,1)}. Draw the network and identify all matches of the feed-forward loop motif. Determine the frequency values for each of the three different frequency concepts.

REFERENCES

1. Y. Artzy-Randrup, S. J. Fleishman, N. Ben-Tal, and L. Stone. Comment on "network motifs: simple building blocks of complex networks" and "superfamilies of evolved and designed networks." *Science*, 305:1107c, 2004.

2. M. M. Babu, N. M. Luscombe, M. Gerstein, and S. A. Teichmann. Structure and evolution of transcriptional regulatory networks. *Current Opinion in Structural Biology*, 14:283–291, 2004.

3. A.-L. Barabási and Z. N. Oltvai. Network biology: Understanding the cell's functional organization. *Nature Reviews Genetics*, 5:101–113, 2004.

4. V. Batagelj and A. Mrvar. Pajek — analysis and visualization of large networks. In M. Jünger and P. Mutzel, editors, *Graph Drawing Software*, pp. 77–103. Springer-Verlog, New York, 2004.

5. J. Berg and M. Lässig. Local graph alignment and motif search in biological networks. *Proceedings of the National Academy of Sciences*, 101:14689–14694, 2004.

6. G. C. Conant and A. Wagner. Convergent evolution of gene circuits. *Nature Genetics*, 34:264–266, 2003.

7. R. Dobrin, Q. K. Beg, A.-L. Barabási, and Z. N. Oltvai. Aggregation of topological motifs in the *Escherichia coli* transcriptional regulatory network. *BMC Bioinformatics*, 5:10, 2004.

8. FANMOD. http://www.minet.uni-jena.de/~wernicke/motifs/index.html.

9. T. Fruchterman and E. Reingold. Graph drawing by force-directed placement. *Software — Practice and Experience*, 21:1129–1164, 1991.

10. M. R. Garey and D. S. Johnson. *Computers and Intractability: A Guide to the Theory of NP-Completeness*. W. H. Freeman and Company, New York, 1979.

11. F. Harary and E. M. Palmer. *Graphical Enumeration*. Academic Press, New York, 1973.

12. M. Himsolt. Graphlet: Design and implementation of a graph editor. *Software — Practice and Experience*, 30:1303–1324, 2000.

13. P. J. Ingram, M. P. Stumpf, and J. Stark. Network motifs: Structure does not determine function. *BMC Genomics*, 7:108, 2006.

14. M. Kanehisa and P. Bork. Bioinformatics in the post-sequence era. *Nature Genetics*, 33:305–310, 2003.

15. N. Kashtan, S. Itzkovitz, R. Milo, and U. Alon. Mfinder tool guide. Technical report, Department of Molecular Cell Biology and Computer Science & Applied Mathematics, Weizman Institute of Science, 2002.

16. N. Kashtan, S. Itzkovitz, R. Milo, and U. Alon. Efficient sampling algorithm for estimating subgraph concentrations and detecting network motifs. *Bioinformatics*, 20:1746–1758, 2004.

17. N. Kashtan, S. Itzkovitz, R. Milo, and U. Alon. Topological generalizations of network motifs. *Physical Review E*, 70:031909, 2004.

18. H. Kitano. Systems Biology: A Brief Overview. *Science*, 295:1662–1664, 2002.

19. C. Klukas, F. Schreiber, and H. Schwöbbermeyer. Coordinated perspectives and enhanced force-directed layout for the analysis of network motifs. In K. Misue, K. Sugiyama, and J. Tanaka, editors, *Asia Pacific Symposium on Information Visualisation (APVIS2006)*, volume 60 of *CRPIT*, pp. 39–48, ACS, 2006.

20. T. I. Lee, N. J. Rinaldi, F. Robert, D. T. Odom, Z. Bar-Joseph, G. K. Gerber, N. M. Hannett, C. T. Harbison, C. M. Thompson, I. Simon, J. Zeitlinger, E. G. Jennings, H. L. Murray, D. B. Gordon, B. Ren, J. J. Wyrick, J.-B. Tagne, T. L. Volkert, E. Fraenkel, D. K. Gifford, and R. A. Young. Transcriptional regulatory networks in *Saccharomyces cerevisiae*. *Science*, 298:799–804, 2002.

21. S. Mangan and U. Alon. Structure and function of the feed-forward loop network motif. *Proceedings of the National Academy of Sciences*, 100:11980–11985, 2003.

22. S. Mangan, S. Itzkovitz, A. Zaslaver, and U. Alon. The incoherent feed-forward loop accelerates the response-time of the gal system of *Escherichia coli*. *Journal of Molecular Biology*, 356:1073–1081, 2006.

23. S. Mangan, A. Zaslaver, and U. Alon. The coherent feedforward loop serves as a sign-sensitive delay element in transcription networks. *Journal of Molecular Biology*, 334:197–204, 2003.

24. S. Maslov and K. Sneppen. Specificity and stability in topology of protein networks. *Science*, 296:910–913, 2002.

25. S. Maslov, K. Sneppen, and U. Alon. Correlation profiles and motifs in complex networks. In S. Bornholdt and H. G. Schuster, editors, *Handbook of Graphs and Networks: From the Genome to the Internet*, pp. 168–198. Wiley-VCH, Weinheim, Germany, 2003.

26. Brain Connectivity Toolbox. http://www.indiana.edu/~cortex/connectivity_toolbox.html.

27. MAvisto. http://mavisto.ipk-gatersleben.de/index.html.

28. A. Mazurie, S. Bottani, and M. Vergassola. An evolutionary and functional assessment of regulatory network motifs. *Genome Biology*, 6:R35, 2005.

29. B. D. McKay. Practical graph isomorphism. In *Proceedings of the 10th Manitoba Conference on Numerical Mathematics and Computing*, volume 30 of *Congressus Numerantium*, pp. 45–87, 1981.

30. Mfinder. http://www.weizmann.ac.il/mcb/UriAlon/index.html.

31. M. Middendorf, E. Ziv, and C. H. Wiggins. Inferring network mechanisms: The *Drosophila melanogaster* protein interaction network. *Proceedings of the National Academy of Sciences*, 102:3192–3197, 2005.

32. R. Milo, S. Itzkovitz, N. Kashtan, R. Levitt, and U. Alon. Response to comment on "network motifs: simple building blocks of complex networks" and "superfamilies of evolved and designed networks." *Science*, 305:1107d, 2004.

33. R. Milo, S. Itzkovitz, N. Kashtan, R. Levitt, S. Shen-Orr, I. Ayzenshtat, M. Sheffer, and U. Alon. Superfamilies of evolved and designed networks. *Science*, 303:1538–1542, 2004.

34. R. Milo, S. Shen-Orr, S. Itzkovitz, N. Kashtan, D. Chklovskii, and U. Alon. Network motifs: Simple building blocks of complex networks. *Science*, 298:824–827, 2002.

35. H. S. Moon, J. Bhak, K. H. Lee, and D. Lee. Architecture of basic building blocks in protein and domain structural interaction networks. *Bioinformatics*, 21:1479–1486, 2005.

36. R. J. Prill, P. A. Iglesias, and A. Levchenko. Dynamic properties of network motifs contribute to biological network organization. *PLoS Biology*, 3:e343, 2005.

37. Large transcriptional regulatory network of yeast. http://www.mrc-lmb.cam.ac.uk/genomes/madanm/tfcomb/.

38. N. Pržulj, D. G. Corneil, and I. Jurisica. Modeling interactome: Scale-free or geometric? *Bioinformatics*, 20:3508–3515, 2004.

39. RegulonDB. http://regulondb.ccg.unam.mx/index.html.

40. M. Reigl, U. Alon, and D. B. Chklovskii. Search for computational modules in the *C. elegans* brain. *BMC Biology*, 2:25, 2004.

41. J. J. Rice, A. Kershenbaum, and G. Stolovitzky. Lasting impressions: Motifs in protein–protein maps may provide footprints of evolutionary events. *Proceedings of the National Academy of Sciences*, 102:3173–3174, 2005.

42. S. Sakata, Y. Komatsu, and T. Yamamori. Local design principles of mammalian cortical networks. *Neuroscience Research*, 51:309–315, 2005.

43. F. Schreiber and H. Schwöbbermeyer. Frequency concepts and pattern detection for the analysis of motifs in networks. *Transactions on Computational Systems Biology*, 3 (LNBI 3737):89–104, 2005.

44. F. Schreiber and H. Schwöbbermeyer. MAVisto: A tool for the exploration of network motifs. *Bioinformatics*, 21:3572–3574, 2005.

45. S. Shen-Orr, R. Milo, S. Mangan, and U. Alon. Network motifs in the transcriptional regulation network of *Escherichia coli*. *Nature Genetics*, 31:64–68, 2002.

46. O. Sporns and R. Kötter. Motifs in brain networks. *PLoS Biology*, 2:e369, 2004.

47. Transcriptional regulatory networks in *Saccharomyces cerevisiae*. http://jura.wi.mit.edu/young/public/regulatory-index.html.

48. M. E. Wall, M. J. Dunlop, and W. S. Hlavacek. Multiple functions of a feed-forward-loop gene circuit. *Journal of Molecular Biology*, 349:501–514, 2005.

49. S. Wernicke. A faster algorithm for detecting network motifs. In *Proceedings of the 5th Workshop on Algorithms in Bioinformatics (WABI '05)*, volume 3692 of *LNBI*, pp. 165–177. Springer, 2005.

50. S. Wernicke and F. Rasche. FANMOD: A tool for fast network motif detection. *Bioinformatics*, 22:1152–1153, 2006.

51. S. Wuchty, Z. N. Oltvai, and A.-L. Barabási. Evolutionary conservation of motif constituents in the yeast protein interaction network. *Nature Genetics*, 35:176–179, 2003.

52. E. Yeger-Lotem, S. Sattath, N. Kashtan, S. Itzkovitz, R. Milo, R. Y. Pinter, U. Alon, and H. Margalit. Network motifs in integrated cellular networks of transcription-regulation and protein–protein interaction. *Proceedings of the National Academy of Sciences*, 101:5934–5939, 2004.

53. L. V. Zhang, O. D. King, S. L. Wong, D. S. Goldberg, A. H. Y. Tong, G. Lesage, B. Andrews, H. Bussey, C. Boone, and F. P. Rot. Motifs, themes and thematic maps of an integrated saccharomyces cerevisiae interaction network. *Journal of Biology*, 4:Epub, 2005.

6

NETWORK CLUSTERING

Balabhaskar Balasundaram and Sergiy Butenko

6.1 INTRODUCTION

Clustering can be loosely defined as the process of grouping objects into sets called *clusters*, so that each cluster consists of elements that are similar in some way. The similarity criterion can be defined in several different ways, depending on applications of interest and the objectives that the clustering aims to achieve. For example, in *distance-based clustering* (see Fig. 6.1) two or more elements belong to the same cluster if they are close with respect to a given distance metric. On the contrary, in *conceptual clustering*, which can be traced back to Aristotle and his work on classifying plants and animals, the similarity of elements is based on descriptive concepts.

Clustering is used for multiple purposes, including finding "natural" clusters (modules) and describing their properties, classifying the data, and detecting unusual data objects (outliers). In addition, treating a cluster or one of its elements as a single representative unit allows us to achieve *data reduction*.

Network clustering, which is the subject of this chapter, deals with clustering the data represented as a network, or a graph. Indeed, many data types can be conveniently modeled using graphs. This process is sometimes called *link analysis*. Data points are represented by vertices, and an edge exists if two data points are similar or related in a certain way. It is important to note that the similarity criterion used to construct the network model of a data set is based on *pairwise relations*, while the similarity criterion used to define a cluster refers to all elements in the cluster and needs to

Analysis of Biological Networks, Edited by Björn H. Junker and Falk Schreiber
Copyright © 2008 John Wiley & Sons, Inc.

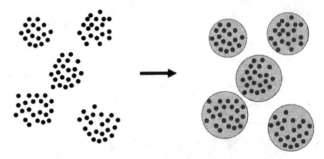

FIGURE 6.1   An illustration of distance-based clustering.

be satisfied by the cluster as a whole and not just pairs of its elements. In order to avoid confusion, from now on we will use the term *"cohesiveness"* when referring to the cluster similarity. Clearly, the definition of similarity (or dissimilarity) used to construct the network is determined by the nature of the data and based on the cohesiveness we expect in the resulting clusters.

In general, network clustering approaches can be used to perform both distance-based and conceptual clustering. In the distance-based clustering, the vertices of the graph correspond to the data points, and edges are added if the points are close enough based on some cutoff value. Alternately, the distances could just be used to weight the edges of a complete graph representing the data set. The following examples illustrate the use of networks in conceptual clustering. Database networks are often constructed by first designating a field as *matching field*, then vertices representing records in the database are connected by an edge if the two matching fields are "close." In *protein interaction networks*, the proteins are represented by vertices, and a pair is connected by an edge if they are known to interact. In *gene coexpression networks*, genes are vertices, and an edge indicates that the pair of genes (end points) are coexpressed over some cutoff value, based on microarray experiments.

It is not surprising that clustering concepts have been fundamental to data analysis, data reduction, and classification. Efficient data organization and retrieval that results from clustering has impacted every field of science and engineering that requires management of massive amounts of data. Cluster analysis techniques and algorithms in the areas of statistics and information sciences are well documented in several excellent textbooks [6,33,40,41,62]. Some recent surveys on cluster analysis for biological data, primarily using statistical methods, can be found in Refs. [42,59]. However, we are unaware of any text devoted to *network* clustering, which draws from several rich and diverse fields of study such as graph theory, mathematical programming, and theoretical computer science. The aim of this chapter is to provide an introduction to various clustering models and algorithms for networks modeled as simple, undirected graphs that exist in the literature. The basic concepts presented here are simple enough to be understood without any special background. Simple algorithms are presented whenever possible, and if more background is required apart from the basic graph theoretic ideas, we refer to the original literature and survey the results. We hope this chapter serves as a starting point to the readers for exploring this interesting area of research.

6.2 NOTATIONS AND DEFINITIONS

Chapter 2 provides an introduction to basic concepts in graph theory. In this section, we give some definitions that will be needed subsequently as well as describe the notations that we require. These definitions can be found in any introductory graph theory textbook, and for further reading we recommend the texts by Diestel [20] or West [66].

Let $G = (V, E)$ be a simple, finite, undirected graph. We consider such graphs only in this chapter, and we use n and m to denote the number of vertices and edges in G. Denote $\bar{G} = (V, \bar{E})$ and $G[S]$, the complement graph of G and the subgraph of G induced by $S \subseteq V$, respectively (see Chapter 2 for definitions). We denote $N(v)$, the set of neighbors of a vertex v in G. Note that $v \notin N(v)$, and $N[v] = N(v) \cup \{v\}$ is called a closed neighborhood. The *degree* of a vertex v is $\deg(v) = |N(v)|$, the cardinality of its neighborhood. The shortest distance (in terms of number of edges) between any two vertices $u, v \in V$ in G is denoted by $d(u, v)$. Then, the *diameter* of G is defined as $\operatorname{diam}(G) = \max_{u,v \in V} d(u, v)$. When G is not connected, $d(u, v)$ is ∞ for u and v in different components and so is the diameter of G. The *edge connectivity* $\kappa'(G)$ of a graph is the minimum number of edges that must be removed to disconnect the graph. Similarly, the *vertex connectivity* (or just connectivity) $\kappa(G)$ of a graph is the minimum number of vertices whose removal results in a disconnected or trivial graph. In the graph in Fig. 6.2, $\kappa(G) = 2$ (e.g., removal of vertices 3 and 5 would disconnect the graph) and $\kappa'(G) = 2$ (e.g., removal of edges $(9,11)$ and $(10,12)$ would disconnect the graph).

A *clique* C is a subset of vertices such that an edge exists between every pair of vertices in C, that is, the induced subgraph $G[C]$ is complete. A subset of vertices I is called an *independent set* (also called a *stable set*) if for every pair of vertices

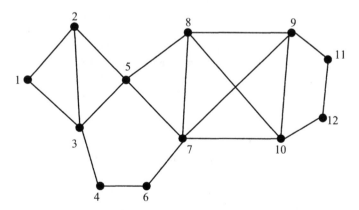

FIGURE 6.2   An example graph.

in I, (i, j) is not an edge, that is, $G[I]$ is edgeless. An independent set (or clique) is *maximal* if it is not a subset of any larger independent set (clique), and *maximum* if there are no larger independent sets (cliques) in the graph. For example, in Fig. 6.2, $I = \{3, 7, 11\}$ is a maximal independent set as there is no larger independent set containing it (we can verify this by adding each vertex outside I to I and it will no longer be an independent set). The set $\{1, 4, 5, 10, 11\}$ is a maximum independent set, one of the largest cardinality in the graph. Similarly $\{1, 2, 3\}$ is a maximal clique and $\{7, 8, 9, 10\}$ is the maximum clique. Note that C is a clique in G if and only if C is an independent set in the complement graph $\bar{G}$. The *clique number* $\omega(G)$ and the independence number $\alpha(G)$ are the cardinalities of a maximum clique and independent set in G, respectively. *Maximal* independent sets and cliques can be found easily using simple algorithms commonly referred to as "greedy" algorithms. We explain such an algorithm for finding a maximal independent set. Since we know that after adding a vertex v to the set, we cannot add any of its neighbors, it is intuitive (or greedy!) to add a vertex of minimum degree in the graph so that we remove fewer vertices leaving a larger graph as we are generally interested in finding large independent sets (or if possible a maximum independent set). The greedy maximal independent set algorithm is presented in the following, and Fig. 6.3 illustrates the progress of this algorithm.

> **greedy_maximal_independent_set_algorithm** (graph $G = (V, E)$)
>   initialize $I := \emptyset$;
>   while $V \neq \emptyset$ {
>   pick a vertex v of minimum degree in $G[V]$;
> $I := I \cup \{v\}$;
> $V := V \setminus N[v]$;
>   }

The greedy maximal clique algorithm works similarly, but with one difference. Since after adding a vertex v, we can only consider neighbors of v to be included, we pick v to be a vertex of maximum degree in the graph in an attempt to find a large clique. This algorithm is part of a clustering approach discussed in detail in Section 6.4. If suitable vertex weights exist, which are determined *a priori* or dynamically during the course of an algorithm, the weights could be used to guide the vertex selection instead of vertex degrees, but in a similar fashion (See Exercise 1).

A *dominating set* $D \subseteq V$ is a set of vertices such that every vertex in the graph is either in this set or has a neighbor in this set. A dominating set is said to be *minimal* if it contains no proper subset that is dominating and it is said to be a *minimum* dominating set if it is of the smallest cardinality. Cardinality of a *minimum* dominating set is called the *domination number*, $\gamma(G)$, of a graph. For example, $D = \{7, 11, 3\}$ is a minimal and minimum dominating set of the graph in Fig. 6.2. A *connected dominating set* is one in which the subgraph induced by the dominating set is connected and an *independent dominating set* is one in which the dominating set is also independent. In Section 6.5, we describe clustering approaches based on these two models.

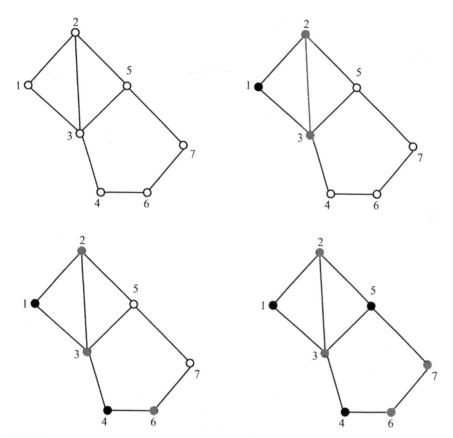

FIGURE 6.3   Greedy maximal independent set algorithm progress. Black vertex is added to *I* and gray vertices are the neighbors considered removed along with the black vertices. If a tie exists between many vertices of minimum degree, we choose the one with the smallest index.

Algorithms and complexity: It is important to acknowledge the fact that many models used for clustering networks are believed to be computationally "intractable," or "NP-hard." Even after decades of research, no efficient algorithm is known for a large class of such intractable problems. However, an efficient algorithm for any such problem implies existence of an efficient algorithms for *all* such problems! An efficient algorithm being one that runs in time (measured as the number of fundamental operations done by the algorithm) that is a fixed polynomial function of the input size. A problem is said be "tractable" if such a polynomial-time algorithm is known. The known algorithms for intractable problems, on the contrary, take time that is exponential in input size. Consequently these algorithms are extremely time consuming on large networks. The "theory of NP-completeness" is a rich field of study in theoretical computer science that studies the tractability of problems and classifies them broadly as "tractable" or "intractable." While there are several clustering models that are tractable, many interesting ones are not. The intractable problems are often approached using heuristic or approximate methods. The basic difference between approximation algorithms and heuristics is the following. Approximation algorithms provide performance guarantees on the quality of the solution returned by the algorithm. An algorithm for a minimization problem with approximation ratio α (> 1) returns a solution with an objective value that is no larger than α times the minimum objective value. Heuristics, on the contrary, provide no such guarantee and are usually evaluated empirically based on their performance on established benchmark testing instances for the problem they are designed to solve. We will describe simple greedy heuristics and approximation algorithms for hard clustering problems in this chapter. An excellent introduction to complexity theory can be found in Refs. [27,50], whereas the area of approximation algorithms is described well in Refs. [7,36,64]. For the application of meta-heuristics in solving combinatorial optimization problems, see Ref. [2].

6.3 NETWORK CLUSTERING PROBLEM

Given a graph $G^0 = (V^0, E^0)$, the *clustering problem* is to find subsets (not necessarily disjoint) $\{V_1^0, \ldots, V_r^0\}$ of V^0 such that $V^0 = \bigcup_{i=1}^r V_i^0$, where each subset is a *cluster* modeled by structures such as cliques or other distance and diameter-based models. The model used as a cluster represents the cohesiveness required of the cluster. The clustering models can be classified by the constraints on relations between clusters (clusters may be disjoint or overlapping) and the objective function used to achieve the goal of clustering (minimizing the number of clusters or maximizing the cohesiveness). When the clusters are required to be disjoint, $\{V_1^0, \ldots, V_r^0\}$ is a *cluster-partition* and when they are allowed to overlap, it is a *cluster-cover*. The first approach is called the *exclusive clustering*, whereas the second *overlapping clustering*.

For a given G^0, assuming that there is a measure of cohesiveness of the cluster that can be varied, we can define two types of optimization problems:

Type I: Minimize the number of clusters while ensuring that every cluster formed has cohesiveness over a prescribed threshold;

Type II: Maximize the cohesiveness of each cluster formed, while ensuring that the number of clusters that result is under a prescribed number K (the last requirement may be relaxed by setting $K = \infty$).

As an example of Type I clustering, consider the problem of clustering an incomplete graph with cliques used as clusters and the objective of minimizing the number of clusters. Alternately, assume that G^0 has non-negative edge weights w_e, $e \in E^0$. For a cluster V_i^0, let E_i^0 denote the edges in the subgraph induced by V_i^0. Treating w as a dissimilarity measure (distance), $w(E_i^0) = \sum_{e \in E_i^0} w_e$ or $\max_{e \in E_i^0} w_e$ are meaningful measures of cohesiveness that can be used to formulate the corresponding Type II clustering problems. Henceforth, we will refer to problems from the literature as Type I and Type II based on their objective.

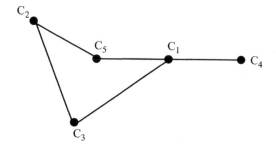

FIGURE 6.4   Abstracted version of the example graph.

After performing clustering, we can abstract the graph G^0 to a graph $G^1 = (V^1, E^1)$ as follows: there exists a vertex $v_i^1 \in V^1$ for every subset (cluster) V_i^0 and there exists an edge between v_i^1, v_j^1 if and only if there exist $x^0 \in V_i^0$ and $y^0 \in V_j^0$ such that $(x^0, y^0) \in E^0$. In other words, if any two vertices from different clusters had an edge between them in the original graph G^0, then the clusters containing them are made adjacent in G^1. We can recursively cluster the abstracted graph G^1 in a similar fashion to obtain a multilevel hierarchy. This process is called the *hierarchical clustering*. Consider the example graph in Fig. 6.2, the following subsets form clusters in this graph: $C_1 = \{7, 8, 9, 10\}$, $C_2 = \{1, 2, 3\}$, $C_3 = \{4, 6\}$, $C_4 = \{11, 12\}$, $C_5 = \{5\}$. In Section 6.4, we will describe how this clustering was accomplished. However, given the clusters of the example graph in Fig. 6.2, call it G^0, we can construct an abstracted graph G^1 as shown in Fig. 6.4. In G^1, C_1 and C_4 are adjacent since $9 \in C_1$ and $11 \in C_4$ are adjacent in G^0. Other edges are also added in a similar fashion.

The remainder of this chapter is organized as follows. We describe two broad approaches to clustering and discuss Type I and Type II models in each category. In Section 6.4, we present a Type I and Type II model, each based on cliques that are popular in clustering biological networks. In Section 6.5, we discuss center-based models that are popular in clustering wireless networks, but have strong potential for use in biological networks. For all the models discussed, simple greedy approaches that are easy to understand are described. We then conclude by pointing to more sophisticated models and solution approaches and some general issues that need to be considered before using clustering techniques.

6.4   CLIQUE-BASED CLUSTERING

A clique is a natural choice for a highly cohesive cluster. Cliques have minimum possible diameter, maximum connectivity, and maximum possible degree for each vertex—respectively interpreted as reachability, robustness, and familiarity, the best situation in terms of structural properties that have important physical meaning in a cluster.

Given an arbitrary graph, a Type I approach tries to partition it into (or cover it using) minimum number of cliques. Type II approaches usually work with a weighted complete graph and hence every partition of the vertex set is a clique partition. The objective here is to maximize cohesiveness (by minimizing maximum edge dissimilarity) within the clusters.

6.4.1 Minimum Clique Partitioning

Type I clique partitioning and clique covering problems are both NP-hard [27]. Consequently, exact approaches to solve these problems that exist in the literature are computationally ineffective for large graphs. Heuristic approaches are preferred for large graphs for this reason.

Before proceeding further, we should note that clique-partitioning and clique-covering problems are closely related. In fact, the minimum number of clusters produced in clique covering and partitioning are the same. Denote covering optimum by c and partition optimum by p. Since every clique partition is also a cover, $p \geq c$. Let $\{V_1^0, \ldots, V_c^0\}$ be an optimal clique cover. Any vertex v present in multiple clusters causing overlaps can be removed from all but one of the clusters to which it belongs, leaving the resulting cover with the same number of clusters and one less overlap. Repeating this as many times as necessary would result in a clique partition with the same number of clusters c. Thus, we can conclude that $p = c$. We are now ready to describe a simple heuristic for clique partitioning:

> **greedy_clique_partitioning_algorithm** (graph $G = (V, E)$)
>   initialize $i := 0; Q := \emptyset$;
>   while $V \setminus Q \neq \emptyset$ {
> $i := i + 1$;
> $C_i := \emptyset$;
> $V' := V \setminus Q$;
>   while $V' \neq \emptyset$ {
>   pick a vertex v of maximum degree in $G[V']$;
> $C_i := C_i \cup \{v\}$;
> $V' := V' \cap N(v)$;
>   }
> $Q := Q \cup C_i$;
>   }

The *greedy clique partitioning algorithm* finds maximal cliques in a greedy fashion starting from a vertex of maximum degree. Each maximal clique found is then fixed as a cluster and removed from the graph. This is repeated to find the next cluster until all vertices belong to some cluster. When the algorithm begins, as no clusters are known, the set Q that collects the clusters found is initialized to an empty set. The outer while-loop checks whether there are vertices remaining in the graph that have not been assigned to any cluster, and proceeds if they exist. When $i = 1$, $V' = V$ (the original graph) and the first cluster C_1 is found by the inner while-loop. First a maximum degree vertex, say v, is found and added to C_1. Since C_1 has to be a clique, all the non-neighbors of v are removed when V' is reset to $V' \cap N(v)$, restricting us to look only at neighbors of v. If the resulting V' is empty, then we stop the inner while-loop, otherwise we repeat this process to find the next vertex of maximum degree in the newly found "residual graph" $G[V']$ to be added to C_1. During the progress of the inner while-loop, all non-neighbors for each newly added vertex are removed from V', thereby ensuring that when the inner while-loop terminates, C_1 is a clique and it is added to the collection of clusters Q. The algorithm is greedy in its approach as we always give preference to a vertex of maximum degree in the residual graph $G[V']$ to be included in a cluster (clique). Then we check if $V \setminus Q$ is not empty, meaning that there are vertices still in the graph not assigned to any cluster. For the next iteration $i = 2$, vertices in C_1 are removed since $V' := V \setminus Q$. The algorithm then proceeds in a similar fashion. The inner while-loop by itself is the greedy algorithm for finding a maximal clique in $G[V']$. Table 6.1 reports the progress of this algorithm on the example graph shown in Fig. 6.2, and Fig. 6.5 illustrates the result of this algorithm.

Extensive work has been done in the recent past to understand the structure and function of proteins based on protein–protein interaction maps of organisms. Clustering protein interaction networks (PINs) using cliques has formed the basis for several studies that attempt to decompose the PIN into functional modules and protein complexes. Protein complexes are groups of proteins that interact with each other at the same time and place, while functional modules are groups of proteins that are known to have pairwise interactions by binding with each other to participate

TABLE 6.1   The Progress of Greedy Clique Partitioning Algorithm on the Example Graph, Fig. 6.2.

iter. i	Q	C_i	V'	v^a
1	$\varnothing$	$\varnothing$	V^b	7
		$\{7\}$	$\{5, 6, 8, 9, 10\}$	8
		$\{7, 8\}$	$\{5, 9, 10\}$	9^c
		$\{7, 8, 9\}$	$\{10\}$	10
		$C_1 = \{7, 8, 9, 10\}$	$\varnothing$	
2	C_1	$\varnothing$	$\{1, 2, 3, 4, 5, 6, 11, 12\}$	3
		$\{3\}$	$\{1, 2, 4, 5\}$	2
		$\{2, 3\}$	$\{1, 5\}$	1^c
		$C_2 = \{1, 2, 3\}$	$\varnothing$	
3	$C_1 \cup C_2$	$\varnothing$	$\{4, 5, 6, 11, 12\}$	4^c
		$\{4\}$	$\{6\}$	6
		$C_3 = \{4, 6\}$	$\varnothing$	
4	$C_1 \cup C_2 \cup C_3$	$\varnothing$	$\{5, 11, 12\}$	11^c
		$\{11\}$	$\{12\}$	12
		$C_4 = \{11, 12\}$	$\varnothing$	
5	$C_1 \cup C_2 \cup C_3 \cup C_4$	$\varnothing$	$\{5\}$	5
		$C_5 = \{5\}$	$\varnothing$	

[a]v is the vertex of maximum degree in $G[V']$.
[b]$V = \{1, 2, 3, 4, 5, 6, 7, 8, 9, 10, 11, 12\}$.
[c]Means a tie was broken between many vertices having the maximum degree in $G[V']$ by choosing the vertex with smallest index.

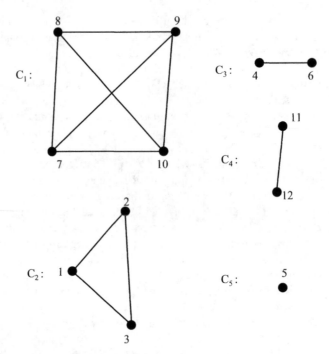

FIGURE 6.5   Result of the greedy clique partitioning algorithm on the graph in Fig. 6.2.

in different cellular processes at different times [63]. Spirin and Mirny [63] use cliques and other high density subgraphs to identify protein complexes (splicing machinery, transcription factors, etc.) and functional modules (signalling cascades, cell-cycle regulation, etc.) in *Saccharomyces cerevisiae*. Gagneur et al. [26] introduce a notion of modular decomposition of PINs using cliques that not only identifies a hierarchical relationship, but in addition introduces a labeling of modules (as "series, parallel, and prime") that results in logical rules to combine proteins into functional complexes. Bader et al. [9] use logistic regression based methodology incorporating screening statistics and network structure in gaining confidence from high-throughput proteomic data containing significant background noise. They propose constructing a high-confidence network of interactions by merging proteomics data with gene expression data. They also observe that cliques in the high-confidence network are consistent with those found in the original network, and the distance-related properties of the original network are not altered in the high-confidence network.

6.4.2   Min–Max k-Clustering

The *min–max k-clustering problem* is a Type II clique partitioning problem with min–max objective. Consider a weighted complete graph $G = (V, E)$ with weights $w_{e_1} \leq w_{e_2} \leq \cdots \leq w_{e_m}$, where $m = n(n-1)/2$. The problem is to partition the graph into no more than k cliques such that the maximum weight of an edge between two vertices inside a clique is minimized. In other words, if $V_1, \ldots, V_k$ is the clique partition, then we wish to minimize $\max_{i=1\ldots,k} \max_{u,v \in V_i} w_{uv}$. This problem is NP-hard and it is NP-hard to approximate within a factor less than two, even if the edge weights obey *triangle inequality* (i.e., for every distinct triple $i, j, k \in V$, $w_{ij} + w_{jk} \geq w_{ik}$). The best possible approximation algorithms (returning a solution that is no larger than twice the minimum) for this problem with edge weights obeying triangle inequality are available in Refs. [30,38].

The weight w_{ij} between two vertices i and j can be thought of as a measure of dissimilarity—larger w_{ij} means more dissimilar i and j are. The problem then tries to cluster the graph into at most k cliques such that the maximum dissimilarity between any two vertices inside a clique is minimized. Given any graph $G' = (V', E')$, the required edge weighted complete graph G can be obtained in different ways using meaningful measures of dissimilarity. The weight w_{ij} could be $d(i, j)$, the shortest distance between i and j in G'. Other appropriate measures are $\kappa(i, j)$ and $\kappa'(i, j)$, which denote, respectively, the minimum number of vertices and edges that need to be removed from G' to disconnect i and j [28]. Since these are measures of similarity (since larger value for either indicates the two vertices are "strongly" connected to each other), we could obtain the required weights as $w_{ij} = |V'| - \kappa(i, j)$ or $|E'| - \kappa'(i, j)$. These structural properties used for weighting can all be computed efficiently in polynomial time [5] and are applicable to protein interaction networks. One could also consider using statistical correlation between pairs of vertices in weighting the edge between them. Since correlation is a similarity measure, we could use one minus the correlation coefficient between pairs of vertices as a dissimilarity measure to weight the edges. This is especially applicable in gene coexpression networks.

The *bottleneck graph* of a weighted graph $G = (V, E)$ is defined for a given number c as follows: $G(c) = (V, E_c)$ where $E_c = \{e \in E : w_e \leq c\}$. The bottleneck graph $G(c)$ contains only those edges with weight at most c. Figure 6.6 illustrates the concept of bottleneck graphs. This notion has been predominantly used as a

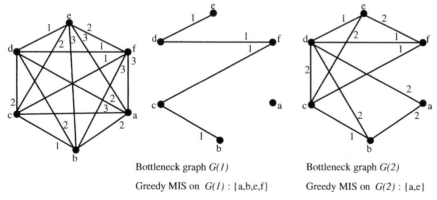

Bottleneck graph G(1)

Greedy MIS on G(1) : {a,b,e,f}

Bottleneck graph G(2)

Greedy MIS on G(2) : {a,e}

FIGURE 6.6 An example complete weighted graph G and its bottleneck graphs $G(1)$ and $G(2)$ for weights 1 and 2, respectively. MIS found using the greedy algorithm with ties between minimum degree vertices broken by choosing vertices in alphabetical order.

procedure to reveal the hierarchy in G. For simplicity assume that the edges are sorted and indexed so that $w_{e_1} \leq w_{e_2} \leq \cdots \leq w_{e_m}$. Note that $G(0)$ is an edgeless graph and $G(w_{e_m})$ is the original graph G. As c varies in the range w_{e_1} to w_{e_m}, we can observe how the components appear and merge as c increases, enabling us to develop a dendrogram or a hierarchical tree structure. Similar approaches are discussed by Girvan and Newman [28] to detect community structure in social and biological networks.

Next we present a bottleneck heuristic for the min–max k-clustering problem proposed in Ref. [38]:

bottleneck_min-max_k-clustering_algorithm (graph $G = (V, E)$, sorted edges $w_{e_1} \leq w_{e_2} \leq \cdots \leq w_{e_m}$)

        initialize $i := 0, stop := false$;
        while $stop = false$ {
           $i := i + 1$;
           $G_b := bottleneck(w_{e_i})$;
           $I := MIS(G_b)$;
           if $|I| \leq k$ {
              return I;
              $stop := true$;
           }
        }

The procedure $bottleneck(w_{e_i})$ returns the bottleneck graph $G(w_{e_i})$, and $MIS(G_b)$ is an arbitrary procedure for finding a maximal independent set (MIS) in G, as illustrated in Fig. 6.6. One of the simplest procedures to finding a MIS in a graph G is the greedy approach described in Section 6.2.

Consider the algorithm during some iteration i. If there exists a MIS in $G(w_{e_i})$ of size more than k, then we cannot partition $G(w_{e_i})$ into k cliques as the vertices in the MIS must belong to different cliques. This also means there is no clique partition with at most k cliques in G with maximum edge weight in all cliques under w_{e_i}. Hence, we know that the optimum answer to our problem is at least $w_{e_{i+1}}$ (this observation is critical to show that the solution returned is no larger than twice the minimum possible when the edge weights satisfy the triangle inequality). Since our objective is to not have more than k cliques, we proceed to the next iteration and consider the next bottleneck graph $G(w_{e_{i+1}})$. On the contrary, if in iteration i the MIS we find in $G(w_{e_i})$ is of size less than or equal to k, we terminate and return the MIS found to create clusters. Note that although we found an MIS of size at most k, it does not imply that there are no independent sets in $G(w_{e_i})$ of size larger than k. This algorithm will actually be optimal if we manage to find a maximum independent set (one of largest size) in every iteration. However, this problem is NP-hard and we have to restrict ourselves to finding MIS using heuristic approaches such as the greedy approach described earlier to have a polynomial time algorithm.

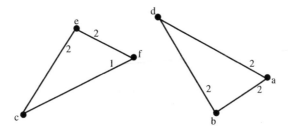

FIGURE 6.7 The output of the bottleneck min–max two-clustering algorithm on the graph shown in Fig. 6.6.

Without loss of generality, let $I = \{1, \ldots, p\}$ be the output of the above algorithm terminating in iteration i with the bottleneck graph $G(w_{e_i})$. We can form p clusters $V_1, \ldots, V_p$ by taking each vertex in I and its neighbors in $G(w_{e_i})$. Note that the resulting clusters could overlap. In order to obtain a partition, we can create these clusters sequentially from 1 to p while ensuring that only neighbors that are not already in any previously formed cluster are included for the cluster currently being formed. In other words, the cluster cover $V_1, \ldots, V_p$ are the closed neighborhoods of $1, \ldots, p$ in the last bottleneck graph $G(w_{e_i})$. To obtain a cluster partition, vertices included in $V_1, \ldots, V_{l-1}$ are not included in V_l for $l = 2, \ldots, p$. $G[V_1], \ldots, G[V_p]$ is a partition of G into p cliques and $p \leq k$. For any edge in any clique $G[V_l]$, it is either between l and a neighbor of l in $G(w_{e_i})$ or it is between two neighbors of l in $G(w_{e_i})$. In the first case, the edge weight is at most w_{e_i} and in the second case, the edge weight is at most $2w_{e_i}$ as the edge weights satisfy the triangle inequality. This is true for all clusters and hence the maximum weight of any edge in the resulting clusters is at most $2w_{e_i}$. Since we found an independent set of size more than k in iteration $i - 1$ (which is the reason we proceeded to iteration i), as noted earlier the optimum answer $w^* \geq w_{e_i}$ and our solution $2w_{e_i}$ is at most $2w^*$, that is, $w^* \leq 2w_{e_i} \leq 2w^*$. Given the intractability of obtaining better approximation, this is likely the best we can do in terms of a guaranteed quality solution. Figure 6.6 showing the bottleneck graphs also shows the MIS found using a greedy algorithm with ties between minimum degree vertices broken according to the alphabetical order of their labels. Figure 6.7 shows the clustering output of the bottleneck min–max k-clustering algorithm with $k = 2$ on the graph shown in Fig. 6.6.

6.5   CENTER-BASED CLUSTERING

In center-based clustering models, the elements of a cluster are determined based on their similarity with the cluster's center or *cluster-head*. The center-based clustering algorithms usually consist of two steps. First, an optimization procedure is used to determine the cluster-heads, which are then used to form clusters around them. Popular approaches such as *k-means clustering* used in clustering biological networks fall under this category. However, k-means and its variants are Type II approaches as k is fixed. We present here some Type I approaches as well as a Type II approach suitable for biological networks.

6.5.1 Clustering with Dominating Sets

Recall the definition of a dominating set from Section 6.2. Minimum dominating set and related problems provide a modeling tool for center-based clustering of Type I. The use of dominating sets in clustering has been quite popular especially in telecommunications [10,18]. Clustering here is accomplished by finding a small dominating set. Since the minimum dominating set problem is NP-hard [27], heuristic approaches and approximation algorithms are used to find a small dominating set.

If D denotes a dominating set, then for each $v \in D$ the closed neighborhood $N[v]$ forms a cluster. By the definition of domination, every vertex not in the dominating set has a neighbor in it and hence is assigned to some cluster. Each v in D is called a *cluster-head* and the number of clusters that result is exactly the size of the dominating set. By minimizing the size of the dominating set, we minimize the number of clusters produced resulting in a Type I clustering problem. This approach results in a cluster cover since the resulting clusters need not be disjoint. Clearly, each cluster has diameter at most two, as every vertex in the cluster is adjacent to its cluster-head and the cluster-head is "similar" to all the other vertices in its cluster. However, the neighbors of the cluster-head may be poorly connected among themselves. Furthermore, some postprocessing may be required as a cluster formed in this fashion from an arbitrary dominating set could completely contain another cluster.

Clustering with dominating sets is especially suited for clustering protein interaction networks to reveal groups of proteins that interact through a central protein, which could be identified as a cluster-head in this method. We will now point out some simple approaches to obtain different types of dominating sets.

Independent Dominating Sets   Recall the definition of independent domination from Section 6.2. Note that a maximal independent set I is also a minimal dominating set (e.g., $\{3, 7, 11\}$ in Fig. 6.2 is a maximal independent set, which is also a minimal dominating set). Indeed, every vertex outside I has a neighbor inside (otherwise it can be added to this set contradicting its maximality) making it a dominating set. Furthermore, if there exists $I' \subset I$ that is dominating, then there exists some vertex $v \in I \setminus I'$ that is adjacent to some vertex in I'. This contradiction to independence of I indicates that I is a minimal dominating set.

Hence finding a maximal independent set results also in a minimal *independent* dominating set that can be used in clustering the graph as described in the introduction. Here, no cluster formed can contain another cluster completely, as the cluster-heads are independent and different. However, for two cluster-heads v_1, v_2, $N(v_1)$ could be completely contained in $N(v_2)$. Neither minimality nor independence are affected by this property. The resulting cluster cover can be easily converted to a partition by observing that vertices common to multiple clusters are simply neighbors of multiple cluster-heads and can be excluded from all clusters but one.

The greedy algorithm for minimal independent dominating sets below proceeds by adding a maximum degree vertex to the current independent set and then deleting that vertex along with its neighbors. Note that this algorithm is greedy because it adds a maximum degree vertex so that a larger number of vertices are removed in each iteration, yielding a small independent dominating set. This is repeated until no more vertices exist. The progress of this algorithm is detailed in Table 6.2. The result of this algorithm on the graph in Fig. 6.2 is illustrated in Fig. 6.8.

> **greedy_minimal_independent_dominating_set_algorithm** (graph $G = (V, E)$)
> initialize $I := \emptyset$, $V' := V$;
> while $V' \neq \emptyset$ {
>   pick a vertex v of maximum degree in $G[V']$;
> $I := I \cup \{v\}$;
> $V' := V' \setminus N[v]$;
> }

Connected Dominating Sets   In some situations, it might be necessary to ensure that the cluster-heads themselves are connected in addition to being dominating, and not independent as in the previous model. A connected dominating set (CDS) is a dominating set D such that $G[D]$ is connected. Finding a minimum CDS is also a NP-hard problem [27], but approximation algorithms for the problem exist [31]. Naturally, G is assumed to be connected for this problem.

The following algorithm is a greedy vertex elimination type heuristic for finding CDS from Ref. [15]. In this heuristic, we pick a vertex of *minimum* degree u in the graph and delete it, if deletion does not disconnect the graph. If it does, then the vertex is fixed (added to set F) to be in the CDS. Upon deletion, if u has no neighbor in the fixed vertices, a vertex of *maximum* degree in $G[D]$ that is also in the neighborhood of u is fixed ensuring that u is dominated. Thus, in every iteration D is connected and is a dominating set in G. The algorithm terminates when all the vertices left in D are fixed ($D = F$) and that is the output CDS of the algorithm. See Table 6.3 for the progress of the algorithm on the example graph and the result is shown in Fig. 6.9. The resulting clusters are all of diameter at most two, as vertices in each cluster are

TABLE 6.2   The Progress of Greedy Minimal Independent Dominating set Algorithm on the Example Graph, Fig. 6.2.

iter.	I	V'	v^a
0	$\emptyset$	V^b	7
1	$\{7\}$	$\{1,2,3,4,11,12\}$	3
2	$\{7,3\}$	$\{11,12\}$	11^c
3	$\{7,3,11\}$	$\emptyset$	

[a] v is the vertex of maximum degree in $G[V']$.
[b] $V = \{1, 2, 3, 4, 5, 6, 7, 8, 9, 10, 11, 12\}$.
[c] Means a tie was broken between many vertices having the maximum degree in $G[V']$ by choosing the vertex with smallest index.

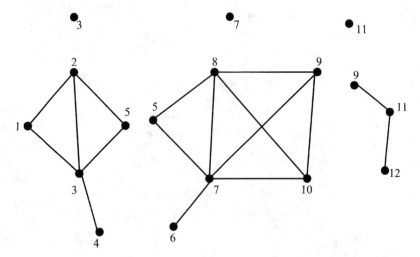

FIGURE 6.8 Result of the greedy minimal independent dominating set algorithm on the graph in Fig. 6.2. The minimal independent dominating set found is {7, 3, 11}. The figure also shows the resulting cluster cover.

adjacent to their cluster-head and the cluster-heads form a CDS.

```
greedy_CDS_algorithm (connected graph G = (V, E))
       initialize D := V; F := Ø;
       while D \ F ≠ Ø {
           pick u ∈ D \ F with minimum degree in G[D];
           if G[D \ {u}] is disconnected
               F := F ∪ {u};
           else {
               D := D \ {u};
               if N(u) ∩ F = Ø {
                   pick w ∈ N(u) ∩ D with maximum degree in G[D];
                   F := F ∪ {w};
               }
           }
       }
```

An alternate constructive approach to CDS uses spanning trees. A spanning tree in $G = (V, E)$ is a subgraph $G' = (V, E')$ that contains all the vertices V and G' is a tree. See Chapter 2 for additional definitions. All the nonleaf (inner) vertices of a spanning tree can be used as a CDS. Larger the number of leaves, smaller the CDS. The problem is hence related to the problem of finding a spanning tree with maximum number of leaves, which is also NP-hard [27]. However, approximation algorithms with guaranteed performance ratios can be found in Refs. [47,48,61] for this problem.

TABLE 6.3   The Progress of Greedy Connected Dominating Set Algorithm on the Example Graph, Fig. 6.2.

D	F	u^a	$N(u) \cap F$	$N(u) \cap D$	w^b
V^c	$\emptyset$	1^d	$= \emptyset$	$\{2, 3\}$	3
$V \setminus \{1\}$	$\{3\}$	2^d	$\neq \emptyset$	-	-
$V \setminus \{1, 2\}$	$\{3\}$	4^d	$\neq \emptyset$	-	-
$V \setminus \{1, 2, 4\}$	$\{3\}$	6	$= \emptyset$	$\{7\}$	7
$V \setminus \{1, 2, 4, 6\}$	$\{3, 7\}$	11^d	$= \emptyset$	$\{9, 12\}$	9
$\{3, 5, 7, 8, 9, 10, 12\}$	$\{3, 7, 9\}$	12	$= \emptyset$	$\{10\}$	10
$\{3, 5, 7, 8, 9, 10\}$	$\{3, 7, 9, 10\}$	5^e	-	-	-
$\{3, 5, 7, 8, 9, 10\}$	$\{3, 7, 9, 10, 5\}$	8	$\neq \emptyset$	-	-
$\{3, 5, 7, 9, 10\}$	$\{3, 7, 9, 10, 5\}$				

$^a u$ is the vertex of minimum degree in $G[D]$ in $D \setminus F$.
$^b w$ is the vertex of maximum degree in $G[D]$ in $N(u) \cap D$.
$^c V = \{1, 2, 3, 4, 5, 6, 7, 8, 9, 10, 11, 12\}$.
^{d}Means a tie was broken between many vertices by choosing the vertex with smallest index.
^{e}Indicates that $G[D \setminus \{u\}]$ is disconnected.

6.5.2 k-Center Clustering

k-Center clustering is a variant of the well-known k-means clustering approach. Several variants of *k-means clustering* have been widely used in clustering data, including biological data [12,16,21,25,32,46,54,56,67]. In these approaches, we seek

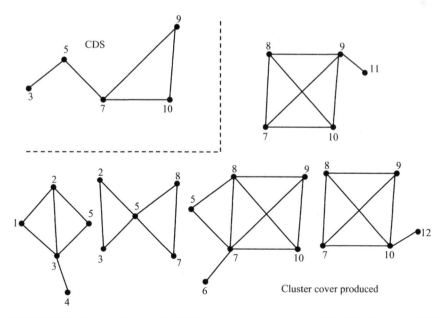

FIGURE 6.9   Result of the greedy CDS algorithm on the graph in Fig. 6.2. The found CDS $\{3, 5, 7, 9, 10\}$ is shown with the resulting cluster cover.

to identify k-cluster-heads (k clusters) such that some objective that measures the dissimilarity (distance) of the members of a cluster to the cluster-head is minimized. The objective could be to minimize the maximum distance of any vertex to the cluster-heads, or the total distance of all vertices to the cluster-heads. Other objectives include minimizing mean and variance of intracluster distance to cluster-head over all clusters. Different choice of dissimilarity measures and objectives yield different clustering problems and often, different clustering solutions. The traditional k-means clustering deals with clustering points in the n-dimensional Euclidean space where the dissimilarity measure is the Euclidean distance and the objective is to minimize the mean squared distance to cluster centroid (the geometric centroid of points in the cluster).

The *k-center problem* is a Type-II center-based clustering model with a min–max objective that is similar to the above approaches. Here the objective is to minimize the maximum distance of any vertex to the cluster-heads, where the distance of a vertex to the cluster-heads is the distance to the closest cluster-head. Formally, the problem is defined on a complete edge weighted graph $G = (V, E)$ with non-negative edge weights w_e, $e \in E$. These weights can be assumed to be a distance or dissimilarity measure. In Section 6.4.2, we have already discussed several ways to weight the edges. The problem is to find a subset of at most k vertices S such that the cost of the k-center given by $w(S) = \max_{i \in V} \min_{j \in S} w_{ij}$ is minimum. It can be interpreted as follows: for each $i \in V$, its distance to S is its distance to a closest vertex in S given by $\min_{j \in S} w_{ij}$. This distance to S is clearly zero for all vertices $i \in S$. The measure we wish to minimize is the distance of the farthest vertex in $V \setminus S$ from S. Given a k-center S, clusters can be formed as follows: Let $V_i = \{v \in V \setminus S : w_{iv} \leq w(S)\} \cup \{i\}$ for $i \in S$, then $V_1, \ldots, V_{|S|}$ cover G. If we additionally require that the distances obey the triangle inequality, then each $G[V_i]$ is a clique with the distance between every pair of vertices being no more than $2w(S)$. This leads to a Type II optimization problem where we are allowed to use up to k clusters and we wish to minimize the dissimilarity in each cluster that is formed (by minimizing $w(S)$). This problem is NP-hard and it is NP-hard to approximate within a factor less than 2, even if the edge weights satisfy the triangle inequality. We present here an approximation algorithm from Ref. [37] that provides a k-center S such that $w(S^*) \leq w(S) \leq 2w(S^*)$, where S^* is an optimum k-center. As it was the case with the bottleneck algorithm for the min–max k-clustering presented in Section 6.4.2, the approximation ratio result is guaranteed only for the case when the triangle inequality holds.

> **bottleneck_k-center_algorithm** (graph $G = (V, E)$, sorted edges $w_{e_1} \leq w_{e_2} \leq \cdots \leq w_{e_m}$)
>
>   Initialize $i := 0, stop := false$;
>   While $stop = false$ {
> $i := i + 1$;
> $G_b := bottleneck(w_{e_i})$;
> $I := MIS(G_b^2)$;
>   If $|I| \leq k$ {
>   Return I;

```
                stop := true;
        }
}
```

The *kth power of a graph* $G = (V, E)$ is a graph $G^k = (V, E^k)$ where $E^k = \{(i, j) : i, j \in V, i < j, d(i, j) \leq k\}$. In addition to edges in E, G^k contains edges between pairs of vertices that have a shortest path of length at most k between them in G. The special case of square graphs is used in this algorithm.

As before, the procedure *bottleneck*(w_{e_i}) returns the bottleneck graph $G(w_{e_i})$, and the procedure $MIS(G_b^2)$ returns any maximal independent set in the square graph of G_b. The above algorithm uses the following observations. If there exists a k-center S of cost w_{e_i} (since the cost is always a weight of some edge in G), then set S actually forms a dominating set in the bottleneck graph $G(w_{e_i})$ and vice versa. On the contrary, if we know that the minimum dominating set in $G(w_{e_i})$ has size $k + 1$ or more, then no k-center of cost at most w_{e_i} exists, and hence the optimum cost we seek is at least $w_{e_{i+1}}$. Since the minimum dominating set size is NP-hard to find, we make use of the following facts. Consider some arbitrary graph G' and let I be an MIS in the square of G'. Then we know that, distance in G' between every pair of vertices in I is at least three. In order to dominate a vertex v in I, we should add either v or a neighbor u of v to any dominating set in G'. But u cannot dominate any other vertex in I as they are all at least distance three apart (if u dominates another vertex v' in I, then there is path via u of length two between v and v'). Hence any minimum dominating set in G' is at least as large as I. In our algorithm, if we find an MIS in the square of the bottleneck graph $G(w_{e_i})^2$ of size at least $k + 1$, then we know that no dominating set of size k exists in $G(w_{e_i})$, and thus we know no k-center of cost at most w_{e_i} exists in G. Thus, we proceed in the algorithm until we find an MIS of size at most k and terminate. The approach is clearly a heuristic as we described in the previous bottleneck algorithm. The approximation ratio follows from the fact that if we terminate in iteration i, then the optimum is at least as large as w_{e_i}, that is, $w(S^*) \geq w_{e_i}$. Since I is an MIS in the square of the bottleneck graph $G(w_{e_i})^2$, every vertex outside I has some neighbor in $G(w_{e_i})^2$ inside I. Thus, we know that every vertex v outside I is at most two steps away in $G(w_{e_i})$ from some vertex u inside I, and the direct edge (u, v) would have weight at most $2w_{e_i}$ by triangle inequality. Hence, $w(I) \leq 2w_{e_i} \leq 2w(S^*)$. This algorithm on the graph from Fig. 6.6 terminates in one step when $k = 3$. To find I, we used the greedy maximal independent set algorithm mentioned in Section 6.2. The results are illustrated in Fig. 6.10.

6.6 CONCLUSION

As pointed out earlier, this chapter is meant to be a starting point for readers interested in network clustering, emphasizing only on the basic ideas and simple algorithms and models that require a limited background. Numerous other models exist for clustering, such as several variants of clique-based clustering [19,44,51], graph

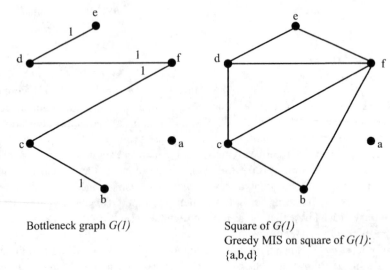

Bottleneck graph *G(1)*

Square of *G(1)*
Greedy MIS on square of *G(1)*:
{a,b,d}

FIGURE 6.10   The bottleneck $G(1)$ of the graph from Fig. 6.6, and the square of $G(1)$. The set I is the output of the bottleneck k-center algorithm with $k = 3$. The corresponding clustering is $\{a\}, \{b, c\}, \{d, e, f\}$; the objective value is 1 (which happens to be optimal).

partitioning models [23,24,49], min-cut clustering [43], and connectivity-based clustering [34,35,58,60] to name a few. In fact, some of these approaches have been used effectively to cluster DNA microarray expression data [19,34,44]. However, more sophisticated approaches that are used in solving such problems are involved and require a rigorous background in optimization and algorithms. Exact approaches for solving clustering problems of moderate sizes are very much possible given the state of the art computational facilities and robust commercial software packages that are available, especially for mathematical programming. For solving large-scale instances, since most problems discussed here are computationally intractable, meta-heuristic approaches such as simulated annealing [1], tabu search [29], or GRASP [22,53] offer an attractive option .

It is also often observed that the models we have discussed in this chapter are too restrictive for use on real-life data resulting in a large number of clusters. One can use the notion of a distance-k neighborhood of a vertex v to remedy this situation. The distance-k neighborhood of a vertex v is $N_k(v)$ defined to be all vertices that are at distance k or less from v excluding v, that is, $N_k(v) = \{u \in V : 1 \le d(v, u) \le k\}$. These neighborhoods can be found easily using *breadth first search algorithm* introduced in Chapter 2. Some empirical properties of the distance-k neighborhood of vertices in real-life networks including biological networks are studied in Ref. [39]. This notion itself has been used to identify molecular complexes by starting with a seed vertex, and adding vertices in distant neighborhoods if the vertex weights are over some threshold [8]. The vertex weights themselves are based on *k-cores* (subgraphs of minimum degree at least k) in the neighborhood of a vertex. *k*-Cores were introduced in social network analysis [57] to identify dense regions of the network, and "resemble" cliques if k is large enough in relation to the size of the k-core found. Moreover, models that relax the notion of cliques [11,13,14] and dominating sets [17,45,52] based on distance-k neighborhoods also exist (see Exercises 3 and 4). Alternately, an edge density based relaxation of cliques called *quasi-cliques* have also been studied [3,4]. These relaxations of the basic models we have discussed are more robust in clustering real-life data containing a significant percentage of errors.

An important dilemma that practitioners are faced with while using clustering techniques is the following. Very rarely does real-life data present a unique clustering solution. Firstly because, deciding which model best represents clusters in the data is difficult, and requires experimentation with different models. It is often better to employ multiple models for clustering the same network as each could provide different insights into the data. Secondly, even under a particular clustering model, several alternative optimal solutions could exist for the clustering problem. These are also of interest in most clustering applications. Therefore, care must be taken in the selection of clustering models, as well as solution approaches, especially given the computational intractability of most clustering problems. These issues are clearly in addition to the general issues associated with the clustering problem such as interpretation of clusters and what they represent.

Appreciation for the efficiency of clustering and categorizing dates back to Aristotle and his work on classifying plants and animals. But it is wise to remember that it was Aristotle who said "*the whole is more than the sum of its parts.*"

6.7 SUMMARY

This chapter discusses the clustering problem and the basic types of clustering problems that can be formulated. Several popular network clustering models are introduced and classified according to the two types. Simple algorithms are presented and explained in detail for solving such problems. Many of the studied models have been popular in clustering biological networks such as protein interaction networks and gene coexpression networks. Models from other fields of study that have relevant biological properties are also introduced. This simple presentation should provide a good understanding of the basic concepts and hopefully encourage the reader to consult other literature on clustering techniques.

6.8 EXERCISES

1. Given a graph $G = (V, E)$ with non-negative vertex weights $w_i, i \in V$, the *maximum weighted clique problem* is to find a clique such that the sum of the weights of vertices in the clique is maximized. Develop a "weight-greedy" heuristic for the problem. What happens to your algorithm when all the weights are unity?

2. One way to weight the vertices of scale-free networks is to use the "cliquish-ness" of the neighborhood of each vertex [65] (see Ref. [8] for an alternative approach). For a vertex v, let m_v denote the number of edges in $G[N(v)]$, the induced graph of the neighborhood. Then the weights are defined as, $w_v = 2m_v/[\deg(v)(\deg(v) - 1)]$ when $\deg(v) \geq 2$ and zero otherwise. Use this approach to weight the vertices of the graph in Fig. 6.2 and apply the algorithm you developed in Exercise 2.

3. Several clique relaxations have been studied in social network analysis as the clique model is very restrictive in its definition [55]. One such relaxation is a *k-clique*, which is a subset of vertices such that the shortest distance between any two vertices in a k-clique is at most k in the graph G. Develop a heuristic for finding a maximal k-clique given k and graph G. What happens to your algorithm when $k = 1$?

   (*Hint*: Recall the definitions of the distance-k neighborhood and power graphs.)

4. In protein interaction networks, it is often meaningful to include proteins that are at distance 2 or 3 from a central protein in a cluster. Consider the following definition of a distance-k dominating set. A set D is said to be *k-dominating* in the graph G, if every vertex u that is not in D has at least one vertex v in D, which is at distance no more than k in G, that is, $d(u, v) \leq k$. Develop a center-based clustering algorithm that uses k-dominating sets.

5. Consider the graph in Fig. 6.2. Construct a weighted complete graph on the same vertex set with the shortest distance between pairs of vertices as edge weights. Apply the heuristic presented in Section 6.4.2 to this graph and in-terpret the resulting clusters.

6. Develop a heuristic for the following Type II clustering problem.

   min–max k-clique clustering problem: Given a connected graph $G = (V, E)$ and a fixed p, partition V into subsets $V_1, \dots, V_p$ such that V_i is a k_i-clique (defined in Exercise 3) for $i = 1, \dots, p$ and $\max\limits_{i=1,\dots,p} k_i$ is minimized.

REFERENCES

1. E. Aarts and J. Korst. *Simulated Annealing and Boltzmann Machines*. John Wiley & Sons, Chichester, 1989.

2. E. L. Aarts and J. K. Lenstra, editors. *Local Search in Combinatorial Optimization*. Princeton University Press, Princeton, 2003.

3. J. Abello, P. M. Pardalos, and M. G. C. Resende. On maximum clique problems in very large graphs. In J. Abello and J. Vitter, editors, *External Memory Algorithms and Visualization*, volume 50 of *DIMACS Series on Discrete Mathematics and Theoretical Computer Science*, pp. 119–130. American Mathematical Society, Providence, RI, 1999.

4. J. Abello, M. G. C. Resende, and S. Sudarsky. Massive quasi-clique detection. *Lecture Notes in Computer Science*, 2286:598–612, 2002.

5. R. K. Ahuja, T. L. Magnanti, and J. B. Orlin. *Network Flows: Theory, Algorithms, and Applications*. Prentice-Hall, Upper Saddle River, NJ, 1993.

6. M. R. Anderberg. *Cluster Analysis for Applications*. Academic Press, New York, 1973.

7. G. Ausiello, M. Protasi, A. Marchetti-Spaccamela, G. Gambosi, P. Crescenzi, and V. Kann. *Complexity and Approximation: Combinatorial Optimization Problems and Their Approximability Properties*. Springer-Verlag, New York, 1999.

8. G. D. Bader and C. W. V. Hogue. An automated method for finding molecular complexes in large protein interaction networks. *BMC Bioinformatics*, 4(2), 2003.

9. J. S. Bader, A. Chaudhuri, J. M. Rothberg, and J. Chant. Gaining confidence in high-throughput protein interaction networks. *Nature Biotechnology*, 22(1):78–85, 2004.

10. B. Balasundaram and S. Butenko. Graph domination, coloring and cliques in telecommunications. In M. G. C. Resende and P. M. Pardalos, editors, *Handbook of Optimization in Telecommunications*, pp. 865–890. Springer Science + Business Media, New York, 2006.

11. B. Balasundaram, S. Butenko, and S. Trukhanov. Novel approaches for analyzing biological networks. *Journal of Combinatorial Optimization*, 10(1):23–39, 2005.

12. K. Birnbaum, D. E. Shasha, J. Y. Wang, J. W. Jung, G. M. Lambert, D. W. Galbraith, and P. N. Benfey. A gene expression map of the arabidopsis root. *Science*, 302:1956–1960, 2003.

13. J.-M. Bourjolly, G. Laporte, and G. Pesant. Heuristics for finding k-clubs in an undirected graph. *Computers & Operations Research*, 27:559–569, 2000.

14. J.-M. Bourjolly, G. Laporte, and G. Pesant. An exact algorithm for the maximum k-club problem in an undirected graph. *European Journal of Operational Research*, 138:21–28, 2002.

15. S. Butenko, X. Cheng, C. A. S Oliveira, and P. M. Pardalos. A new heuristic for the minimum connected dominating set problem on ad hoc wireless networks. In R. Murphey and P. M. Pardalos, editors, *Cooperative Control and Optimization*, pp. 61–73. Kluwer Academic Publisher, Dordrecht, 2004.

16. S. E. Calvano, W. Xiao, D. R. Richards, R. M. Felciano, H. V. Baker, R. J. Cho, R. O. Chen, B. H. Brownstein, J. P. Cobb, S. K. Tschoeke, C. Miller-Graziano, L. L. Moldawer, M. N. Mindrinos, R. W. Davis, R. G. Tompkins, and S. F. Lowry. A network-based analysis of systemic inflammation in humans. *Nature*, 437:1032–1037, 2005.

17. G. J. Chang and G. L. Nemhauser. The k-domination and k-stability problems on sun-free chordal graphs. *SIAM Journal on Algebraic and Discrete Methods*, 5:332–345, 1984.

18. Y. P. Chen, A. L. Liestman, and J. Liu. Clustering algorithms for ad hoc wireless networks. In Y. Pan and Y. Xiao, editors, *Ad Hoc and Sensor Networks*. Nova Science Publishers, New York, 2004.

19. E. J. Chesler and M. A. Langston. Combinatorial genetic regulatory network analysis tools for high throughput transcriptomic data. Technical Report ut-cs-06-575, CS Technical Reports, University of Tennessee, 2006.

20. R. Diestel. *Graph Theory*. Springer-Verlag, Berlin, 1997.

21. F. Duan and H. Zhang. Correcting the loss of cell-cycle synchrony in clustering analysis of microarray data using weights. *Bioinformatics*, 20:1766–1771, 2004.

22. T. A. Feo and M. G. C. Resende. Greedy randomized adaptive search procedures. *Journal of Global Optimization*, 6:109–133, 1995.

23. C. E. Ferreira, A. Martin, C. C. de Souza, R. Weismantel, and L. A. Wolsey. Formulations and valid inequalities for the node capacitated graph partitioning problem. *Mathematical Programming*, 74(3):247–266, 1996.

24. C. E. Ferreira, A. Martin, C. C. de Souza, R. Weismantel, and L. A. Wolsey. The node capacitated graph partitioning problem: A computational study. *Mathematical Programming*, 81:229–256, 1998.

25. J. S. Fetrow, M. J. Palumbo, and G. Berg. Patterns, structures, and amino acid frequencies in structural building blocks, a protein secondary structure classification scheme. *Proteins: Structure, Function, and Bioinformatics*, 27(2):249–271, 1997.

26. J. Gagneur, R. Krause, T. Bouwmeester, and G. Casari. Modular decomposition of protein-protein interaction networks. *Genome Biology*, 5(8):R57.1–R57.12, 2004.

27. M. R. Garey and D. S. Johnson. *Computers and Intractability: A Guide to the Theory of NP-completeness*. W.H. Freeman and Company, New York, 1979.

28. M. Girvan and M. E. J. Newman. Community structure in social and biological networks. *Proceedings of the National Academy of Sciences*, 99(12):7821–7826, 2002.

29. F. Glover and M. Laguna. *Tabu Search*. Kluwer Academic Publishers, Dordrecht, 1997.

30. T. F. Gonzalez. Clustering to minimize the maximum intercluster distance. *Theoretical Computer Science*, 38(2-3):293–306, 1985.

31. S. Guha and S. Khuller. Approximation algorithms for connected dominating sets. *Algorithmica*, 20:374–387, 1998.

32. V. Guralnik and G. Karypis. A scalable algorithm for clustering protein sequences. Workshop on Data Mining in Bioinformatics, 2001.

33. J. A. Hartigan. *Clustering Algorithms*. John Wiley and Sons, New York, 1975.

34. E. Hartuv, A. Schmitt, J. Lange, S. Meier-Ewert, H. Lehrachs, and R. Shamir. An algorithm for clustering cDNAs for gene expression analysis. In *RECOMB*, pp. 188–197, 1999.

35. E. Hartuv and R. Shamir. A clustering algorithm based on graph connectivity. *Information Processing Letters*, 76(4–6):175–181, 2000.

36. D. S. Hochbaum. *Approximation Algorithms for NP-hard Problems*. PWS Publishing Company, 1997.

37. D. S. Hochbaum and D. B. Shmoys. A best possible heuristic for the *k*-center problem. *Mathematics of Operations Research*, 10:180–184, 1985.

38. D. S. Hochbaum and D. B. Shmoys. A unified approach to approximation algorithms for bottleneck problems. *Journal of the ACM*, 33(3):533–550, 1986.

39. P. Holme. Core-periphery organization of complex networks. *Physical Review E*, 72:046111–1–046111–4, 2005.

40. A. K. Jain and R. C. Dubes. *Algorithms for Clustering Data*. Prentice-Hall, Upper Saddle River, NJ, 1988.

41. M. Jambu and M. O. Lebeaux. *Cluster Analysis and Data Analysis*. North-Holland, New York, 1983.

42. D. Jiang, C. Tang, and A. Zhang. Cluster analysis for gene expression data: A survey. *IEEE Transactions on Knowledge & Data Engineering*, 16:1370–1386, 2004.

43. E. L. Johnson, A. Mehrotra, and G. L. Nemhauser. Min-cut clustering. *Mathematical Programming*, 62:133–152, 1993.

44. G. Kochenberger, F. Glover, B. Alidaee, and H. Wang. Clustering of microarray data via clique partitioning. *Journal of Combinatorial Optimization*, 10(1):77–92, 2005.

45. S. Kutten and D. Peleg. Fast distributed construction of small k-dominating sets and applications. *Journal of Algorithms*, 28(1):40–66, 1998.

46. K. Lee, J. Sim, and J. Lee. Study of protein–protein interaction using conformational space annealing. *Proteins: Structure, Function, and Bioinformatics*, 60(2):257–262, 2005.

47. H. Lu and R. Ravi. The power of local optimization: Approximation algorithms for maximum-leaf spanning tree. In *Proceedings of the 30th Annual Allerton Conference on Communication, Control and Computing*, pp. 533–542, 1992.

48. H. Lu and R. Ravi. Approximating maximum leaf spanning trees in almost linear time. *Journal of Algorithms*, 29(1):132–141, 1998.

49. A. Mehrotra and M. A. Trick. Cliques and clustering: A combinatorial approach. *Operations Research Letters*, 22:1–12, 1998.

50. C. H. Papadimitriou. *Computational Complexity*. Addison-Wesley, Reading, MA, 1994.

51. X. Peng, M. A. Langston, A. M. Saxton, N. E. Baldwin, and J. R. Snoddy. Detecting network motifs in gene co-expression networks. In *Proceedings of the International Conference for the Critical Assessment of Microarray Data Analysis*, 2004.

52. L. D. Penso and V. C. Barbosa. A distributed algorithm to find k-dominating sets. *Discrete Applied Mathematics*, 141:243–253, 2004.

53. M. G. C. Resende and C. C. Ribeiro. Greedy randomized adaptive search procedures. In F. Glover and G. Kochenberger, editors, *Handbook of Metaheuristics*. Kluwer Academic Publishers, Dordrecht, 2003.

54. K. G. Le Roch, Y. Zhou, P. L. Blair, M. Grainger, J. K. Moch, J. D. Haynes, P. De la Vega, A. A. Holder, S. Batalov, D. J. Carucci, and E. A. Winzeler. Discovery of gene function by expression profiling of the malaria parasite life cycle. *Science*, 301:1503–1508, 2003.

55. J. Scott. *Social Network Analysis: A Handbook*. Sage Publications, London, 2000.

56. E. Segal, M. Shapira, A. Regev, D. Pe'er, D. Botstein, D. Koller, and N. Friedman. Module networks: identifying regulatory modules and their condition-specific regulators from gene expression data. *Nature Genetics*, 34:166–176, 2003.

57. S. B. Seidman. Network structure and minimum degree. *Social Networks*, 5:269–287, 1983.

58. R. Shamir and R. Sharan. Algorithmic approaches to clustering gene expression data. In T.Jiang, T. Smith, Y. Xu, and M. Q. Zhang, editors, *Current Topics in Computational Molecular Biology*, pp. 269–300. MIT Press, 2002.

59. W. Shannon, R. Culverhouse, and J. Duncan. Analyzing microarray data using cluster analysis. *Pharmacogenomics*, 4:41–52, 2003.

60. R. Sharan, R. Elkon, and R. Shamir. Cluster analysis and its applications to gene expression data. *Ernst Schering workshop on Bioinformatics and Genome Analysis*, pp. 83–108. Springer-Verlag, 2002.

61. R. Solis-Oba. 2-approximation algorithm for finding a spanning tree with maximum number of leaves. In *ESA'98: Proceedings of the 6th Annual European Symposium on Algorithms*, pp. 441–452, Springer-Verlag, New York, 1998.

62. H. Spath. *Cluster Analysis Algorithms*. Ellis Horwood, Chichester, 1980.

63. V. Spirin and L. A. Mirny. Protein complexes and functional modules in molecular networks. *Proceedings of the National Academy of Sciences*, 100(21):12123–12128, 2003.

64. V. V. Vazirani. *Approximation Algorithms*. Springer-Verlag, Berlin, 2001.

65. D. Watts and S. Strogatz. Collective dynamics of "small-world" networks. *Nature*, 393:440–442, 1998.

66. D. West. *Introduction to Graph Theory*. Prentice-Hall, Upper Saddle River, NJ, 2001.

67. W. Zhong, G. Altun, R. Harrison, P.C. Tai, and Y. Pan. Improved K-means clustering algorithm for exploring local protein sequence motifs representing common structural property. *IEEE Transactions on NanoBioscience*, 4(3):255–265, 2005.

7

PETRI NETS

INA KOCH AND MONIKA HEINER[1]

7.1 INTRODUCTION

Over the last few years, the great effort in genomics as well as the progressive development of high-throughput technologies resulted in a high amount of qualitative data, outnumbering the amount of quantitative data. This lack of quantitative data makes the application of quantitative methods for modeling and analysis of biochemical networks difficult and in many cases even infeasible. Often the measurement of quantitative data *in vivo* as well as *in vitro* is very complicated or not possible due to experimental limits or ethical reasons. At the same time many qualitative data are not considered in research. In order to get some information about the system behavior using the huge amount of qualitative data, methods for their analysis have been derived, which are mainly based on the incidence matrix (stoichiometric matrix) of the underlying net graph, see Chapter 1.

Petri nets are special graphs, which have been developed to easily model and analyze mathematically exactly systems with concurrent processes. The basic ideas of qualitative Petri nets have been introduced in 1962 by Carl Adam Petri in his dissertation [22] to describe and simulate networks of causally related, discrete actions. Since that time, many theorems and algorithms have been developed and implemented to analyze such systems [20,33]. Qualitative Petri nets have been extended by various

---

[1] Both authors equally contributed to this chapter. This chapter has been partly supported by the Federal German Ministry of Education and Research (BMBF), BCB project 0312705D (Ina Koch).

Analysis of Biological Networks, Edited by Björn H. Junker and Falk Schreiber

notions of time to describe quantitatively — among others — stochastic [2] and continuous system behavior [6]. Nowadays, applications of Petri net theory comprise the analysis of technical systems, administrative systems, and others.

First applications of Petri nets to biochemical systems were published in 1993 [24] and 1994 [11,25]. Meanwhile, metabolic networks [12,17,21,37], signal transduction networks [8,17,26], gene regulatory networks [17,32,35], and combinations of them [4,16,18,30] have been successfully modeled and analyzed using various classes of Petri nets, qualitative as well as quantitative ones. Thus, in contrast to other qualitative methods, Petri nets provide a unique mathematical description method for different abstraction levels. In addition, Petri nets allow to combine all these different abstraction levels within one model.

Petri nets have an executable graphical representation, supporting the intuitive understanding of the system modeled and the communication between experimentally and theoretically working scientists. There are movable objects, the animation of which visualizes possible flows through the network. Model animation helps to experience the network behavior and allows to test whether the model actually does behave in the desired manner.

One of the great advantages of Petri net theory is that it comprises the definition of structural and behavioral properties. For their determination, various mathematically sound analysis techniques have been developed. Petri net theory allows to formulate model validation criteria, which increase the confidence in the model. So the whole modeling process is an iterative procedure of model design, model animation, and model analysis.

Moreover, Petri nets support the integration of qualitative as well as quantitative methods by serving as a mathematically unifying description. Until now, there are two communities, one developing and using qualitative methods, and the other using quantitative methods as ordinary differential equations (ODEs) or stochastic approaches. The combination of the qualitative and quantitative world is a significant step toward integrated analysis of biochemical systems.

Finally, there exist many reliable public-domain software tools as editors, animators, and analyzers, see Section 7.5.

In the following sections, we introduce Petri nets as the so-called *place/transition nets*, which represent the basic Petri net class. Here, no time dependencies and therefore neither continuous nor stochastic behavior are considered. We explain how to use Petri net theory for modeling of biochemical systems as an iterative process of editing, animating, and analyzing the system. The most important analysis techniques are defined and explained using one running example, the combined glycolysis and pentose phosphate pathway in erythrocytes. We focus on the analysis of qualitative properties of the total system behavior. Afterwards, we sketch the derivation and evaluation of a related quantitative model given as continuous Petri net. We additionally summarize other published case studies. For the own examples, we provide the corresponding Petri nets and analysis results in the supplementary material on the book's web page.

7.2 QUALITATIVE MODELING

7.2.1 The Model

For graphical representation, biologists use *hypergraphs*, already introduced and discussed in Chapter 2. For illustration, see our running example in Fig. 7.5. Petri nets refine this type of representation by replacing each hyperarc by a transition to describe a biochemical reaction. Petri nets are directed, bipartite, attributed graphs. *Bipartite* means that they consist of two types of nodes (vertices), which are called in our context places $P = \{p_1, \ldots, p_m\}$ and transitions $T = \{t_1, \ldots, t_n\}$, and directed arcs (edges), which connect only nodes of different type. Arcs are weighted by natural numbers, whereby the arc weight (attribute) may be read as the multiplicity of the arc. Consequently, the arc weight of 0 stands for the absence of an arc. The arc weight 1 is the default value and is usually not drawn explicitly. Please note, instead of the classical graph-theoretic terms *vertex* and *edges* we use here the terms *nodes* and *arcs*, which are in the given context more popular.

Places typically model passive system elements such as conditions, states, or chemical compounds. We distinguish between primary compounds as metabolites, proteins, or protein complexes, and secondary (auxiliary) compounds as *ADT, ATP*, and so on. Transitions generally stand for active system elements such as events or chemical reactions (e.g., the enzyme-catalyzed conversion from one metabolite to another), complex forming, or de-/phosphorylation. In graphical representations, places are depicted as circles, transitions as rectangles, and arcs as arrows. The arcs in the Petri net describe the *causal relation* between active and passive system elements. They connect an event (a transition) with its preconditions (preplaces), which must be fulfilled to trigger this event, and with its postconditions (postplaces), which will be fulfilled when the event takes place.

The fulfillment of a condition is characterized by tokens, which are the dynamic elements and represent movable objects residing in places. Principally, a place in a discrete Petri net may carry any integer number of tokens, indicating different degrees of fulfillment. Tokens correspond usually to molecules, moles, concentration levels, or gene expression levels, depending on the chosen abstraction level. A given distribution of tokens over all places describes a certain system state and is called a *marking m* of the Petri net. Accordingly, the *initial marking m_0* describes the system state before any system behavior took place. The following definition summarizes this informal introduction.

Definition 7.1 (Petri net)   A Petri net is a quadruple $\mathcal{N} = (P, T, f, m_0)$, where

- P and T are finite, nonempty, and disjoint sets. P is the set of places. T is the set of transitions.
- $f : ((P \times T) \cup (T \times P)) \rightarrow \mathbb{N}_0$ defines the set of directed arcs, weighted by non-negative integers.
- $m_0 : P \rightarrow \mathbb{N}_0$ gives the initial marking.

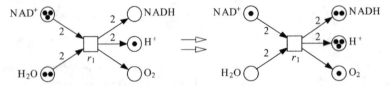

FIGURE 7.1 Petri net model of a single chemical reaction (light-induced phosphorylation), given by its stoichiometric equation $2NAD^+ + 2H_2O \rightarrow 2NADH + 2H^+ + O_2$, and two of its states, linked by a single firing of r_1.

Each marking is defined by the given token situation in all places $m \in \mathbb{N}_0^{|P|}$, whereby $|P|$ denotes the number of places in the Petri net. $m(p)$ yields the number of tokens on place p in the marking m. A place p with $m(p) = 0$ is called *clean (empty, unmarked)* in m, otherwise it is called *marked*. A set of places is called clean, if all its places are clean, otherwise marked.

Furthermore, we introduce the following notations. The preset (preplaces or pretransitions) of a node $x \in P \cup T$ is defined as ${}^\bullet x := \{y \in P \cup T | f(y, x) \neq 0\}$, and its postset (postplaces or posttransitions) as $x^\bullet := \{y \in P \cup T | f(x, y) \neq 0\}$. We extend both notions to set of nodes $X \subseteq P \cup T$ and define the set of all prenodes ${}^\bullet X := \bigcup_{x \in X} {}^\bullet x$, and the set of all postnodes $X^\bullet := \bigcup_{x \in X} x^\bullet$. Fig. 7.8 provides examples for these notations. A node (place or transition) $x \in P \cup T$ is called a *boundary node*, if ${}^\bullet x = \emptyset$ or $x^\bullet = \emptyset$. Transitions without preplaces (postplaces) are called *input (output) transitions* and are drawn as flat rectangles. Input (output) transitions realize the necessary compound supply (removal) into (out of) the system. Analogously, we define *input (output) places*, representing the input (output) compounds of a biochemical network. Thus, input and output nodes model the interface between the network under consideration and its environment.

Having introduced the structure of a Petri net, let us now turn to its execution. To bring a Petri net to life we need the firing rule, which defines the dynamic behavior of a Petri net. The firing rule consists of two parts: the precondition and the actual firing behavior. If all preplaces of a transition are sufficiently marked with tokens, that is, at least with the number of the corresponding arc weights, this transition is *enabled* (has *concession*) and may *fire* (*occur*). If a transition fires, tokens are removed from all its preplaces and added to all its postplaces, each according to the corresponding arc weights, compare Fig. 7.1. NAD^+ and H_2O are the preplaces of the transition r_1, and $NADH$, H^+, and O_2 form the postplaces. Two moles of NAD^+ and two moles of H_2O are necessary to produce two moles of $NADH$, two moles of H^+, and one *mole* of O_2. Here, as generally in metabolic networks, the arc weights reflect the corresponding stoichiometric factors.

Definition 7.2 (Firing rule)   Let $\mathcal{N} = (P, T, f, m_0)$ be a Petri net.

- A transition t is *enabled* in a marking m, written as $m[t\rangle$, if
$\forall p \in {}^\bullet t : f(p, t) \leq m(p)$.

- A transition t, which is enabled in m, may fire. When t in m *fires*, a new marking m' is reached, written as $m[t\rangle m'$, with
  $$\forall p \in P : m'(p) = m(p) - f(p, t) + f(t, p).$$
- The firing itself is timeless and atomic.

Please note, we consider a time-free model. So the firing rule does not take time into consideration, as for example, how often a transition fires or how long a transition needs for firing. The term *atomic* means that the firing itself cannot be decomposed into smaller parts at the chosen abstraction level. If a transition fires, the removal of tokens from the preplaces and the production of tokens in the postplaces takes place at once, not consuming time. Thus, there is no system state in between.

All markings (system states), which can be reached from a certain marking m by any transition firing sequence of arbitrary length, form the *set of reachable markings* $[m\rangle$. The set of markings $[m_0\rangle$, reachable from the initial marking, is also called the *state space* of a given system.

The Petri net structure reflects the biochemical topology, whereas the Petri net behavior, produced by the repeated firing of transitions, describes the set of all *partial order* sequences of chemical reactions from the input to the output compounds of a given network, respecting the given stoichiometric relations. Moreover, the same modeling idea can be applied on a more abstract level, where stoichiometric details are not known or do not matter, resulting into a partial order description of causal relations of the basic (re)actions involved. Petri nets use a partial order description to handle concurrent (independent) actions. Consider, for example, the small system of three reactions in Fig. 7.2. It is not clear, which of the two transitions r_2 and r_3 will fire first, that is, they are *unordered* with regard to their occurrence. Contrary, the pairs (r_1, r_2) and (r_1, r_3) are each *ordered*, that is, r_1 has to fire prior to r_2 and r_3. After the firing of r_1, the two transitions r_2 and r_3 can fire concurrently (independently). Considering all transition pairs of of a given net, we have usually ordered as well as unordered pairs. Thus, we get a partial order. Fig. 7.3 summarizes the basic branching net structures, all biochemical systems are made of, in terms of the Petri net modeling principle. Please note, the branching degree can be greater than two.

Petri nets have been applied to different types of biological networks — metabolic networks, signal transduction networks, and gene regulatory networks, see Sections 7.1 and 7.6. What is the difference between metabolic and signaling

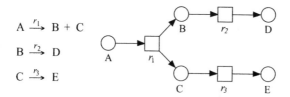

$A \xrightarrow{r_1} B + C$

$B \xrightarrow{r_2} D$

$C \xrightarrow{r_3} E$

FIGURE 7.2   The partial order description principle in a Petri net model. The transition r_1 has to fire prior to r_2 and r_3. Then, r_2 and r_3 can fire concurrently (independently). The transition pairs (r_1, r_2) and (r_1, r_3) are each ordered, the pair (r_2, r_3) is unordered.

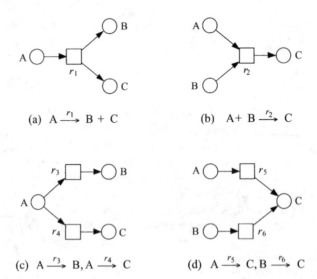

(a) $A \xrightarrow{r_1} B + C$ (b) $A + B \xrightarrow{r_2} C$

(c) $A \xrightarrow{r_3} B, A \xrightarrow{r_4} C$ (d) $A \xrightarrow{r_5} C, B \xrightarrow{r_6} C$

FIGURE 7.3   The basic branching structures, all biochemical systems are made of. Please note, the branching degree can be greater than two. The first (second) column exemplifies forward (backward) branching of nodes. Branching transitions relate to concurrent system behavior, and branching places to alternative system behavior. The essential difference between (a) and (c), or (b) and (d), respectively, cannot be expressed unambiguously using monochromatic graphs, that is, graphs with one node type. For example, we get identical graph structures for (a) and (c) by replacing the transitions r_1, r_3, and r_4 by arcs connecting directly the transition's preplace and postplace.

pathways from the point of view of modeling? The main difference is that metabolic networks are related to substance flows (mass flows), which are determined by the stoichiometry of the underlying chemical reaction equations. The arc weights correspond to the stoichiometric factors. Thus, the network can be completely characterized by the set of chemical stoichiometric reactions. In signaling networks, we have a flow of information in the form of signals without stoichiometry. The signal flow is realized by activation and deactivation of proteins or protein complexes building signal cascades. Here, we usually use an arc weight of one. Accordingly, two basic structures of metabolic and signal transduction networks can be found, compare Fig. 7.4.

In order to get readable Petri net representations, we take advantage of widely used short-hand notations. Not influencing the underlying graph structure, their only purpose is to support clearer and better understandable figures.

Read arcs (*test arcs*) are represented by bidirectional arcs or by arcs ending in a black dot instead of an arrow, see Fig. 7.13. They shortly stand for two inverse, unidirectional arcs. Usually, signal transduction does not involve the immediate resetting of the triggering signal(s). Therefore, such situations are modeled by read arcs. Similarly, enzyme reactions are also modeled by read arcs, if any, because they are catalytic reactions, that is, there is no consumption of the enzyme.

Logical nodes (*fusion nodes*) are colored in gray and serve as links connecting distributed Petri net components. They are often used for secondary compounds,

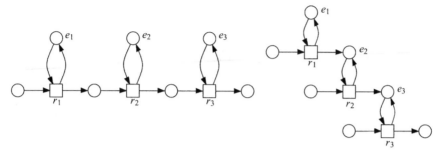

FIGURE 7.4 The essential structural difference between metabolic networks (left) and signal transduction networks (right), expressed in terms of Petri nets. Metabolic networks are related to substance flows (mass flows). In signal transduction networks we have a flow of information by activation and deactivation of proteins or protein complexes, building signal cascades.

which are involved in many reactions such that their adjacent arcs would make the model unreadable. Accordingly, these secondary compounds are also called *ubiquitous*, because they are found everywhere in the cell.

Hierarchical nodes are used to exploit hierarchical structuring techniques. Transition-bordered subnets are abstracted by *macro transitions* represented by two centrically nested squares. This notation is often applied to abstract both directions of reversible reactions or to abstract linear reaction sequences.

We distinguish between the terms *pathway* and *network* in such a way that a pathway represents a very special, often functionally related smaller part of a whole network. Graph-theoretically, a pathway can be any subnetwork. It does not have to be linear; it can have bifurcations. Generally, all forward/backward branching structures as given in Fig. 7.3 are possible within a pathway. A network describes a special more or less complicated cell behavior or even the whole model of a cell. It can consist of several pathways as our running example, which involves the glycolysis and pentose phosphate pathway.

Running example: For illustrating the introduced terms we use as running example a simplified model of the combined glycolysis (G) and pentose phosphate pathway (PPP), in the following written as G-PPP, in erythrocytes (red blood cells). This pathway, see Fig. 7.5, is well understood and can be found in many articles and textbooks such as Ref. [1]. Its Petri net version in Ref. [25] serves as basis of our model, see Fig. 7.6. It is a simplified variation of the whole G-PPP, which nevertheless preserves representative biological behavior. This helps us to focus on the Petri net modeling and validation aspects.

Glycolysis is the main energy-conversion pathway in many organisms. Through a sequence of reactions, glucose is metabolized to pyruvate with the concomitant production of ATP. Pyruvate can be further anaerobically processed (i.e., without oxygen) to lactate. At the beginning, glucose is converted into glucose 6-phosphate, which can form fructose 6-phosphate following the glycolysis, or enter the PP pathway by the production of ribulose 5-phosphate. In the glycolytic pathway fructose 6-phosphate is converted into fructose 1,6-biphosphate producing then

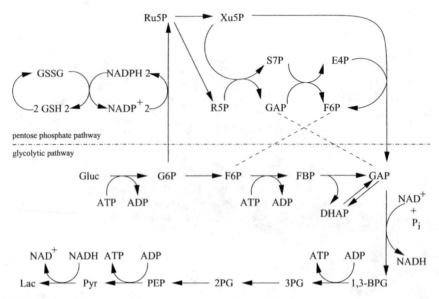

FIGURE 7.5 The hypergraph representation of the combined glycolysis and pentose phosphate pathway in erythrocytes, based on Ref. [25]. The horizontal line separates the glycolysis from the PPP. The two dashed lines indicate the involvement of *GAP* and *F6P* in both pathways.

dihydroxyacetone phosphate and glyceraldehyde 3-phosphate, which can be interconverted rapidly and reversible. Glyceraldehyde 3-phosphate — also a product of the PP pathway — is converted through the formation of 1,3 biphosphoglycerate, 3-phosphoglycerate, 2-phosphoglycerate, and phosphoenolpyruvate into pyruvate, which gives lactate by lactic acid fermentation.

The PP pathway generates NADPH to provide the needed four electrons in the reductive (i.e., gains electrons) biosynthesis (e.g., the synthesis of fatty acids), and produces five-carbon sugars. It consists of two phases, the oxidative (loses electrons) generation of NADPH and the nonoxidative interconversion of sugars. Glucose 6-phosphate is converted into ribulose 5-phosphate using necessarily NADP$^+$, which is generated from glutathione disulfide (the oxidized glutathione) and NADPH by a characteristic reaction for erythrocytes. Ribulose 5-phosphate forms ribose 5-phosphate and xylulose 5-phosphate. In the nonoxidative stage (i.e., without gaining or losing electrons), the pathway performs the interconversion into three-carbon sugars (here glyceraldehyde 3-phosphate), seven-carbon sugars (here sedoheptulose 7-phosphate), six-carbon sugars (here fructose 6-phosphate), and four-carbon sugars (here erythrose 4-phosphate).

The PP pathway and glycolysis are linked by the metabolites glyceraldehyde 3-phosphate and fructose 6-phosphate. All reactions are catalyzed by special enzymes, which are not given in the hypergraph representation, but in the Petri net model, compare Table 7.2.

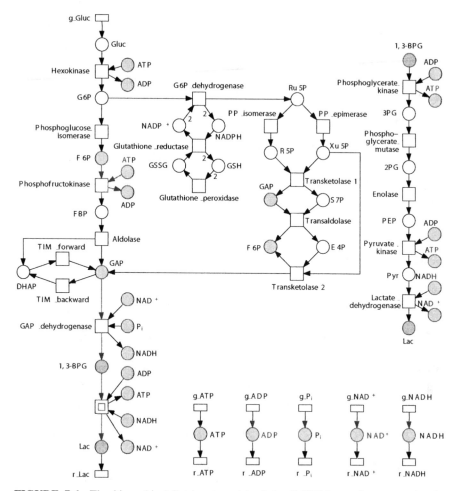

FIGURE 7.6 The hierarchical Petri net model of the G-PPP in erythrocytes, whereby both hierarchy levels are drawn in one picture. The hierarchical node, indicated by a macro transition and drawn as two centrically nested squares, stands for the connected component, shown on the upper right side. Both hierarchy levels are connected by the dark gray colored places. These places are special logical places, automatically generated and colored by the hierarchy manager of the Petri net drawing tool. The light gray colored nodes represent logical places, connecting distributed net parts. The places $NADP^+$ and $GSSG$ get an initial marking, which will be motivated in Section 7.3.2. The interface to the environment is modeled by the input transition g_Gluc and the output transition r_Lac. Additionally, all secondary compounds get input as well as output transitions, modeling the assumption that they are available in sufficient amount. Thus an open system has been modeled.

The Petri net model of the G-PPP in Fig. 7.6 was developed straight-forwardly by replacing the hyperarc representation in Fig. 7.5. Places represent the primary and secondary compounds (see Table 7.1), and transitions stand for the chemical reactions for interconversion of these compounds (see Table 7.2). Transitions are named after the enzyme, catalyzing this reaction.

We rearrange the direction of the substance flow, compare the Petri net model in Fig. 7.6 with the biological hypergraph representation in Fig. 7.5, such that the main substance flow goes top down. We use logical places for seven compounds (*ADP*, *ATP*, *F6P*, *GAP*, *NAD*$^+$, *NADH*, *P$_i$*). There is only one reversible reaction, modeled by two transitions, one for the forward (*TIM_forward*) and the other for the backward (*TIM_backward*) reaction. We apply one hierarchical transition to hide five linear reactions of the glycolytic part on a lower net level, which are depicted on the upper right side of Fig. 7.6.

Contrary to Fig. 7.5, we model explicitly the interface to the environment by the input transition *g_Gluc* and the output transition *r_Lac*. Additionally, all secondary compounds get input as well as output transitions, modeling the assumption that they are available in sufficient amount. Input/output transitions are drawn as flat rectangles, whereby *g* and *r* stand for *generate* and *remove*, respectively. Later we will replace this style of *open environment modeling* by a closed style, see Section 7.3.2.

Now we are interested in examining the model in order to increase our confidence in it. Therefore, we want to validate the model. *Model validation* aims at checking the constructed model for consistency and correct reflection of the known behavior of the modeled system in reality. One initial approach might be model animation, that is, playing the token game, to observe some possible behavior. However, model animation has the disadvantage of not allowing definite conclusions about any system properties. Contrary, *model analysis* derives definitive statements on behavioral system properties by considering all possible behavior. In the following, we introduce those model properties in terms of Petri net terminology, which are expected to be meaningful for biochemical network analysis, and apply them to our running example. Please note, all following definitions in this chapter refer to a Petri net as introduced in the definitions 7.1 and 7.2.

7.2.2 The Behavioral Properties

For a model-based analysis of the system behavior, the behavioral properties of interest of the real system have to be mapped on properties of the model of the system. To accomplish this task, we have to understand first the various notions of behavioral properties provided by the applied modeling language. In Petri net theory, there are three orthogonal (independent) behavioral properties, *liveness*, *reversibility*, and *boundedness*; see Ref. [20] for eight tiny Petri nets to confirm the orthogonality of these properties.

Liveness is an important expected property of a biochemical Petri net. It should exhibit an infinite net behavior in the sense that no part of the network will ever stop working, if a sufficient amount of input compounds enters the net. An unintended interruption of the substance or signal flows is likely to indicate a modeling error. A

TABLE 7.1   Compounds of the G-PPP and their abbreviations

Abbr.	Compound name	Abbr.	Compound name
1,3-BPG	1,3-Biphosphoglycerate	NAD^+	Nicotinamide adenine dinucleotide
2PG	2-Phosphoglycerate		(oxidized form)
3PG	3-Phosphoglycerate	*NADH*	Nicotinamide adenine dinucleotide
ADP	Adenosine diphosphate		(reduced form)
ATP	Adenosine triphosphate	$NADP^+$	Nicotinamide adenine dinucleotide
DHAP	Dihydroxyacetone phosphate		phosphate (oxidized form)
E4P	Erythrose 4-phosphate	*NADPH*	Nicotinamide adenine dinucleotide
F6P	Fructose 6-phosphate		Phosphate (reduced form)
FBP	Fructose 1,6-biphosphate	*PEP*	Phosphoenolpyruvate
G6P	Glucose 6-phosphate	P_i	Inorganic orthophosphate
GAP	Glyeraldehyde 3-phosphate	*Pyr*	Pyruvate
Gluc	Glucose	*R5P*	Ribose 5-phosphate
GSH	Reduced glutathione	*Ru5P*	Ribulose 5-phosphate
GSSG	Oxidized glutathione	*S7P*	Sedoheptulose 7-phosphate
Lac	Lactate	*Xu5P*	Xylulose 6-phosphate

TABLE 7.2   Transitions and the corresponding enzyme names

Transition name	Full enzyme name
Aldolase	Aldolase
Enolase	Enolase
G6P_dehydrogenase	Glucose 6-phosphate dehydrogenase
GAP_dehydrogenase	Glyeraldehyde 3-phosphate dehydrogenase
Glutathione_peroxidase	Glutathione peroxidase
Glutathione_reductase	Glutathione reductase
Hexokinase	Hexokinase
Lactate_dehydrogenase	Lactate dehydrogenase
Phosphofructokinase	Phosphofructokinase
Phosphoglucose_isomerase	Phosphoglucose isomerase
Phosphoglycerate_kinase	Phosphoglycerate kinase
Phosphoglycerate_mutase	Phosphoglycerate mutase
PP_epimerase	Phosphopentose epimerase
PP_isomerase	Phosphopentose isomerase
Pyruvate_kinase	Pyruvate kinase
TIM_backward	Triose phosphate isomerase
TIM_forward	Triose phosphate isomerase
Transaldolase	Transaldolase
Transketolase1	Transketolase
Transketolase2	Transketolase

transition is called live, if forever it will eventually be enabled again, independently of what happened in the past. A transition is called dead, if it will never be enabled again, independently of what will happen in the future. Otherwise the transition is called nonlive. With other words, a transition is live, if it is not dead in the whole state space.

Definition 7.3 (Liveness of a transition)

- A transition t is *dead in the marking m*, if it is not enabled in any marking m' reachable from m:
$$\nexists\, m' \in [m\rangle : m'[t\rangle.$$
- A transition t is *live*, if it is not dead in any marking reachable from m_0.

Definition 7.4 (Liveness of a Petri net)

- A marking m is *dead*, if there is no transition, which is enabled in m.
- A Petri net is *deadlock-free* (weakly live), if there are no reachable dead markings.
- A Petri net is *live* (strongly live), if each transition is live.

Strong liveness includes weak liveness, but not vice versa. Consider Fig. 7.7 for an example illustrating the various liveness notions. The transitions r_1 and r_2 can never fire; so they are dead at the initial marking. The transitions r_3, r_4, and r_7 can only fire finitely often; so they are nonlive. Also the transitions r_5 and r_6 are nonlive; they can fire arbitrarily often, but only as long as the place set $\{B, C, D\}$ still carries a token. All these transitions $r_3 - r_7$ become dead, as soon as the place set $\{B, C, D\}$ got clean by firing of r_7. The remaining transitions r_8 and r_9 are live; during the repeated firing of r_5 and r_6, there is always the chance that in the future a token will reach the rightmost cycle, initiating the infinite firing of r_8 and r_9. Consequently, there is no state reachable, where no transition is enabled (dead marking); so the Petri net is deadlock-free (weakly live).

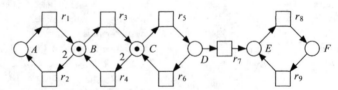

FIGURE 7.7   A Petri net to illustrate the three orthogonal behavioral properties. The transitions r_1 and r_2 in the leftmost cycle are dead at the initial marking. The transitions r_8 and r_9 in the rightmost cycle are live. All other transitions are nonlive. So the Petri net is weakly live, because not all transitions are live. The Petri net is not reversible, because the token decrease by firing of r_4 cannot be reverted. The place A is 0-bounded, place B is 1-bounded and all other places are 2-bounded, so the Petri net is 2-bounded.

Reversibility is a property, which relates to the possible paths between all reachable system states. It gives us information, whether the system contains irreversible system behavior. A reversible system is likely to be more robust against disturbances than a system with irreversible system behavior, because it can reinitialize itself. Reversibility is important, if we want to ensure that each reachable marking can be reached from each arbitrary other reachable marking.

Definition 7.5 (Reversibility)   A Petri net is *reversible*, if the initial marking can be reached again from each reachable marking: $\forall m \in [m_0\rangle : m_0 \in [m\rangle$.

This excludes an irreversible system behavior like unexpected burning out or accumulation of certain compounds. Reversibility ensures *reproducibility* of a reachable marking, including the initial one. The Petri net in Fig. 7.7 is not reversible. There is no transition, reverting the token decrease by the firing of r_4, so the initial marking is not reachable anymore as soon as r_4 has fired.

Boundedness is a property, which indicates, whether there could be unlimited compound accumulations, represented by an infinite token number in the corresponding places. Boundedness is essential for any steady state network analysis. If the maximal token number in all places is finite, that is, bounded by a positive integer number, the Petri net is bounded. So the number of reachable states is finite. The practical analyzability of bounded Petri nets is much larger than of unbounded ones, which exhibit an infinite number of reachable states.

Definition 7.6 (Boundedness)

- A place p is *k-bounded* (bounded for short), if there exists a positive integer number k, which represents an upper bound for the number of tokens on this place in all reachable markings of the Petri net:

$$\exists k \in \mathbb{N}_0 : \forall m \in [m_0\rangle : m(p) \le k.$$

- A Petri net is *k-bounded* (bounded for short), if all its places are *k*-bounded.
- If the Petri net is bounded in every initial marking, it is called to be *structurally bounded*.

Whereas reversibility represents a property, which gives information on the net behavior with respect to the repeated reachability of all states, boundedness makes a statement on the finiteness or infinity of the number of reachable states in dependence on the limited or unlimited token amount in the places.

In the Petri net in Fig. 7.7 the place A is 0-bounded, place B is 1-bounded, and all other places are 2-bounded, so the Petri net as a whole is 2-bounded. The Petri net of our running example, see Fig. 7.6, is unbounded. Many biological network models are unbounded due to the modeling style of input and output transitions, performing the compound exchange with the surroundings. Input transitions do not have preplaces, that is, no preconditions. Therefore, they are always enabled and impose an infinite number of tokens into the system, which leads to an infinite number of markings. We can model our running example also as a bounded Petri net, which will be explained in Section 7.3.2.

Another behavioral property of general interest relates to the situation, in which two transitions are enabled, but the firing of one transition disables the other one. Consider again Fig. 7.7: one token in place D will enable both its posttransitions r_6 and r_7. But only one of the enabled transitions is actually able to fire. Such a situation is called a *dynamic conflict*. The occurrence of dynamic conflicts indicates alternative (branching) system behavior, whereby the decision between these alternatives is taken nondeterministically.

Besides these four general properties, it often has to be checked, whether a given (sub)marking m' of special interest is *reachable* from a marking m, that is, if $m' \in [m\rangle$. For example, in Fig. 7.7, it might be of interest whether a marking is reachable, where the places E and F are marked at the same time. We will see in Section 7.3.5 how to express reachability properties using temporal logics.

Up to now, the conclusions on the behavioral properties have been derived by informal reasoning. This may work for small Petri nets but not for our running example. So let us turn to analysis techniques to decide behavioral properties in an algorithmic (i.e., automatically computable) way.

7.3 QUALITATIVE ANALYSIS

Basically, two different ways of qualitative net analysis can be distinguished, first, the static analysis, which considers the whole state space without constructing it, and second, the dynamic analysis, which does construct the full or partial state space for deciding the properties of interest.

Static analysis techniques are introduced in the following three subsections: *Structural analysis*, *Invariant analysis*, and *MCT-sets*. Most of the considered properties depend on the graph structure only. Some properties also take into account the initial marking.

Afterwards, two subsections discuss dynamic analysis techniques. It is common sense to distinguish between *general properties*, which can be applied to any system without considering its special functionality, and *special properties*, which do reflect the intended functionality. We introduce here the basic ideas. There exist several sophisticated improvements, which are, however, beyond the scope of this introduction.

7.3.1 Structural Analysis

The following structural properties are elementary graph properties and reflect the modeling approach. They can be read as preliminary consistency checks. Additionally, certain combinations of structural properties allow conclusions on behavioral properties, which is, however, not discussed in all details in this material.

- A Petri net is *ordinary*, if all arc weights are equal to 1. This includes homogeneity (see next bullet). A nonordinary Petri net cannot be live and 1-bounded at the same time.

- A Petri net is *homogeneous*, if all outgoing arcs of a given place have the same multiplicity.

- A Petri net is *pure*, if there are no two nodes, connected in both directions. This excludes read arcs.

- A Petri net is *conservative*, if all transitions fire token-preservingly, that is, all transitions add exactly as many tokens to their postplaces as they subtract from their preplaces. A conservative Petri net is structurally bounded.

- A Petri net is *connected*, if it holds for all pairs of nodes a and b that there is an undirected path from a to b. So the direction of arcs is ignored here. Disconnected parts of a Petri net cannot influence each other, so they may be treated separately. In the following, we consider only connected Petri nets.

- A Petri net is *strongly connected*, if it holds for all pairs of nodes a and b that there is a directed path from a to b. Strongly connectedness involves connectedness and the absence of boundary nodes. It is a necessary condition for a Petri net to be live and bounded at the same instant.

- A Petri net is *free of boundary nodes*, if there are no transitions without pre-/postplaces and no places without pre-/posttransitions. Such a self-contained, that is, closed system needs a nonclean initial marking to become live. A Petri net with input transitions is unbounded.

A Petri net is *free of static conflicts*, if there are no two transitions sharing a preplace. Transitions, sharing a preplace, might have to compete, if the tokens on a shared preplace are limited. Therefore, they are said to be in a static conflict. For example, place B in Fig. 7.8 is a shared place. Its posttransitions r_2 and r_3 compete for the tokens on B, therefore r_2 and r_3 are in a static conflict. It depends on the token situation whether a dynamic conflict occurs or not. A static conflict is generally given, if a compound is involved in several reactions, compare Fig. 7.3 case (c), which is the case especially for multifunctional proteins and often for secondary compounds. Static conflicts indicate situations, where dynamic conflicts, that is, nondeterministic choices may occur in the system behavior. Such nondeterministic choices are impossible, if there are no static conflicts in a Petri net.

All these structural properties above are not influenced by the initial marking. Most of these properties can be decided locally in the graph structure. Only connectedness and strongly connectedness have to consider the global graph structure, which is done using standard graph algorithms. Furthermore, the following advanced structural Petri net properties are of interest, which have to be decided by combinatorial algorithms.

A nonempty set of places $D \subseteq P$ is called *structural deadlock (co-trap)*, if every transition, which fires tokens onto a place in this structural deadlock set, also has a preplace in this set, that is, ${}^\bullet D \subseteq D^\bullet$ (the set of pretransitions is contained in the set of posttransitions). Thus, pretransitions of a structural deadlock cannot fire, if the place set is clean. Therefore, a structural deadlock cannot get tokens again, as soon as it is

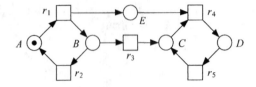

structural deadlock: $^{\bullet}\{A, B\} \subseteq \{A, B\}^{\bullet}$ trap: $\{C, D, E\}^{\bullet} \subseteq {}^{\bullet}\{C, D, E\}$

pretransitions: $^{\bullet}\{A, B\} = \{r_1, r_2\}$ posttransitions: $\{C, D, E\}^{\bullet} = \{r_4, r_5\}$

posttransitions: $\{A, B\}^{\bullet} = \{r_1, r_2, r_3\}$ pretransitions: $^{\bullet}\{C, D, E\} = \{r_1, r_3, r_4, r_5\}$

FIGURE 7.8 The token on place A can rotate in the left cycle by repeated firing of r_1 and r_2. Each round produces an additional token on place E, making this place unbounded. This cycle can be terminated by firing of transition r_3, which brings the circulating token from the left to the right side of the Petri net. The place set $\{A, B\}$ cannot get tokens again as soon as it got clean. Thus, it is a (proper) structural deadlock. Contrary, the place set $\{C, D, E\}$ cannot become clean again as soon as it got a token. The repeated firing of r_4 and r_5 reduces the total token number, but cannot remove all of them. Thus, the place set $\{C, D, E\}$ is a (proper) trap.

clean, and then all its posttransitions $t \in D^{\bullet}$ are dead. A Petri net without structural deadlocks is live, while a system in a dead state has a clean structural deadlock. Generally, a clean structural deadlock is not desired in biological systems.

A nonempty set of places $Q \subseteq P$ is called *trap*, if every transition, which subtracts tokens from a place of the trap set, also has a postplace in this set, that is, $Q^{\bullet} \subseteq {}^{\bullet}Q$ (the set of posttransitions is contained in the set of pretransitions). Thus, posttransitions of a trap always return tokens to the place set. Therefore, once a trap contains tokens, it cannot become clean again. There can be a decrease of the total token amount within a trap, but not down to zero. In biological systems, a trap indicates irreversible compound deposition. It depends on the application, if this effect is desired or should not occur.

An input place p establishes a structural deadlock $D = \{p\}$ on its own, and an output place q a trap $Q = \{q\}$. If D and D' are structural deadlocks (traps), then $D \cup D'$ is also a structural deadlock (trap). If each transition has a preplace, then $P^{\bullet} = T$, and if each transition has a postplace, then ${}^{\bullet}P = T$. Therefore, in a net without boundary transitions, the whole set of places is a structural deadlock as well as a trap. As we will see in Section 7.3.2, there are special token-preserving place sets, characterized by the so-called p-invariants, which are deadlocks and traps at the same time. For those deadlocks (traps), which do not correspond to p-invariants, we introduce the notion *proper deadlock (proper trap)*. See also Fig. 7.8 for an example to illustrate these two notions.

Running example: Our open model (Fig. 7.6) is not ordinary, but homogeneous, and pure. It is not conservative, but connected, however, not strongly connected. It has boundary transitions, especially input transitions, therefore it is unbounded. There are three static conflicts concerning primary compounds and several static conflicts involving secondary compounds. For example, the transitions *Phosphoclucose_isomerase* and *G6P_dehydrogenase* compete for the tokens on *G6P*. Therefore, depending on the nondeterministic conflict decision, a token will continue the glycolytic pathways or enter the PPP. There are neither proper structural deadlocks nor proper traps.

7.3.2 Invariant Analysis

System invariants are well-established concepts in mathematical reasoning. In biochemical network analysis, invariants can be biologically interpreted, and therefore provide additional insights into the network behavior. Invariants give information about structural composition principles, and the possible net behavior. Additionally, system inconsistencies can be detected by invariant analysis. Therefore, model validation should include a check of all invariants for their biological plausibility.

Besides the Petri net invariants [15], various types of system invariants have been defined independently. The Petri net *place-invariants* (p-*invariants*) correspond to the known notion of moieties [19], which represent compound conservation relations. The Petri net *transition-invariants* (t-*invariants*) correspond to the *elementary modes* [28], which have been introduced to describe the steady state behavior of metabolic networks. In Refs [8,9] it is shown that exactly the same analysis principles can be applied to signal transduction networks. In order to reduce the generating system of the solution space, further subsets of t-invariants as *extreme pathways* [29] and *generic pathways* [14] have been proposed. However, all these related concepts will not be explained here, because of space limitations.

To introduce invariants, we need the notion of the incidence matrix, which for metabolic networks coincides with the stoichiometric matrix. Fig. 7.9 gives an example for the incidence matrix. The reactions (transitions) index the columns, and the compounds (places) index the rows of the matrix structure. A matrix element indicates the token change of a compound (defined by the row) by the firing of a transition (defined by the column). Thus, both the incidence as well as the stoichiometric matrix define a bipartite graph. The net structure of a pure Petri net (which excludes read arcs) is fully represented by the incidence matrix. Using the incidence matrix, we define two homogeneous linear equation systems, the solution of which are the p- and t-invariants. The following definition recalls the essential basic terms.

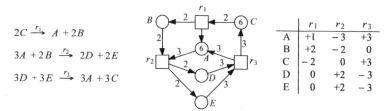

FIGURE 7.9 A stoichiometric equation system with the corresponding Petri net and its incidence (stoichiometric) matrix. The given initial marking makes the system live.

Definition 7.7 (p-invariants, t-invariants)   Let $\mathcal{N} = (P, T, f, m_0)$ be a Petri net.

- The incidence matrix of $\mathcal{N}$ is a matrix $C : P \times T \rightarrow \mathbb{Z}$, indexed by P and T, such that $C(p, t) = f(t, p) - f(p, t)$.
- A *place vector (transition vector)* is a vector $x : P \rightarrow \mathbb{Z}$, indexed by P ($y : T \rightarrow \mathbb{Z}$, indexed by T).
- A place vector (transition vector) is called a p-invariant (t-invariant), if it is a non-trivial non-negative integer solution of the homogeneous linear equation system $x \cdot C = 0$ ($C \cdot y = 0$).
- The set of nodes corresponding to an invariant's non-zero entries are called the *support* of this invariant x, written as $supp(x)$.
- An invariant x is called *minimal*, if it does not contain any other invariant z, that is, $\nexists$ invariant $z : supp(z) \subset supp(x)$, and the greatest common divisor of all entries of x is 1.
- A Petri net is covered by p-invariants — CPI, (covered by t-invariants — CTI), if every place (transition) belongs to a p-invariant (t-invariant).

The set of minimal (p- or t-) invariants builds a unique generating system for all invariants. All possible invariants x can be computed as non-negative linear combinations of x_i, the minimal ones: $n \cdot x = \sum(a_i \cdot x_i)$, with $n, a_i \in \mathbb{N}_0$, that is, the allowed operations are addition, multiplication by a natural number, and division by a common divisor. A minimal t-invariant (p-invariant) defines a connected subnet, consisting of its support, its pre- and postplaces (pre- and posttransitions), and all arcs in between. The computation of invariants requires only structural reasoning; the state space does not need to be generated. Therefore, the state space explosion problem, see Section 7.3.4, does not apply here. However, the number of minimal invariants can grow exponentially with the size of the Petri net.

Now, let us return to the example in Fig. 7.9. Two types of invariants can be derived from the incidence matrix, see Definition 7.7, third bullet. The following two homogeneous linear equation systems have to be solved to compute the p- and t-invariants.

$$
\begin{array}{rrrrl}
x_1 & +2x_2 & -2x_3 & & = 0 \\
-3x_1 & -2x_2 & & +2x_4 +2x_5 & = 0 \\
3x_1 & & +3x_3 & -3x_4 -3x_5 & = 0
\end{array}
$$

$$
\begin{array}{rrrl}
y_1 & -3y_2 & +3y_3 & = 0 \\
2y_1 & -2y_2 & & = 0 \\
-2y_1 & & +3y_3 & = 0 \\
& +2y_2 & -3y_3 & = 0 \\
& +2y_2 & -3y_3 & = 0
\end{array}
$$

There exist four solutions for the minimal p-invariants, and one solution for the minimal t-invariant. The solutions are written as vectors with a length of the number of places (that is, as place vectors), and as vectors with a length of the number of transitions (that is, as transitions vectors), respectively. The solutions for the p-invariants are $(2, 0, 1, 0, 3)$, $(0, 1, 1, 0, 1)$, $(2, 0, 1, 3, 0)$, and $(0, 1, 1, 1, 0)$, and for the t-invariants $(3, 3, 2)$. The supports, defined by the nonzero entries of these vectors, are then for the p-invariants $\{A, C, E\}$, $\{B, C, E\}$, $\{A, C, D\}$, and $\{B, C, D\}$, and for the t-invariants $\{r_1, r_2, r_3\}$. So the Petri net is CPI and CTI.

A *p-invariant* x is technically a place vector, standing for a set of places, over which the weighted sum of tokens is constant independently of any firing. The weight for each place is given by the positive integer number in the solution vector. That means, any two reachable markings m_1, m_2 hold $x \cdot m_1 = x \cdot m_2$. So p-invariants represent token-preserving sets of places. Their supports are structural deadlocks and traps at the same time. (But caution: not every place set, which is structural deadlock and trap at the same time, is also a p-invariant.) Therefore, they need tokens in the initial marking to allow liveness. A place belonging to a p-invariant is bounded, and a Petri net, which is CPI, is structurally bounded.

In the context of metabolic networks, p-invariants reflect compound conservations, while in signal transduction networks p-invariants often correspond to the different (inactive, active) states of a given compound (protein or protein complex).

A *t-invariant* is technically a transition vector, defining a multiset of transitions. A multiset is a set, which may contain an element in multiple copies, that is, a transition can occur more than one times as specified by the integer number in the solution vector. This multiset of transitions has altogether a zero effect on the marking, that is, after all of them have fired, a given marking is reproduced. A t-invariant is called *realizable*, if such behavior is actually possible due to the sufficient number of tokens in a reachable marking.

Obvious t-invariants are called *trivial t-invariants*. They consist, for example, of the two transitions representing the forward and backward directions of reversible reactions, or of the generating and removing of a compound, often used to model the system environment. A t-invariant has two biological interpretations.

1. The entries of a t-invariant represent a multiset of transitions, which reproduce a given marking by their partially ordered firing. That means that they occur basically one after the other. The partial order sequence of the firing events of the t-invariant's transitions may contribute to a deeper understanding of the system behavior.

2. The entries of a t-invariant may also be read as the relative firing rates of transitions, all of them occurring permanently and concurrently. This activity level corresponds to the steady state behavior.

Independently of the interpretation, the net representations of minimal t-invariants stand for minimal self-contained subnetworks with an enclosed biological meaning. In metabolic networks, minimal t-invariants describe minimal sets of enzymes, which are necessary for the network function at steady state. In signal transduction networks, the signal response behavior may be reflected by nonminimal t-invariants.

It is a crucial question, whether the Petri net is covered by t-invariants (CTI). This property ensures that every transition participates in a t-invariant. That means that every chemical reaction in the system may occur as part of the basic behavior of the Petri net. However, to ensure that every chemical reaction can actually contribute to the system behavior, it is essential that the net is covered by realizable t-invariants. The CTI property is a necessary condition for bounded Petri nets to be live.

Running example: Minimal invariants are often sparse vectors, meaning that typically many entries are zero. To avoid annoying notations, we give the invariants in the style of a multiset notation by enumerating only all nonzero entries of their vectors in a suitable order. Each entry is specified by the corresponding node name, multiplied by the entry value, if larger than 1, indicating the multiple presence of the given node element in the multiset. There are two p-invariants:

$$x_1 = (2 \cdot GSSG, GSH),$$
$$x_2 = (NADP^+, NADPH).$$

Both represent obvious compound conservations. The weight 2 in x_1 counts a token on GSSG twice. The weighted token sum in both p-invariants is 2, because

$$x_1 \cdot m_0 = 2 \cdot m_0(GSSG), \text{ and}$$
$$x_2 \cdot m_0 = m_0(NADP^+).$$

There are eight minimal t-invariants, covering the Petri net. As expected, there are six trivial t-invariants, one for the reversible reaction, and five for the input/output transitions of the secondary compounds:

$$y_1 = (TIM\_backward, TIM\_forward),$$
$$y_2 = (g\_ATP, r\_ATP),$$
$$y_3 = (g\_ADP, r\_ADP),$$
$$y_4 = (g\_P_i, r\_P_i),$$
$$y_5 = (g\_NAD^+, r\_NAD^+),$$
$$y_6 = (g\_NADH, r\_NADH).$$

There are two nontrivial minimal t-invariants, one describing the lactate formation by the glycolysis only

$$y_7 = (\ g\_Gluc, 2 \cdot g\_ADP, 2 \cdot g\_P_i,$$

$$Hexokinase, Phosphoclucose\_isomerase, Phosphofructokinase,$$

$$Aldolase, TIM\_forward,$$

$$2 \cdot GAP\_dehydrogenase, 2 \cdot Phosphoglycerate\_kinase,$$

$$2 \cdot Phosphoglycerate\_mutase, 2 \cdot Enolase,$$

$$2 \cdot Pyruvate\_kinase, 2 \cdot Lactate\_dehydrogenase,$$

$$2 \cdot r\_Lac, 2 \cdot r\_ATP\ ),$$

and the other, describing the lactate formation by including also the PPP

$$y_8 = (\ 3 \cdot g\_Gluc, 5 \cdot g\_ADP, 5 \cdot g\_P_i,$$

$$3 \cdot Hexokinase, 3 \cdot G6P\_dehydrogenase,$$

$$6 \cdot Glutathione\_reductase, 6 \cdot Glutathione\_peroxidase,$$

$$PP\_isomerase, 2 \cdot PP\_epimerase,$$

$$Transketolase1, Transaldolase, Transketolase2,$$

$$2 \cdot Phosphofructokinase, 2 \cdot Aldolase, 2 \cdot TIM\_forward,$$

$$5 \cdot GAP\_dehydrogenase, 5 \cdot Phosphoglycerate\_kinase,$$

$$5 \cdot Phosphoglycerate\_mutase, 5 \cdot Enolase,$$

$$5 \cdot Pyruvate\_kinase, 5 \cdot Lactate\_dehydrogenase,$$

$$5 \cdot r\_Lac, 5 \cdot r\_ATP\ ).$$

The net representations of these two nontrivial t-invariants generate the essential partial order behavior of the modeled system. Reading the entries of y_7 and y_8 as multisets of transitions, we can check the nontrivial minimal t-invariants for realizability in the given initial marking, which then involves the realizability of the trivial t-invariants. To follow the flow of entering tokens, the transitions have to fire basically in the order as given in the t-invariant descriptions above. Doing so, we observe the intermediate need of secondary compounds, ATP and NAD^+, which are released afterwards. Moreover, the given initial marking limits the resources in $NADP^+$ and $GSSG$, causing a sequential firing of the reactions $G6P\_dehydrogenase$, $Glutathione\_reductase$, and $Glutathione\_peroxidase$ in the PPP.

The nontrivial t-invariant entries, read as relative firing rates, indicate the higher activity level of the PPP-induced compound flow in the steady state. The transition $TIM\_backward$ does not belong to a nontrivial t-invariant, therefore, it does not contribute to the steady state behavior of the two subnetworks, defined by the non-trivial t-invariants. The input and output transitions of a t-invariant, easily recognizable due to the adopted naming convention, specify the total equation of the corresponding subnetwork, which are for y_7 and y_8, respectively:

$$Gluc + 2 \cdot ADP + 2 \cdot P_i \rightarrow 2 \cdot Lac + 2 \cdot ATP$$
$$3 \cdot Gluc + 5 \cdot ADP + 5 \cdot P_i \rightarrow 5 \cdot Lac + 5 \cdot ATP.$$

Environment modeling: We employ the nontrivial minimal t-invariants to construct a closed system, representing the steady- state behavior by a bounded model. Boundedness usually ensures a higher degree of practical analyzability, because all system states (i.e., all token distributions) may principally be enumerated and evaluated, see Section 7.3.4. To make a Petri net model bounded we have to ensure that the input of tokens into the net as well as the output of tokens from the net are controlled such that an infinite accumulation of tokens in places will be avoided. This requires two adaptations: removing the boundary transitions and controlling the dynamic conflicts.

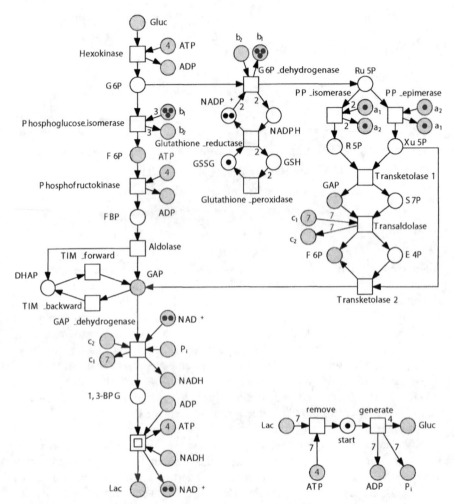

FIGURE 7.10   The closed Petri net model of the G-PPP, using the second environment modeling style. The artificial transitions *generate* and *remove*, and the artificial place *start* build the environment net component, given on the lower right side. This net component corresponds to the inverse total equation of the open network. Furthermore, the artificial places $a_1, a_2, b_1, b_2, c_1, c_2$ have been added to control the three dynamic conflicts according to the steady state ratio of the involved transitions. These control places come in pairs, forming p-invariants. The number 7 in place c_1 specifies the amount of tokens residing in this place in the initial marking. The computation of all supplementary net components relies on the nontrivial minimal t-invariants. This self-contained Petri net is structurally bounded, live and reversible. It represents a closed system.

To replace the boundary transitions, we complement the network by a net component, modeling that environment behavior necessary to keep the network in its steady state, compare Fig. 7.10. This environment behavior is determined by the inverse total equation of the whole network, which we get by summation of the total equations of the two nontrivial minimal t-invariants, given above.

$$4 \cdot Gluc + 7 \cdot ADP + 7 \cdot P_i \rightarrow 7 \cdot Lac + 7 \cdot ATP.$$

Technically this is done by two artificial transitions *remove* and *generate* (separated by an artificial place *start*), which generate the necessary tokens for the system to bring it to live, and remove exceeding tokens from the system to avoid an infinite token accumulation.

To control those dynamic conflicts, which result into different token ratios, we introduce supplementary control places. The place pairs (a_1, a_2), (b_1, b_2), and (c_1, c_2) regulate the token flow in the dynamic conflicts between the transitions

- *PP_isomerase and PP_epimerase (Ru5P)*,
- *Phosphoglucose_isomerase and G6P_dehydrogenase (G6P)*, and
- *Transaldolase and GAP_dehydrogenase (GAP)*, respectively.

Each pair of control places forms a p-invariant. The arc weights and the necessary initial marking for each place pair are computed by help of the minimal t-invariants. Let us consider the control places c_1 and c_2, regulating the conflict, induced by the shared place *GAP*. There are three posttransitions, competing for the tokens on *GAP*. The transition *TIM_backward* has not to be considered, because the cycle via the reversible reaction does not change the token amount. The transition *GAP_dehydrogenase* participates in the t-invariant y_7 twice, and in the t-invariant y_8 five times, making together 7. Contrary, the transition *Transaldolase* has to fire only once, according to t-invariant y_8. In summary, these two transitions have to fire in a ratio 7:1, which is enforced by the arc weights, connecting c_1 and c_2 with the transitions in conflict, and the initial marking for c_1 and c_2.

This kind of environment behavior reflects explicit assumptions about the quantitative ratio of input/output compounds, while in the open system model no assumptions about the the quantitative ratio of input/output compounds are made.

There are no transitions without preplaces anymore. Therefore, we have a chance to get a bounded model. Opposite to the open model, see Fig. 7.6, the closed model, see Fig. 7.10, is strongly connected. Furthermore, it is CPI, therefore structurally bounded. It still is CTI, which for bounded models is a necessary condition to be live. The environment component removes seven tokens from *ATP*, so the closed model is not homogeneous anymore. All other structural properties are the same as for the open model.

7.3.3   MCT-Sets

In order to support the examination of t-invariants for their biological meaning, transitions can be classified into *maximal common transition sets (MCT-sets)* according to their common occurrence in the minimal t-invariants. This is especially helpful, if the amount of t-invariants is too large to be explored manually.

MCT-sets are defined over the supports of the minimal t-invariants. Supports are sets, which can technically be read as vectors over Booleans, which allows the access to the ith entry by indexing.

Definition 7.8 (Maximal common transition sets (MCT-Sets))   Let X denote the set of all (nontrivial) minimal t-invariants x of a given Petri net.

- Two transitions t_i and t_j belong to the same MCT-set, if they participate in exactly the same minimal t-invariants, that is, $\forall x \in X, \forall i, j \in \{1, \ldots, n\}$: $supp(x)(i) = supp(x)(j)$.
  Equally, we can define the following.
- A transition set $A \subseteq T$ is called an MCT-set, if
  $\forall x \in X : \quad A \subseteq supp(x) \vee A \cap supp(x) = \emptyset.$

The support-oriented classification according to the first bullet in the definition above establishes an equivalence relation in the transition set T, leading to a partition of T. The equivalence classes A are the MCT-sets, defining disjunctive subnets, which are not necessarily connected. These subnets represent a possible structural decomposition of large biochemical networks into rather small subnets, the decomposition being based on statically decidable properties only. MCT-sets can be read as the smallest biologically meaningful functional units. They can serve as building blocks of the whole network.

Running example: Considering the two nontrivial minimal t-invariants y_7 and y_8 of the open model (see Fig. 7.6), we find four MCT-sets. The first set contains the common region (intersection) of both t-invariants, comprising almost the whole glycolytic pathway

$$
\begin{aligned}
A = \; & supp(y_7) \cap supp(y_8) \\
= \; & \{ \, g\_Gluc, g\_ADP, g\_P_i, Hexokinase, Phosphofructokinase, Aldolase, \\
& TIM\_forward, GAP\_dehydrogenase, Phosphoglycerate\_kinase, \\
& Phosphoglycerate\_mutase, Enolase, Pyruvate\_kinase, \\
& Lactate\_dehydrogenase, r\_Lac, r\_ATP \, \}.
\end{aligned}
$$

The next two sets contain those transitions, which are specific to one of the two t-invariants. The specific region of the t-invariant y_7 belongs to the glycolytic pathway

$$
\begin{aligned}
B = \; & supp(y_7) - supp(y_8) \\
x = \; & \{ \, Phosphoglucose\_isomerase \, \},
\end{aligned}
$$

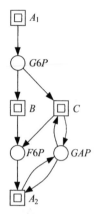

FIGURE 7.11   The coarse net structure of the Petri net given in Fig. 7.6, according to the structuring principle inherent in the minimal t-invariants. Each macro transition stands for a connected subnet defined by a set of transitions, occurring together in all nontrivial minimal t-invariants. Each elementary (loop-free) macro transition sequence in the coarse net structure corresponds to a nontrivial minimal t-invariant of the whole network. There are two such sequences (A_1, B, A_2) and (A_1, C, A_2), sharing the beginning and the end. The places shown in the coarse net structure are the boundary places of the subnets, building the interface between the subnets. Only the primary compound flow is represented here.

and the specific region of the t-invariant y_8 includes the PPP

$$C = \ supp(y_8) - supp(y_7)$$
$$= \{\ G6P\_dehydrogenase,$$
$$Glutathione\_reductase, Glutathione\_peroxididase$$
$$PP\_isomerase, PP\_epimerase,$$
$$Transketolase1, Transaldolase, Transketolase2\ \}.$$

All remaining transitions

$$D = \ \{\bigcup_{i=1,6} supp(y_i)\} - supp(y_7) - supp(y_8)$$
$$= \{\ g\_ATP, g\_NAD^+, g\_NADH, TIM\_backward, r\_ADP, r\_P_i\ \}$$

are only part of trivial t-invariants. This means that these transitions do not contribute to the steady state behavior of the nontrivial t-invariants.

Thus, the main building blocks of the Petri net, and by this way of the underlying biochemical network, are represented by the first three MCT-sets, each defining a connected subnetwork. The two subnets, describing the two pathways, are defined by the union of the first MCT-set with the second or third one, respectively. However, if we neglect the arc connections established by secondary compounds, the MCT-set A

breaks down into two subsets

$$A_1 = \{g\_Gluc, Hexokinase\},$$
$$A_2 = A - A_1,$$

which are each connected subsets according to the *primary compound flow*. We obtain the coarse network structure as given in Fig. 7.11, highlighting the structuring principle inherent in the nontrivial minimal t-invariants. Each elementary (loop-free) macro transition sequence in the coarse net structure corresponds to a nontrivial minimal t-invariant of the whole network. There are two such sequences (A_1, B, A_2) and (A_1, C, A_2), sharing the beginning and the end.

In the closed model, we have only two t-invariants, a trivial one of the reversible reaction, and the other covering all remaining transitions. A net decomposition into building blocks in this case is not necessary, because the structure is obvious.

7.3.4   Dynamic Analysis of General Properties

Boundedness has already been decided statically. However, in order to decide liveness and reversibility, we generally have to compute a data structure, describing the whole system behavior. The easiest way to do this is constructing a dedicated graph, the *reachability graph*. The nodes of a reachability graph represent all possible states (markings) of the Petri net. The arcs in between are labeled by single transitions, the firing of which causes the related state change.

Definition 7.9 (Reachability graph)   Let $\mathcal{N} = (P, T, f, m_0)$ be a Petri net. The reachability graph of $\mathcal{N}$ is the graph $\mathcal{RG}(\mathcal{N}) = (V_\mathcal{N}, E_\mathcal{N})$, where

- $V_\mathcal{N} := [m_0\rangle$ is the set of nodes,
- $E_\mathcal{N} := \{(m, t, m') \mid m, m' \in [m_0\rangle, t \in T : m[t\rangle m'\}$ is the set of arcs.

The reachability graph is finite for bounded Petri nets only, for example, see Fig. 7.12. In state s_1, the transitions r_1 and r_2 are enabled, but only one of them can fire. They are in a dynamic conflict. The firing of one transition disables the other one. Constructing the reachability graph, both alternatives are considered. Contrary, in state s_2, the transitions r_3 and r_4 are enabled concurrently. They can fire in any order. The firing of one transition does not disable the other one. There are two interleaving sequences of r_3 and r_4, leading from state s_2 to state s_5. Thus, a branching node in the reachability graph, that is, a node with several successors, means either alternative or concurrent behavior. The difference is not locally decidable in the reachability graph anymore. The transitions r_4 and r_5 may be read as reversible reactions, producing several two-states loops.

Altogether, the reachability graph gives a concise representation of all possible single step firing sequences. Consequently, concurrent system behavior is described by enumerating all permutations of concurrent transitions, shortly called *interleaving*

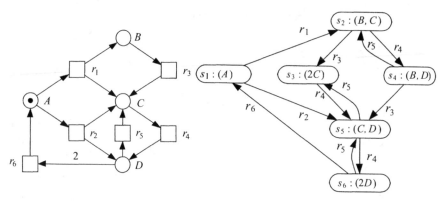

FIGURE 7.12 A Petri net and its reachability graph. The six states, forming the nodes of the reachability graph, are given in a multiset notation. The arcs are labeled by the transition, the firing of which causes the related state change.

firing sequences. Therefore, the reachability graph represents the interleaving semantics. The much more challenging partial order (true concurrency) semantics is beyond the scope of the introduction presented here.

Reachability graphs tend to be huge, because they comprise all possible system states (markings). The state space grows rapidly for two reasons: concurrency is resolved by all interleaving sequences, and many tokens within a p-invariant can distribute themselves rather arbitrarily. The state space explosion motivates the static analyses, as presented in the three preceding subsections. If we succeed in constructing the complete reachability graph, we are able to decide the behavioral Petri net properties, introduced in Section 7.2.2. Please note, then the following used notation *"iff"* stands shortly for *"if and only if."*

1. A Petri net is k-bounded, iff there is no node in the reachability graph with a token number larger than k in any place.
2. A Petri net is reversible, iff the reachability graph is strongly connected.
3. A Petri net is deadlock-free, iff the reachability graph does not contain terminal nodes, that is, nodes without outgoing arcs.
4. In order to decide liveness, we partition the reachability graph into maximal sets of strongly connected nodes, which we call strongly connected components (SCC). An SCC is called terminal, if no other SCC is reachable in the partitioned graph. A transition is live, iff it appears in all terminal SCCs of the partitioned reachability graph. A Petri net is live, which includes deadlock freedom, iff this holds for all transitions.

The occurrence of dynamic conflicts is checked at best during the construction of the reachability graph, because branching nodes do not necessarily mean alternative system behavior. The reachability of a given marking m' is tested by constructing a (shortest) path (i.e., a sequence of directly connected nodes), leading from m_0 to m'. The construction of such a path does not succeed, if the node m' does not exist.

Let us return to the example in Fig. 7.12. The reachability graph is finite, so the Petri net is bounded. A closer look at all states reveals 2-boundedness. The reachability graph is strongly connected, so the Petri net is reversible. All transitions appear in this strongly connected reachability graph at least once, so the Petri net is live. There are dynamic conflicts.

Running example: The reachability graph is finite for the closed system only, whereby its size depends on the given initial marking. We observe a slight growth in the state space by increasing the initial token numbers in *ATP* and *NAD$^+$*, because this allows a higher concurrency degree within the network. To get liveness and reversibility, the minimal token numbers in these places of the initial marking are (4 *ATP*, 2 *NAD$^+$*) or (5 *ATP*, *NAD$^+$*), respectively. Additionally, we make the following observations. There are no dynamic conflicts concerning ATP, that is, no restriction of the concurrency degree, if $m_0(ATP) > 5$. There is no further increase of the size of the reachability graph (42.576 states, 204.172 arcs), if $m_0(ATP) > 7$ and/or $m_0(NAD^+) > 7$.

While liveness and reversibility of an unbounded model are generally decidable, there are no efficient algorithms known, and thus no tools exist for this task. There are several theorems of static analysis approaches (some of them are mentioned in Sections 7.3.1 and 7.3.2), which sometimes help, but not for our running example.

7.3.5 Dynamic Analysis of Special Properties

To validate the model, it is often of interest to prove — besides the general properties — additional special properties, which reflect the intended functionality of the network. We have to formulate these special properties in a unambiguous language. Temporal logics, a mathematical mechanism has been proven to be best suited for this purpose. It provides a flexible formalism that considers the validity of logical statements in temporal relations in the sense of *before* and *after*.

The analysis technique, deciding whether a temporal-logic property holds in a model, is called *model checking*, and the tools implementing the algorithms are called *model checkers*. Model checking generally requires boundedness. If the Petri net is 1-bounded, there exists a particularly rich choice of model checkers, exploiting different data structures and algorithms [27].

One of the widely used temporal logics is the *computational tree logic* (CTL). It is called after the data structure used — the computational tree, which we get by unwinding the reachability graph. Therefore, CTL represents a branching time logic with interleaving semantics.

The application of this validation approach needs to understand temporal logics. Here, we restrict ourselves to an informal introduction into CTL. CTL is, as any temporal logic, an extension of a classical (propositional) logic. The atomic (i.e., basic, not more dividable) propositions consist of statements on the current token situation in a given place. To simplify the notation, places are interpreted as (non-negative) integer variables, which allows statements as $Lac = 7$, and $ATP > 7$ (meaning that the place *Lac* carries seven tokens, and *ATP* at least eight tokens). Atomic propositions can be combined to composed propositions using the standard logical operators: ¬ (negation), ∧ (conjunction), ∨ (disjunction), and → (implication), for example, $Lac = 7 \wedge ATP > 7$.

The truth value of such a composed proposition may change by the execution of the Petri net. These temporal relations between propositions are expressed by the additionally available temporal operators. In CTL there are basically four of them (ne**X**t, **F**inally, **G**lobally, **U**ntil), which come in two versions (**E** for Existance, **A** for All), making together eight operators.

Let $\phi_{[1,2]}$ be arbitrary temporal-logic formulae. Then, the following formulae hold in state m,

- **EX** ϕ : if there is a state reachable by one step, where ϕ holds.
- **EF** ϕ : if there is a path, where ϕ holds finally.
- **EG** ϕ : if there is a path, where ϕ holds globally, that is, forever.
- **E** $(\phi_1 \; \mathbf{U} \; \phi_2)$: if there is a path, where ϕ_1 holds as long until ϕ_2 holds finally.

The other four operators, which we get by replacing the Existence operator by the All operator, are defined likewise by extending the requirement *"there is a path"* to *"for all paths holds"*. For a more comprehensive introduction into temporal logics see Ref. [5]. For typical patterns, how to specify biologically relevant properties of biochemical networks using CTL, see Ref. [3].

Running example: We demonstrate this technique by the following samples of meaningful statements, the truth of which can be determined for the closed version by model checking.

- **property 1:** The initially provided *ATP* can be used up in between, that is, there is a reachable state, where the place *ATP* is empty.

$$\mathbf{EF} \; (ATP = 0)$$

- **property 2:** A cyclic behavior concerning the presence/absence of *ATP* is possible forever. Technically spoken, in all states, that is, forever, holds: if the place *ATP* is nonempty, then there is a path, where it becomes empty finally, and vice versa, if the place *ATP* is empty, then there is a path, where it becomes finally nonempty.

$$\mathbf{AG} \; [\; ATP \neq 0 \rightarrow \mathbf{EF} \; (ATP = 0) \;] \wedge [ATP = 0 \rightarrow \mathbf{EF} \; (ATP \neq 0) \;]$$

- **property 3:** The total equation of the network holds forever. For the network without the reversible reaction *TIM_backward* this translates into: in all states, that is, forever holds: starting from a state with four tokens in *Gluc*, seven tokens in *ADP* and seven tokens in P_i (which is the state produced by the transition *generate*), all paths will finally reach a state with seven tokens in *Lac* and at least seven tokens in *ATP* (which is the state enabling the transition *remove*).

$$\mathbf{AG} \, [\, (Gluc = 4 \wedge ADP = 7 \wedge P_i = 7) \rightarrow \mathbf{AF}(Lac = 7 \wedge ATP > 7) \,]$$

For the network with the reversible reaction *TIM_backwards*, we have to weaken the statement, because the two transitions modeling the reversible reaction build cycles. These cycles can — at least structurally — prevent the system from reaching the state, which enables the transition *remove*.

$$\mathbf{AG} \, [\, (Gluc = 4 \wedge ADP = 7 \wedge P_i = 7) \rightarrow \mathbf{EF}(Lac = 7 \wedge ATP > 7) \,]$$

- **property 4:** It is possible to produce lactate without phosphofructokinase (which is producing *FBP*). In other words, it holds forever: if we start in a state, where neither *FBP* nor *Lac* carry a token, then there is a path reaching a state, where *Lac* gets a token, while *FBP* remains empty in all states along this path.

$$\mathbf{AG} \, [\, (FBP = 0 \wedge Lac = 0) \rightarrow \mathbf{E} \, (FBP = 0 \, \mathbf{U} \, Lac > 0) \,]$$

Obviously, it would be of great help to have a dedicated technical language for expressing typical patterns of those kinds of special properties.

7.3.6   Model Validation Criteria

To summarize the preceding validation steps, the model in its two versions has passed the following general-purpose validation criteria.

- validation criterion 1
  - All expected structural properties hold.
  - All expected general behavioral properties hold.
- validation criterion 2
  - CPI (closed system).
  - No minimal p-invariant without biological meaningful interpretation.
  - No known compound conservation without corresponding minimal p-invariant.
- validation criterion 3
  - CTI.
  - No minimal t-invariant without biological interpretation.
  - No known biological behavior without corresponding minimal t-invariant.
- validation criterion 4
  - All expected special behavioral properties, expressed as temporal-logic formulae, hold.

It is worth noting that not all of the validation criteria outlined above are always feasible. For example, it makes sense only to ask for CPI as well as CTI for closed systems (i.e., self-contained systems without boundary nodes). The presented technique to construct a closed model out of an open model adds artificial places, which may produce many artificial p-invariants. For a biological interpretation only those p-invariants have to be considered, which do not include artificial places. In the case of signal transduction networks it depends on the modeling style, whether the essential system behavior can be explained by the discussion of minimal t-invariants only. Finally, the third validation criterion relies on temporal logics as a flexible query language to describe special properties. Thus, it requires seasoned understanding of the network under investigation, combined with the skill to correctly express the expected behavior in temporal logics. In summary, the set of meaningful validation criteria has to be adjusted to the case study on hand.

7.4   QUANTITATIVE MODELING AND ANALYSIS

Having validated the discrete model, the next step could consist of quantitative evaluations taking into account time-related information. Biochemical systems are inherently governed by stochastic laws. However, due to computational limits, the stochastic behavior is often approximated by a continuous one. In a continuous model of a biochemical system, all chemical reactions take place continuously. Moreover, the rates of all the chemical reactions typically depend on the time-dependent, continuous concentrations of the involved compounds. Hence, systems of ordinary differential equations (ODEs) appear to be a natural choice, as commonly used, for example, in the classical metabolic control analysis [10].

However, instead of creating the ODEs from scratch, we derive the continuous model from the discrete Petri net by assigning rate functions to all of the transitions in the network. Doing so, we get a *continuous Petri net*, preserving the structure of the discrete one. In biochemically interpreted continuous Petri nets, the rate functions may apply certain kinetic equation patterns, for example, the mass action or the Michaelis Menten equation pattern, and usually contain various kinetic parameters as dissociation or equilibrium constants. In a continuous Petri net, the marking of a place is no longer an integer, but a positive real number, called token value, which we are going to interpret as the concentration of a given compound. The instantaneous firing of a transition is carried out like a continuous flow, whereby the current firing rate generally depends on the current marking of its preplaces. The following definition summarizes this informal introduction.

Definition 7.10 (Continuous Petri net)   A continuous Petri net is a quintuple $\mathcal{CON} = (P, T, f, v, m_0)$, where

- P and T are finite, nonempty, and disjoint sets. P is the set of continuous places. T is the set of continuous transitions.

- $f : ((P \times T) \cup (T \times P)) \rightarrow \mathbb{R}_0^+$ defines the set of directed arcs, weighted by non-negative real values.
- $v : T \rightarrow H$ assigns to each transition a firing rate function, whereby $H := \bigcup_{t \in T} \{h_t | h_t : \mathbb{R}^{|^\bullet t|} \rightarrow \mathbb{R}\}$ is the set of all firing rate functions, and $v(t) = h_t$ for all transitions $t \in T$.
- $m_0 : P \rightarrow \mathbb{R}_0^+$ gives the initial marking.

Each continuous marking is a place vector $m \in (\mathbb{R}_0^+)^{|P|}$, and $m(p)$ yields again the marking on place p, which is now a real number. A continuous transition t is enabled in m, if $\forall p \in {}^\bullet t : m(p) > 0$. Due to the influence of time, a continuous transition is forced to fire as soon as possible. Corresponding to the concentration-dependent chemical reaction rates, the firing rate of a continuous transition typically depends on the token values of the transition's preplaces. So, we get marking-dependent (i.e., variable) firing rates. Please note, a firing rate may also be negative, in which case

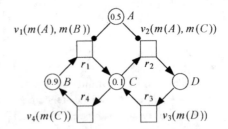

$$\frac{dm(B)}{dt} = -v_1\Big(m(A), m(B)\Big) \qquad\qquad +v_4\Big(m(C)\Big)$$

$$\frac{dm(C)}{dt} = +v_1\Big(m(A), m(B)\Big) \quad -v_2\Big(m(A), m(C)\Big) \quad +v_3\Big(m(D)\Big) \quad -v_4\Big(m(C)\Big)$$

$$\frac{dm(D)}{dt} = \qquad\qquad +v_2\Big(m(A), m(C)\Big) \quad -v_3\Big(m(D)\Big)$$

$$m_0(A) = 0.5, m_0(B) = 0.9, m_0(C) = 0.1, m_0(D) = 0$$

FIGURE 7.13 Example of an abstract continuous Petri net and the ODEs defined by it. To distinguish the discrete and continuous case, continuous nodes are represented in bold face type. Please note, replacing the read arcs by two inverse arcs each would result into a different system of ODEs. The rate functions v_i, assigned to each transition r_i, generally depend on the current marking of the transition's preplaces. These rate functions, which contain various kinetic parameters, may follow certain kinetic laws (e.g., the mass action, Michaelis Menten, etc.), which can be interpreted as equation patterns. Each place, subject to changes, gets its own differential equation, describing the continuous change of its token value by the continuous flows of its adjacent transitions. Each differential equation basically corresponds to a line in the incidence matrix, compare the table-like notation of the ODEs above. Please note that A is a place name, while $m(A)$ refers to the marking on place A. To simplify the notation in the generated ODEs, places are usually interpreted as (non-negative) real variables, which allows to write, e.g., $v_1(A, B)$ instead of $v_1(m(A), m(B))$. The model is self-contained, so it needs a nonclean initial marking.

the chemical reaction takes place in the reverse direction. This feature is commonly used to model reversible reactions by just one transition, where positive firing rates correspond to the forward direction, and negative ones to the backward direction.

Altogether, the semantics of a continuous Petri net is defined by a system of ODEs, whereby one equation describes the continuous change over time on the token value of a given place by the continuous increase of its pretransitions' flow and the continuous decrease of its posttransitions' flow:

$$\frac{dm(p)}{dt} = \sum_{t \in {}^\bullet p} f(t, p) v(t) - \sum_{t \in p^\bullet} f(p, t) v(t).$$

The token values (concentrations) of places that are tested by read arcs only (e.g., A in Fig. 7.13) do not change over time. Therefore, no equations for them are required. Each differential equation basically corresponds to a line in the incidence matrix, whereby now the matrix elements consist of the rate functions multiplied by the arc weight, compare the system of ODEs given in Fig. 7.13. Moreover, as soon as there are transitions with more than one preplaces, we generally get a nonlinear system, which calls for a numerical treatment of the system on hand.

Following this approach, the continuous Petri net becomes the structured description of the corresponding ODEs. Due to the explicit structure and the previous model validation of the underlying discrete structure, we expect to get descriptions, which are less error prone compared with other approaches. In order to simulate the continuous Petri net, exactly the same algorithms are employed as for ODEs in standard notation, that is, numerical differential equation solvers. Thus, we get a powerful combination of qualitative and quantitative models, complementing each other by the appropriate methods.

7.5 TOOL SUPPORT

There is a rich choice of software tools to model and analyze Petri nets of various types. The Petri net Web site [23] supplies all the entry points. In this chapter, the following public domain software tools have been used.

- Snoopy [31] for modeling and animation/simulation of standard (discrete) as well as continuous Petri nets. Snoopy's export feature supports various analysis tools, among them the following ones of this list. There is also an export to SBML, allowing access to tools like Ref. [19] for more detailed evaluations of continuous Petri nets in addition to the standard algorithms of ODE solvers provided by Snoopy.
- Integrated net analyzer (INA) [34] for most of the qualitative analyses.
- The model checker idd-ctl [36] for deciding special properties expressed in temporal logics. However, for 1-bounded models, we recommend to use the model checking kit [27].

7.6   CASE STUDIES

In the following, we sketch four published case studies, one metabolic network, and three signal transduction networks, which might help to get a deeper understanding of the material introduced in this chapter. The Petri nets and the related analysis data are available on the book's web page.

Sucrose-to-starch breakdown in Solanum tuberosum (potato) tubers [12]: This Petri net model (17 places, 25 transitions) describes the main carbon metabolism in potato tubers. It was developed in strong cooperation with experimentally working scientists. The Petri net represents an open system. The invariant analysis gives three minimal p-invariants and 19 minimal t-invariants, 7 of them are trivial ones, reflecting the reversible reactions. The model is covered by the 12 nontrivial t-invariants. The nontrivial invariants represent biologically meaningful subnetworks describing the different possibilities of the sucrose breakdown through invertase or sucrose synthase producing the hexoses, which further result in the starch production, the glycolysis, the ATP consumption, and also the futile cycles. Decomposing the network according to the known biological subnetworks, a systematic validation of the t-invariants was made step by step manually.

This pathway is attractive, because it is not very large, but complicated enough to see the state explosion while computing the reachability graph. Even in a derived simplified model version (18 places, 14 transitions), which is a closed system and 117-bounded, the state space amounts to $3.3 \cdot 10^{10}$, which takes about 1 min of computation time, using the latest issue of our model checking software [36]. The state explosion is caused by the stoichiometric factors up to 29 and the many reversible reactions, occurring concurrently.

The detailed evaluation of the t-invariants discovered a transition, involved only in a trivial t-invariant, which means that this transition has no effect in the steady-state. This has been confirmed by the continuous model. Consequently, this transition could be deleted without changing the steady state behavior.

Apoptosis in mammalian cells [8]: This case study demonstrates the application of Petri net based model validation to a signal transduction pathway using the same mathematical principles as for metabolic networks, but with another, more abstract interpretation. The developed Petri net (37 places and 45 transitions) extends [17]. It represents an open system, modeling basic processes of apoptosis, taking into consideration the pathways induced by the Fas receptor, the TNFR−1 (tumor necrosis factor receptor 1) as well as intrinsic apoptotic stimuli. There are 10 minimal t-invariants, describing the system behavior: four for the Fas-induced pathway, five for the TNFR−1-induced pathway, and one for the apoptotic stimuli induced pathway.

This case study confirms that Petri net based model validation with the focus on t-invariant analysis can also be applied to a signal transduction pathway.

Mating pheromone response pathway in Saccharomyces cerevisiae [26]: In this case study, the well understood signal transduction pathway of pheromone response is modeled and analyzed. The Petri net (42 places, 48 transitions) extends an ODE model [13] and represents an open system.

Here, special net structures are discussed, typically occurring in signal transduction networks. The notion of feasible t-invariants is introduced, which represent self-contained subnets being active under a given input situation. Each of these subnets stands for a signal flow in the network. There are seven feasible t-invariants, which all include the receptor activation, resulting in different responses for the cell, as, for example, changed gene transcriptions or feedback degradations. To support the t-invariant evaluation, the concept of MCT-sets is introduced for the first time. Seven disjunctive MCT-sets with more than one transition are found, each describing a connected subnet. In this study also knockouts were performed confirming experiments known from literature.

Raf-1/MEK/ERK pathway [7]: This case study discusses one of the standard examples used in the systems biology community — the core model of the influence of the Raf-1 kinase inhibitor protein (RKIP) on the extracellular signal regulated Kinase (ERK) signaling pathway. It is considered as a closed system. The qualitative as well as the quantitative model (11 places, 11 transitions) are given and analyzed thoroughly, whereby the qualitative analysis follows basically the outline of this chapter. Moreover, the partial order interpretation of a t-invariant by its so-called infinite partial order run is given.

It is shown that analyses based on the discrete Petri net model of the system can be used to derive the sets of initial concentrations required by the corresponding continuous ordinary differential equation model. All of them result into the same steady-state, and no other initial concentrations produce meaningful steady states.

Further case studies: The following papers also provide case studies with the complete Petri nets being available. This list is meant to illustrate the diversity of the application areas but is far away from being exhaustive. Qualitative (place/transition) Petri nets are used in the first three papers, whereas the remaining four papers apply quantitative (stochastic as well as continuous) Petri nets.

A metabolic network is discussed in Ref. [21], which provides a medium-sized model of the citric acid cycle (Krebs cycle), the second stage in glucose oxidation. The Krebs cycle, which takes the products of glycolysis, is a complex interacting set of nine subreaction networks. An analysis technique is proposed, which resembles the t-invariant analysis, to identify relevant biochemical signaling subcircuits.

The gene regulatory network, underlying the carbon starvation stress response in *E. coli*, is modeled and analyzed in Ref. [35]. The model construction starts at a Boolean graph, where genes are treated as binary switches. Logic minimization automates the construction of a compact qualitative Petri net model, which is by construction 1-bounded. The model checking kit is used to check the model for its ability to correctly switch between the exponential and stationary phases of growth and for the mutually exclusive presence of entities.

The integration of regulatory and metabolic processes into a coherent qualitative modeling framework is demonstrated in Ref [30]. The regulated metabolic network of the biosynthesis of tryptophan in *E. coli* takes into account two types of regulatory feedbacks (inhibitions). The model design exploits, similar to the previous approach, a systematic translation of Boolean graphs into standard Petri nets. The integrated model is validated by dynamic analysis. Three representative initial markings (no/low/high incoming flow of external tryptophan) are considered and the corresponding reachability graphs (which are quite small — 9/66/120 nodes) are evaluated.

A stochastic model to simulate the σ^{32} stress circuit in *E. coli* is given in Ref. [32]. This model of a gene regulatory pathway is used to confirm various hypotheses and to validate experimental results. Opposite to Ref. [30], [35], this model is constructed from scratch, that is, it is not derived from a Boolean graph.

A regulatory network controlling the commitment and sporulation of *Physarum polycephalum* is developed stepwise in Ref. [16]. The resulting stochastic Petri net consistently describes the structure and simulates the dynamics of the molecular network as analyzed by genetic, biochemical, and physiological experiments within a single coherent model. The Petri net is used to simulate the stochastic behavior of wild-type plasmodia as well as the so-called time-resolved somatic complementation (TRSC) experiments.

Hybrid Petri nets allow the combination of discrete and continuous net elements. Thus, discrete as well as continuous behavior can be described within one model. This is used in Ref. [4] to perform a case study on the urea cycle disorder, a genetic disease caused by a deficiency of one enzyme in the metabolic system. The metabolic behavior is described continuously, while the control of gene expression is represented by discrete net elements. The developed hybrid Petri net is used to estimate the regulation both on genomic and metabolic levels.

A generalization of hybrid Petri nets, the hybrid function Petri nets (HFPN) are used in Ref. [17]. Several typical examples are given: a metabolic pathway (the glycolysis), a signaling pathway (the Fas legand induced apoptosis), and three gene regulatory networks (switching mechanism of λ phage, circadian rhythm of *Drosophila melanogaster*, the *lac* operon regulation of *E. coli*). The combination of the *lac* operon gene regulatory mechanism and the glycolytic pathway is elaborated in more detail in Ref. [18]. Five mutants of the *lac* operon are simulated. These papers demonstrate that different abstraction levels, that is, gene regulation, metabolic and signaling pathway, and even combinations of them, can be modeled and analyzed quantitatively using HFPN.

7.7 SUMMARY

In this chapter, we have described the application of Petri net theory to model and analyze biochemical networks, first qualitatively before continuing with quantitative analyses. After introducing the basic Petri net definitions, we demonstrated how to systematically build such a discrete model, which provably reflects the qualitative biological behavior without any knowledge of kinetic parameters. The Petri net animation supports the intuitive understanding, while the analysis techniques promote a profound understanding and thorough validation of a given network.

The techniques have been discussed using a metabolic network as running example. The mathematical concepts can be applied equally to signal transduction networks, gene regulatory networks, or even networks comprising different network types. Thus, Petri nets may serve as common intermediate language providing a unifying framework for different abstraction levels as well as for a pool of interpretation-independent analysis techniques.

There are further promising Petri net concepts, which have been proven to be useful to investigate biochemical networks, but which have not been discussed here due to space limitations. Among them there are the partial order semantics and related analysis techniques, colored Petri nets as short-hand notation for place/transitions nets, and timed, but still discretely treatable Petri nets as interval time Petri nets and stochastic Petri nets.

We strongly advice a two-step technology for the modeling and analysis of biochemical networks in a systematic manner: (1) perform a qualitative (i.e., time-free) modeling and analysis, especially for model validation, increasing the confidence in the model, (2) perform a quantitative (i.e., timed) modeling and analysis, for hopefully reliable predictions of the system behavior. For both steps, we favor the deployment of discrete or continuous Petri nets, respectively, sharing the same net structures for a given case. The quantitative models are derived from the qualitative ones by the addition of quantitative (i.e., kinetic) parameters. Hence, all those models are likely to share some behavioral properties.

7.8 EXERCISES

1. The following subtasks might help to increase the familiarity with the incidence matrix and the related analysis techniques.

   (a) Show that a Petri net with read arcs cannot be described uniquely by the incidence matrix. Which other situations are not fully reflected by the incidence matrix? Give examples.

   (b) A well-defined matrix operation is the transposition, which exchanges rows and columns. If we apply the matrix transposition to the incidence matrix of a Petri net, what happens with the Petri net and its invariants?

   (c) MCT-sets can be computed including or excluding the trivial t-invariants. Give the MCT-sets for the running example, taking into account also the trivial t-invariants. Are there Petri nets, where we get the same results?

2. Consider the following toy example for a system of chemical equations.

$$2\,C + 2\,O_2 \;\rightarrow\; 2\,CO$$
$$2\,C + O_2 \;\rightarrow\; CO_2$$
$$2\,C + CO_2 \;\leftrightarrows\; 2\,CO$$

Derive a corresponding Petri net model and apply all the qualitative analysis techniques, presented in this chapter.

   (a) Which structural properties hold?

   (b) Determine and interpret the p-invariants, t-invariants, and MCT-sets.

   (c) Construct a closed system.

(d) Determine and interpret the p-invariants, t-invariants, and MCT-sets for the closed system. Determine the boundedness degree.

(e) Try to construct the reachability graph for the open system. What is the problem? Construct the reachability graph for the closed system. Decide liveness, reversibility, and existence of dynamic conflicts. Are there concurrent reactions? Check the realizability of the minimal t-invariants.

(f) The infinitely repeated occurrence of the first chemical reaction equation might be translated into
$$\mathbf{AG} \; [\; (C = 2 \wedge O_2 = 2) \rightarrow \mathbf{AF}(CO = 2) \;], \; \text{or}$$
$$\mathbf{AG} \; [\; (C = 2 \wedge O_2 = 2) \rightarrow \mathbf{EF} \, (CO = 2) \;].$$
What is the difference? Which ones holds?
You should be able to solve this task without tool support.

3. Extend the running example of this chapter by the following reactions:

   - $F6P \rightarrow G6P$, which makes the reaction *Phosphoglucose_isomerase* a reversible one,
   - $1,3\text{-}PBG \xrightarrow{DPGM} 2,3\text{-}PG$ and $2,3\text{-}PG \xrightarrow{DPGase} 3PG$, opening a new branch within the macro transition between $1, 3\text{-}BPG$ and $3PG$.

   Adapt step-wise the given Petri net model and apply all the qualitative analysis techniques presented in this chapter, following the outline (a) – (f) as given in the preceding task. To solve this task you will need adequate tool support.

4. While CTL model checking fits particularly for the decision of special properties, it can also be used to decide the general properties introduced in Section 7.2.2. Express for the closed model of the running example the following properties in CTL and check them using an appropriate model checking tool.

   (a) Check the liveness of the transitions *Aldolase* and *Transaldolase*. What had to be done to decide the liveness of the Petri net?

   (b) Check the reversibility of the Petri net.

   (c) Determine the boundedness degree for $Ru5P$ and GAP. Does the answer depends on the chosen initial marking within the range discussed in Section 7.3.4? What had to be done to determine the boundedness degree of the Petri net?

   (d) The control places a_1 and a_2 have been introduced to resolve the dynamic conflict between the two posttransitions of $Ru5P$, which occurs in the open system. How can we check that this dynamic conflict actually disappears in the closed system?

   (e) In Section 7.3.4, it is stated that dynamic conflicts concerning ATP disappear, if $m_0(ATP) > 5$. How can we verify this statement?

   (f) Check the token preservation within the p-invariants $(2 \cdot GSSG, GSH)$, $(NADP^+, NADPH)$, and (c_1, c_2).

5. Visualize the inherent structure of the following differential equation system by deriving the corresponding continuous Petri net. Is the solution unique?

$$\frac{dA_1}{dt} = k_2 \cdot A_2 - k_1 \cdot A_1, \quad \frac{dB_1}{dt} = -k_3 \cdot A_2 \cdot B_1 + k_6 \cdot B_2,$$

$$\frac{dA_2}{dt} = k_1 \cdot A_1 - k_2 \cdot A_2, \quad \frac{dB_2}{dt} = k_3 \cdot A_2 \cdot B_1 - k_4 \cdot A_2 \cdot B_2 + k_5 \cdot B_3 - k_6 \cdot B_2,$$

$$\frac{dB_3}{dt} = k_4 \cdot A_2 \cdot B_2 - k_5 \cdot B_3$$

Which structural and behavioral properties hold for the underlying discrete Petri net? You should be able to solve this task without tool support.

REFERENCES

1. J. M. Berg, J. L. Tymoczko, and L. Stryer. *Biochemistry*. W.H. Freeman and Company, New York, 2002.

2. F. Bause and P. Kritzinger. *Stochastic Petri Nets, an Introduction to the Theory*. Vieweg, Weisbaden, 1996.

3. N. Chabrier-Rivier and F. Fages. Symbolic model checking of biochemical networks. In *Proceedings of the 1st Computational Methods in Systems Biology (CMSB03)*, 2602 volume of LNCS, pp. 149–162. Springer, New York, 2003.

4. M. Chen and R. Hofestädt. Quantitative Petri net model of gene regulated metabolic networks in the cell. *In Silico Biology* 3, 0029, 2003.

5. E. M. Clarke, O. Grumberg, and D. A. Peled. *Model Checking*. MIT Press, Cambridge, MA, 2001.

6. R. David and H. Alla. *Discrete, Continuous, and Hybrid Petri Nets*. Springer 2005.

7. D. Gilbert and M. Heiner. From Petri nets to differential equations – an integrative approach for biochemical network analysis. In *Proceedings of the 27th International Conference on Applications and Theory of Petri Nets and Other Models of Concurrency (ICATPN 06)*, volume 4024 of LNCS, pp. 181–200, Springer, New York, 2006.

8. M. Heiner, I. Koch, and J. Will. Model validation of biological pathways using Petri nets — demonstrated for apoptosis. *Journal of BioSystems*, 75(1–3):15 – 28, 2004.

9. M. Heiner and I. Koch. Petri net based model validation in systems biology. In *Proceedings of the 25th International Conference on Applications and Theory of Petri Nets (ICATPN 04)*, volume 3099 of LNCS, pp. 216–237, Springer 2004.

10. R. Heinrich and T. A. Rapoport. A linear steady-state treatment of enzymatic chains. General properties, control and effector strength. *European Journal of Biochemistry*, 42(1):89–95, 1974.

11. R. Hofestädt. A Petri net application of metabolic processes. *Journal of System Analysis, Modeling and Simulation*, 16:113–122, 1994.

12. I. Koch, B. H. Junker, and M. Heiner, Application of Petri net theory for modeling and validation of the sucrose breakdown pathway in the potato tuber. *Bioinformatics*, (21)7:1219–1226, 2005.

13. B. Kofahl and E. Klipp, Modeling the dynamics of the yeast pheromone pathway. *Yeast*, 21(10):831–850, 2004.

14. A. Larhlimi and A. Bockmayr. Minimal metabolic behaviors and the reversible metabolic space. FU Berlin, DFG-Research Center Matheon. Preprint No.299, 2005.

15. K. Lautenbach. Exact liveness conditions of a Petri net class (in German). GMD Report 82, Bonn, 1973.

16. W. Marwan, A. Sujathab, and C. Starostzik. Reconstructing the regulatory network controlling commitment and sporulation in *Physarum polycephalum* based on hierarchical Petri net modeling and simulation. *Journal of Theoretical Biology*, 236:349–365, 2005.

17. H. Matsuno, Y. Tanaka, H. Aoshima, A. Doi, M. Matsui, and S. Miyano. Biopathways representation and simulation on hybrid functional Petri net. *In Silico Biology*, 3(3):389–404, 2003.

18. H. Matsuno, S. Fujita, A. Doi, M. Nagasaki, and S. Miyano. Towards pathway modeling and simulation. In *Proceedings of the 24th International Conference on Applications and Theory of Petri Nets (ICATPN 03)*, volume 2697 of LNCS, pp. 3–22, Springer 2003.

19. P. Mendes. GEPASI: A software package for modeling the dynamics, steady-states and control of biochemical and other systems. *Computational Applications in Biosciences*, 9:563–571, 1993.

20. T. Murata. Petri nets: Properties, analysis and applications. In *Proceedings of the IEEE*, 77(4):541–80, 1989.

21. J. S. Oliveira, C. G. Bailey, J. B. Jones-Oliveira, D. A. Dixon, D. W. Gull, and M. L. Chandler. A computational model for the identification of biochemical pathways in the Krebs cycle. *Journal of Computational Biology*, (10)1:5782,2003.

22. C. A. Petri. Communication with automata (in German). TU Darmstadt, Ph.D. Thesis, 1962.

23. Petri net home page. http://www.informatik.uni-hamburg.de/TGI/PetriNets/.

24. V. N. Reddy, M. L. Mavrovouniotis, and M. N. Liebman. Petri net representations in metabolic pathways. In *Proceedings of the 2nd International Conference on Intelligent Systems in Molecular Biollogy (ISMB 93)*, pp. 328–336, 1993.

25. V. N. Reddy. Modeling biological pathways: A discrete event systems approach. University of Maryland, College Park, MD, Master's Thesis, May 1994.

26. A. Sackmann, M. Heiner, and I. Koch. Application of Petri net based analysis techniques to signal transduction pathways. *BMC Bioinformatics*, 7:482, 2006.

27. C. Schröter, S. Schwoon, and J. Esparza. The Model Checking Kit. In *Proceedings of the 24th International Conference on Applications and Theory of Petri Nets (ICATPN 03)*, volume 2697 of LNCS, pp. 463–472, Springer 2003.

28. S. Schuster, C. Hilgetag, and R. Schuster. Determining elementary modes of functioning in biochemical reaction networks at steady state. In *Proceedings of the Second Gauss Symposium*, pp. 101–114, 1993.

29. C. H. Schilling, D. Letscher, and B. O. Palsson. Theory for the systemic definition of metabolic pathways and their use in interpreting metabolic function from a pathway-oriented perspective. *Journal of Theoretical Biology*, 203:229–248, 2000.

30. E. Simão, E. Remy, D. Thieffry, and C. Chaouiya. Qualitative modeling of regulated metabolic pathways: application to the tryptophan biosynthesis in *E. coli*. *Bioinformatics* 21(Suppl 2):ii190–ii196, 2005.

31. Snoopy - a software tool to design and animate hierarchical graphs. BTU Cottbus, CS Dep. http://www-dssz.informatik.tu-cottbus.de/software/snoopy.html, 2004.

32. R. Srivastava, M. S. Peterson, and W. E. Bentley. Stochastic kinetic analysis of the *Escherichia coli* stress circuit using σ^{32}-targeted antisense. *Biotechnology and Bioengineering*, 1(75):120–129, 2001.

33. P. H. Starke. *Analysis of Petri net models* (in German). Teubner Verlag, Stuttgart, 1990.

34. P. H. Starke. INA — The Integrated Net Analyzer, Humboldt University, Berlin, http://www2.informatik.hu-berlin.de/ starke/ina.html, 1992.

35. L. J. Steggles, R. Banks, and A. Wipat. Modeling and analyzing genetic networks: From Boolean networks to Petri nets. In *Proceedings of Computational Methods in Systems Biology* (CMSB 06), volume 4210 of LNCS/LNBI, pp. 127–141, Springer 2006.

36. A. Tovchigrechko. Model checking using interval decision diagrams, Department of Computer Science, BTU Cottbus, Ph.D. Thesis, 2006, Submitted.

37. K. Voss, M. Heiner, and I. Koch. steady state analysis of metabolic pathways using Petri nets. *In Silico Biology*, 3(31):367-87, 2003.

PART III

BIOLOGICAL NETWORKS

8

SIGNAL TRANSDUCTION AND GENE REGULATION NETWORKS

Anatolij P. Potapov

8.1 INTRODUCTION

Biological objects exhibit emergent properties that are not readily explainable by the features of their constituent parts: A system is more than just a sum of different elements. In the postgenomics era, there has been an increasingly strong emphasis placed upon a systems biology approach [34]. This approach takes advantage of the large amount of gene sequence information available and the fast progress of high-throughput molecular technologies that open the possibility of large-scale analyses of complex biological molecular systems. It focuses not so much on the individual components themselves but rather on the nature of the links that connect them and the functional states of the networks that result from the assembly of all such links [70]. Functional properties are not in molecules but appear as a result of their coordinated actions.

The term "network" is the hallmark of this philosophy. A biological network is the representation of multiple interactions within a cell; a global view intended to help understand how relationships between molecules dictate cellular behavior. Graphs are used to represent the topology of such a system (Chapter 2) and abstract the

I apologize to those whose work I was not able to mention due to space limitations. I am grateful to E. Wingender, N. Voss, and B. Goemann for many discussions and N. Sasse for help in preparing some figures. This work was supported in part by grant 503568 (COMBIO) within the 6th Framework Programme for Research, Technological Development and Demonstration of the European Commission.

inherent connectivity of many objects within the system, while ignoring their detailed form. Recent breakthroughs in graph theory have provided a new view of the topological design of complex networks, many of which have been found to have a scale-free topology [10,18,50] (Chapters 1–3). These networks exhibit small-world properties [69]: they are compact and display increased clustering (Chapter 3). Moreover, they follow a power-law degree distribution: most components participate in only one or two interactions, but a few participate in dozens and function as hubs. A scale-free topology provides the networks with amazing robustness against random failures [2,3]. The past 5 years have witnessed an unprecedented acceleration in the study of different biological networks. It has been shown that protein interaction networks (Chapter 9) as well as metabolic networks (Chapter 10) from various organisms exhibit a scale-free topology [21,26,44,54,67,68,71].

This chapter provides the reader with the recent progress in our understanding how large cellular regulatory networks are organized and function. It aims at readers not familiar with these subjects and present material without going in many details. More detailed information can be found in recent reviews and books about systems biology and biological networks, such as Refs. [1,5,9,10,18,59]. Here, we discuss on the role of regulatory networks in the evolution and existence of different organisms. Special emphasis is made on signal transduction and gene regulation networks, their particular properties as compared with metabolic, and protein interaction networks and their topological features. The topology of regulatory networks is a kind of skeleton providing a qualitative framework, on which quantitative data can further be superimposed for reasons of quantitative modeling and simulation.

8.2 DECISIVE ROLE OF REGULATORY NETWORKS IN THE EVOLUTION AND EXISTENCE OF ORGANISMS

The existence of living cells relies on numerous highly interconnected interactions and chemical reactions between various types of molecules such as proteins, DNA, RNA, and small metabolites. Actually, various activities of cells are controlled by the action of molecules upon molecules. Among different molecule types, proteins appear as central players. Being the products of gene expression, on the one hand, and playing a key role in the regulation of gene expression, on the other hand, proteins significantly contribute to linking genes to each other and forming multiple regulatory circuits in a cell (Fig. 8.1).

With the realization that in higher organisms only a tiny fraction of DNA is translated into proteins, regulation appears to be a very reasonable functional role for unexpressed DNA. The number of regulators in evolutionary different organisms increases with the size of a genome. Moreover, the number of regulators N_{reg} grows even faster than the number of genes, N_{total}, it regulates: The fraction of regulators increases with the genome size as [59,63].

$$\frac{N_{\text{reg}}}{N_{\text{total}}} \sim N \quad \text{for prokaryotes}$$

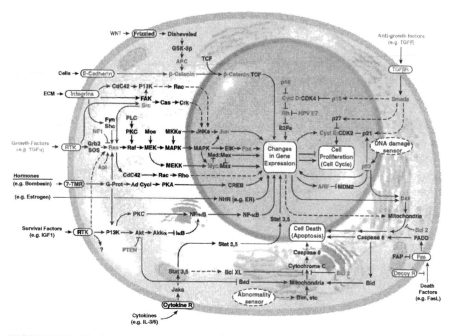

FIGURE 8.1   The integrated regulatory circuit of a mammalian cell. Point arrows are activating reactions; bar-ended arrows are inhibiting ones. Inhibiting arrows in some cases are shown to act on molecules, in other cases they act on reactions. See text for more details. (Reproduced from Ref.[23] with permission from Elsevier.)

$$\frac{N_{\text{reg}}}{N_{\text{total}}} \sim N^{0.3} \quad \text{for eukaryotes}$$

Here, the symbol "$\sim$" means "proportional." That reflects the increasing evolutionary importance of gene regulation and indicates that each added gene and its protein product should be adopted and regulated with respect to all already existing genes. It is the evolution of the gene regulatory networks and not the genes themselves that play the critical role in making organisms different from one another. It is likely that new phenotypes rarely appeared through the introduction of completely new proteins, and innovations mainly occur through the establishment of novel connections between existing or duplicated proteins to generate new regulatory circuits and thereby new regulatory behaviors [14].

Figure 8.1 presents the generalized regulatory circuit of a mammalian cell [23]. Various extracellular ligands, such as hormones, cytokines, growth factors, and so on, approach the cell from outside. The vast majority of them do not enter the cell. Typically, such a ligand binds to the corresponding receptor embedded in the cell membrane, which separates the cell from its external environment. Each such binding triggers a cascade of downstream signal transduction reactions. One receptor activates a limited number of signaling pathways. The targets of signal transduction pathways are metabolic enzymes and transcription factors. The last are proteins that regulate the transcription of genes that are located in the nucleus (shown as the internal circle). Transcription factors can be in two states and transit between active and inactive states. After getting a proper signal and becoming active, transcription factors specifically bind to the regulatory regions of genes and change the level of the expression of these genes. Most signaling molecules themselves are the products of gene expression. For reasons of simplicity, gene expression arrows are not shown in Fig. 8.1, but their presence is supposed. Actually, there are multiple regulatory circuits and intensive communication between the nucleus, where genes are transcribed and mRNAs are synthesized, on the one hand, and the cytoplasm (space between the nucleus and the cell membrane) where mRNAs are translated into proteins, on the other hand. Due to the presence of many cycles, regulatory networks cannot be properly modeled by means of directed acyclic graphs.

8.3 GENE REGULATORY NETWORK AS A SYSTEM OF MANY SUBNETWORKS

A gene regulatory network is a complex set of highly interconnected processes that govern the rate at which different genes in a cell are expressed in time, space, and amplitude. Such a network is commonly displayed by many pairs of proteins and genes, in which the first protein/gene regulates the abundance and/or activity of the second protein/gene. The primary role of proteins in the network is to control the synthesis, activity, and degradation of other proteins, which altogether control the flow of matter and energy through the cell. Networks can be considered as static or dynamic ones [43]. The complexity and the content of a network might change in time and space: it might have spatial and temporal dimensions [4,8,40,57].

Gene regulatory networks refer to a wide range of systems dealing with various aspects of complex interrelationships between genes and their products in a cell. They can be considered over protein interaction networks, signal transduction networks, transcription networks, as well as gene expression, gene coexpression, and gene interaction networks [11,57,62,64]. Each of them is identified in regard to specific physical, chemical, and functional properties, as well as to the level of abstraction used. Fig. 8.2 presents a simplified scheme of intracellular regulation circuits that are displayed by means of a bipartite graph. There is significant overlapping of gene expression and signal transduction parts. Proteins are products of gene expression and at the same time they control the expression of genes. Proteins catalyze metabolic reactions. Most regulatory events consume metabolites and critically depend on the supply of metabolites.

Gene expression is a complex process of converting a particular DNA sequence into the corresponding protein (Chapter 1). It includes transcription of a gene, that is the synthesis of the corresponding RNA, as well as several posttranscriptional events when the RNA is transformed into mRNA and is delivered to places where the mRNA is translated into the corresponding protein chain. All these steps are subject to regulation. Therefore, the list of regulatory networks can further be extended with networks that control RNA splicing, regulate mRNA turnover, and translation. For

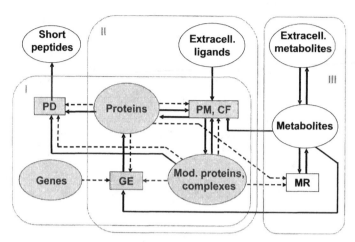

FIGURE 8.2 A simplified scheme of intracellular regulation circuits with gene expression (box I), signal transduction (box II), and metabolic (box III) processes. It has been viewed as a bipartite graph consisting of molecular entities (ellipses) and regulatory events (rectangles). Molecular entities are genes, proteins, modified proteins, their complexes, peptides, and metabolites. Regulatory events are gene expression (GE), protein modification and complex formation (PM, CF), protein degradation (PD) into short peptides, and metabolic reactions (MR). Mass flow is shown with fat arrows. Dashed arrows represent the catalytic action of molecular entities on the corresponding regulatory event; catalysts are themselves not consumed in the corresponding process.

instance, some small 20–25 nucleotide-long double-stranded RNAs are found to play a significant role in posttranscriptional regulation, possibly influencing the stability, compartmentalization, and translation of mRNAs of a diverse range of proteins [15]. Such regulation mediated through the control of mRNA turnover plays an important role in such cellular activities as proliferation, morphogenesis, and apoptosis. In a real cell, all these networks are closely interrelated and together with metabolic networks are integrated within the whole network of the cell.

The group of large molecular networks with well characterized topology includes metabolic networks [27,41,54] and protein interaction networks [44,71] in yeast and some other organisms. In contrast, knowledge on the topological design of more complex networks of regulatory processes is much less complete and mainly limited to transcription networks in relatively simple unicellular organisms—*Escherichia coli* [7,42,61] and yeast [22,40]. Significant advances are made in understanding the robustness in bacterial chemotaxis and in analyzing the developmental gene regulatory networks of *Drosophila* and sea urchin, all reviewed in Ref. [5].

8.4 DATABASES ON GENE REGULATION AND SOFTWARE TOOLS FOR NETWORK ANALYSIS

Recent advances in high-throughput experimental technologies have resulted in a great body of information, which continue to grow exponentially. The collection of such information in a computer-readable form is a prerequisite for making use of these data for analysis. A suitable database structure as well as the quality and consistence of the stored data are of great importance. Presently, many databases on different aspects of gene regulation are available [29,32]. Updated information on them appears every year in the *Nucleic Acids Research* database issue. Representative examples of such databases and knowledge bases containing information about regulatory interactions are aMAZE [38], EcoCyc [33], GeneNet [6], KEGG [31], RegulonDB [56], Reactom [28], TRANSPATH [36,37], and TRANSFAC [46,47]. More such databases can be found at http://www.hsls.pitt.edu/guides/genetics/obrc/enzymes_pathways/ signaling_pathways/.

The aMAZE database [38] focuses on information about genetic regulation, biochemical pathways, signal transductions, and aims on modeling the systems of catalyzed chemical reactions by means of simulation software packages such as GEPASI (http://www.gepasi.org/). KEGG [30,31] and EcoCyc [33] provide rich information on metabolism, metabolic pathways, as well on signal transduction, gene regulation, and cellular processes. GeneNet database [6] provides information on structure and functional organizations of gene networks and metabolic and signal transduction pathways. Reactome [28] is a reach resource on pathway information and reactions in human biology. TRANSPATH [16,36,37] provides encyclopedic information about the intracellular signal transduction pathways and offers molecular details of the signal flow from the cell surface into the nucleus. TRANSPATH and accompanying tools can be used for data visualization and modeling, as well as for the analysis of gene expression data. TRANSFAC [46,47] is the database on many aspects of transcription regulation in eukaryotes. It presents the largest archive of eukaryotic transcription factors, their genomic binding sites, and DNA-binding profiles. Among various tools for analyzing networks, a particular place is taken by Pajek (http://vlado.fmf.uni-lj.si/pub/networks/pajek/) that is a freely downloadable software package for the analysis of really large networks including thousands of vertices and edges. Both unipartite and bipartite networks (see Chapter 2) can be analyzed. A wide range of algorithms for network analysis have been implemented in the package and most run very rapidly. The networks can be displayed in a variety of layouts.

8.5   PECULIARITIES OF SIGNAL TRANSDUCTION NETWORKS

Signal transduction links intracellular processes to extracellular environment and modulates cellular functions in response to various external stimuli. Signal transduction pathways are initiated by the binding of extracellular ligands to receptors and resulting in one or more specific cellular responses. Various extracellular ligands (e.g., hormones, cytokines, growth factors) act on the cell from outside (Fig. 8.1). The vast majority of them do not enter the cell. Typically, such a ligand binds to the corresponding receptor embedded in the cell membrane. This binding triggers a cascade of downstream signal transduction reactions. The final targets of signal transduction pathways are transcription factors and metabolic enzymes. Signal transduction processes rely on a cascade of reversible chemical modification of proteins, as well as on the formation of complexes. The most common type of modification is

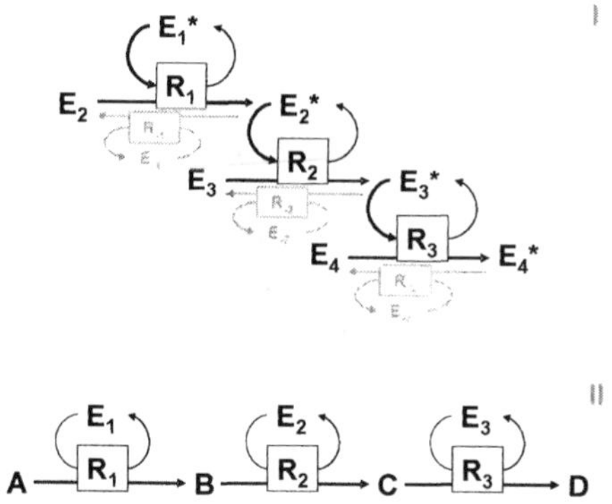

FIGURE 8.3 Generalized topology of a signal transduction path (I) and a metabolic path (II). See text for details.

phosphorylation: the covalent attachment of a phosphate group to a specific amino acid of a target protein. That is catalyzed by specific enzymes, protein kinases. The attachment of negatively charged phosphate group induces spatial reorganization of the target protein and affects its functional activity. The opposite reaction of removing the phosphate group is catalyzed by other enzymes, protein phosphatases that restore the initial functional status of target proteins. The generalized topology of a signal transduction path is shown in Fig. 8.3. In contrast to metabolic paths, signal transduction paths consist of a rather limited mass flow and mainly provide an "information transmission" along a sequence of reactions. That is, one enzyme modulates the activity of another one, which in its turn modulates the activity of the third enzyme, each being not consumed in the reaction it catalyzes. For example, protein E_2^* catalyzes the transformation of E_3 into E_3^* (Fig. 8.3). The functional activity of E_3^* differs from that of E_3. This modification is reversible: another protein, E_{-2}, catalyzes the transformation of E_3^* into the initial form E_3. It is the characteristic of signaling cascades that each key component becomes activated by the previous step to activate the key molecule of the subsequent reaction (Fig. 8.3).

In contrast to protein interaction networks (see also Chapter 9) that refer to the association of protein molecules and are undirected, signal transduction networks refer to reactions and are basically directed. Although signal transduction processes are mediated by protein–protein interactions, these processes show important peculiarities as compared with interactions. The number of different directed graphs with V vertices, $N_{dir}(V)$, is much larger than that of undirected graphs, $N_{undir}(V)$, with the same number of vertices [58]:

$$\frac{N_{dir}(V)}{N_{undir}(V)} = 2^{(V^2-1)/2}$$

Protein interaction networks are "isotropic" in the sense that they do not have a well-defined input and output [12]. In contrast, regulatory and signal transduction networks are "anisotropic" by their definition as the networks whose primary task is to transform a set of inputs into a second set of outputs. By far not all protein–protein interactions are followed by chemical reactions. Therefore, many interactions are not "mirrored" in signal transduction. On the contrary, many components of signal transduction—e.g., steroid hormones, second messengers such as cAMP, Ca^{2+}, diacylglycerol, 3-phosphorylated inositol lipids, stress, UV, irradiation, etc.—are not proteins. Protein–DNA interactions between transcription factors and the regulatory regions of genes must be added to the list as well. All these reactions are not mirrored in the system of protein–protein interactions. Finally, gene regulatory and signal transduction networks have much in common with protein interaction networks. At the same time, they refer to different aspects of cellular activity and display several important differences.

8.6 TOPOLOGY OF SIGNAL TRANSDUCTION NETWORKS

While the topological analysis of metabolic networks is a well-established field, similar approaches have scarcely applied to large signal transduction networks. Despite of a great body of information about many details of different signal transduction pathways, there is modest progress in understanding how they are organized an function as an integral system. That is particularly problematic in regard to signal transduction processes in multicellular organisms of higher eukaryotes. Available information about such signal transduction systems is fragmental and different fragments often refer to different species and cell types. The human organism includes more than 200 different cell types. Altogether, that makes the task of getting the large scale and integral descriptions of various signaling pathways rather problematic.

To overcome these problems and get the first approximation of the properties of signal transduction networks, a genome-wide analysis at the level above the level of species may be useful. In this case, molecules are represented with their ortholog abstractions. The last ignore the species-specific differences between the homologous molecules from various species, and consider all members of a homologous group as one molecular entity. Instead of rather complete mechanistic reactions, that depict the underlying biochemical mechanisms in all their details, the corresponding simple semantic representations are preferred. That excludes the necessity to operate on all reaction details many of which are still unknown. This simplified representation style is familiar to biologists as it is often used in pathway cartoons in review literature description: $X \rightarrow Y$, where X and Y represent signal donors and acceptors, respectively. The use of ortholog abstractions and semantic reactions helps to integrate many information fragments and increase the size of regulatory networks available for analysis. Just as maps of metabolic networks describe the potential pathways that may be used by a cell to accomplish metabolic tasks, these regulatory networks describe potential pathways that can be used in eukaryotic cells to regulate gene expression programs. Of course, such regulatory networks are kind of abstraction and represent the genome-wide properties of the corresponding regulatory systems from those species that were taken into consideration.

Using this approach and information collected from TRANSPATH database on signal transduction [36,37], a signal transduction network consisting of several thousand vertices and edges might be extracted and analyzed. The network is clearly sparse, that is, one molecule has in general only few incoming and outgoing links to other molecules. According to our preliminary results, such a network displays scale-free properties. Its topology follows a power-law degree distribution and shows small-world properties in terms of a network diameter and clustering (Chapters 1, 3, and 6).

Despite the obvious importance of a degree distribution for characterizing the class of a network, this feature does not tell us anything about how vertices are connected to one other: a huge number of combinations can be possible within the same degree statistics. The degree–degree correlation in networks can be evaluated in terms of the assortative mixing that quantifies the extent to which vertices connect preferentially to other vertices with similar characteristics. It can be made by means of the mixing coefficient r, a Pearson correlation coefficient for the degrees of the two vertices on each side of an edge [49,50], which can range within $-1 \leq r \leq 1$. This correlation function is zero for no assortative mixing and positive or negative for assortative or disassortative mixing, respectively. First evaluations on regulatory networks indicate that direct links between high-degree proteins are systematically suppressed, whereas those between a high-degree and low-degree pairs of proteins are favored [45]. If so, high-degree hubs in regulatory networks might be separated from one other by vertices with lower degrees and such low-degree vertices may play rather an important role in sustaining the integrity of the networks and providing communication between their distant parts [19,20].

Complex organization of signaling pathways may significantly contribute to the robustness of cellular functions. Robustness is the ability of a system to maintain its functionalities against external and internal perturbations [35]. Robustness is considered as a ubiquitously observed property of biological systems, which may be a key to understanding cellular complexity. The structure of a signaling network that involves both positive and negative feedback loops may display an amazingly robust behavior. The chemotaxis pathway of bacteria consists of about 12 proteins that receive signals from extracellular environment and transmit the signals to the machinery that controls the movement of the bacterial flagella. That enables bacterial cells to sense gradients of attractants (e.g., food) or repellents (e.g., poison) and to move up or down the gradients. The robustness of such bacterial adaptation relies on a special design of the chemotaxis signaling pathways, that is on the network of positive and negative feedbacks and not on fine-tuning of biochemical parameters [5].

Robustness of cellular functions and the complexity of cellular systems are considered to be intimately linked [35,60]. The primary functions of biological systems are usually robust to wide range of perturbations. However, these systems can show extreme fragility in regard to other apparently small perturbation. This coexistence of extremes in robustness and fragility ("robust yet fragile") is one of the most important properties of highly evolved or designed complexity [13,19,20,60].

8.7 TOPOLOGY OF TRANSCRIPTION NETWORKS

Transcriptional regulation is one of the most fundamental biological control mechanisms of gene expression. This is mediated by transcription factors that are proteins

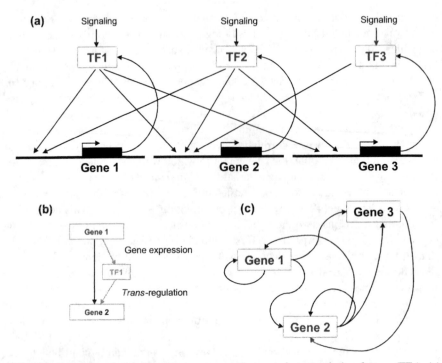

FIGURE 8.4 Interrelationships between genes that encode transcription factors (TFs). (*a*) Each TF binds to specific DNA sequences in the regulatory regions of a gene and, thereby, participates in formation of the transcription–preinitiation complexes near the start sites of transcription. (*b*) The composition of an edge in the network of TF-genes. Each edge represents gene expression and *trans*-regulation. (*c*) The network of TF-genes.

able to recognize and bind specific DNA sequence elements in the regulatory regions of genes (Fig. 8.4). In general, one gene may be regulated by more than one transcription factors, and some transcription factors may control more than one genes. In higher eukaryotes, transcription factors often function in a cooperative manner: a particular set of different transcription factors is necessary to initiate the transcription of a particular gene. The presence of all such factors in a set is required. One gene can be switched on by means of several sets of transcription factors, thereby enabling the transcription of the gene under different circumstances. Each transcription factor can take part in regulation of several genes. The genes encoding transcription factors (TF-genes) are themselves subject to control by other and the same transcription factors. From this, a transcriptional network emerges, which is responsible for controlling essential biological processes such as morphogenesis, cell proliferation, differentiation, homeostasis, and metabolism. Transcriptional networks are often the main targets of signal transduction [46,55,56].

Bacterial transcription networks are extremely flexible in evolution. Transcription factors evolve much faster than their target genes [7,39]. The transcription regulatory networks of unicellular organisms, such as *E. coli* and *S. cerevisiae*, display a scale-free type of their topological organization [22,62a]. While most molecules are engaged in only a few interactions, a few hubs are linked to a significantly higher number of other molecules. In both these networks, the degree distribution follows a power law $P(k) \sim k^{-\gamma}$ with the degree exponent $\gamma \approx 2$ [62a].

The composition of transcription networks in yeast may significantly change with time or environmental conditions. In response to diverse stimuli, yeast transcription factors alter their interactions, thereby rewiring the network [40]. A few transcription factors are present in the network under various conditions and serve as permanent hubs, but most act transiently and appear under certain conditions only. There are at least two different types of regulatory subnetworks: endogenous and exogenous [40]. Endogenous subnetworks are made by constructions, which regulate processes that are intrinsic to the cell (e.g., cell cycle and sporulation). These subnetworks display a multistage architecture and high local interconnectivity. Transcription factor hubs have a relatively small number of targets, many of which are other transcription factors. The hubs are often distant from the "terminal effectors" of processes, being separated by several interactions and reactions. Accordingly, such regulation is expected to be complex and rather slow, that is it requires a relatively long time. In contrast, exogenous subnetworks regulate events that respond to external stimuli (e.g., DNA damage and stress response) and induce a rapid expression of genes. Exogenous subnetworks include few transcription factors. However, these factors control a large number of targets, many of which are the "terminal effectors" that coordinate the response of a cell to stimuli. The ability to dynamically reorganize transcriptional regulatory networks and operate on permanent and temporal transcription factor hubs appears to be an important feature of yeast.

Yeast transcription factors can be further divided into several groups based on their ability to recognize their targets [24]. There are condition-invariant factors that bind essentially the same set of targets under any condition. There are condition-enabled factors that bind targets under certain circumstances only. There are condition-expanded factors that bind additional targets in specific circumstances. In addition to that, there are condition-altered factors that bind different targets under distinct situations. These diverse capacities of transcription factors provide regulatory networks with the necessary flexibility and enable their rewiring in a condition-dependent manner.

A particular transcription network, which is made by those genes only that encode transcription factors (TF-genes), is of special interest. Such a network of TF-genes represents the central core, skeleton of a larger gene network that includes all other target genes. Mammalian network of TF-genes can be represented as a directed graph where vertices are TF-genes and edges are causal links between the genes, each edge combining both gene expression and *trans*-regulation events (Fig. 8.4). Following the very preliminary evaluations, 121 vertices and 212 edges can be identified and positioned in this network [52]. The network provides a genome-wide view above the level of three species: human, mouse, and rat. The architecture of the network is presented in Fig. 8.5. The network is sparse: that is, on average each TF-gene is connected to few other TF-genes only. There is a hierarchy of degree distribution with many

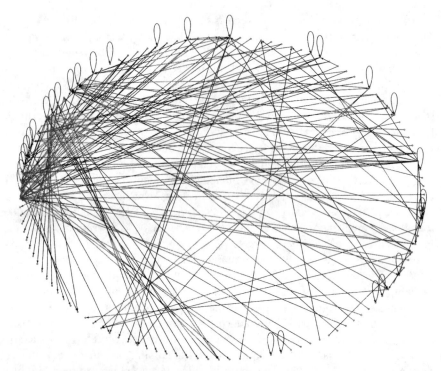

FIGURE 8.5 Mammalian network of transcription factor genes. The network is visualized by means of software Pajek (http://vlado.fmf.uni-lj.si/pub/networks/pajek/).

low-degree vertices and a few high-degree vertices. Many vertices display the presence of self-loops, thereby, indicating the capacity of the corresponding TF-genes for autoregulation.

Within this "complete" network, a special subnetwork centered at *p53* TF-gene can be identified. The *p53* TF-gene encode transcription factor *p53* that is a very important tumor suppressor and play a key role in regulating cell cycle, proliferation, and apoptosis, a process leading cell death. The *p53* subnetwork consists of all TF-genes, which can communicate with *p53* TF-gene, that is, send to and receive from *p53* gene regulatory messages. This subnetwork consists of 44 vertices and 80 edges (Table 8.1).

The problem of extracting such a subnetwork from the whole network might be of general interest for the reader, and *p53* TF-gene subnetwork can be used for that as an illustrative example. The problem can be solved as follows. The "complete" mammalian network of TF-genes is considered as a graph $G = (V, E)$. We are interested in finding its subgraph $G' = (V', E')$ that is centered at vertex v and includes all vertices that communicate with this vertex v. Accordingly, V' and E' are subsets of V and E, respectively (Chapter 2). Let consider vertex *p53* (*p53* TF-gene) as a central vertex v. Then, the subset V' consists of vertex *p53*, all those vertices from which vertex *p53* is reachable and all those vertices that are reachable from vertex

TABLE 8.1   Topological Properties of the Mammalian Network of Transcription Factor Genes. V—Vertices; E—Edges; and ω – Slopes of Straight Lines in Double Logarithmic Plots Calculated by Linear Regression (Figs. 8.7 and 8.8a, Respectively).

Parameter	Complete network	p53 subnetwork
Size		
V	121	44
E	212	80
Shortest path length		
average	2.2	2.4
maximal	5.0	5.0
Degree distribution		
γ_{in}	1.34 ± 0.28	0.56 ± 0.26
γ_{out}	1.78 ± 0.14	1.50 ± 0.16
γ_{inout}	1.46 ± 0.21	0.97 ± 0.21
Average clustering coefficient		
C	0.134	0.241
C/C_{random}	4.6	2.8
Clustering coefficient distribution		
ω	0.66 ± 0.25	1.1 ± 0.24

Note: C_{random} is the average clustering coefficient of a classical random graph with the same number of vertices and edges. Standard deviation is indicated when necessary.
Source: Reproduced from Ref. [52] with permission from JSBi.

p53 (Fig. 8.6). The subset E' is made by all edges the tail and the head of which are both from the subset V'.

There are different ways by means of which the subgraph $G' = (V', E')$ can be extracted from the graph $G = (V, E)$. For instance, the subset V' can be found by using the depth first search (DFS) algorithm (Chapter 2). The complete algorithm for finding the *p53* gene subnetwork is

p53_centered_subnetwork_DFS_algorithm (vertex v)
(1) $U =$ all vertices upstream of *p53* (by means of DFS);
(2) $D =$ all vertices downstream of *p53* (by means of DFS);
(3) $V' =$ union of sets U and D and vertex *p53*;
(4)   Add all edges $\{v, w\} \in E$ to E' if $v, w \in U$;
(5)   Add all edges $\{v, w\} \in E$ to E' if $v, w \in D$;
(6)   Add all edges $\{v, w\} \in E$ to E' if $v \in D$ and $w \in U$

The subsets of vertices found by means of the upstream search and the downstream search might show some overlapping that is due to the presence of cycles in the network G. Therefore, V' is actually the union of these "upstream" and "downstream" subsets (line 3). To avoid bypaths around *p53*, not all edges $\{v, w\} \in E$ are taken into account. An edge $\{v, w\} \in E$ is not part of E' if v is in the "upstream" subset of *p53* and w is in the "downstream" of *p53* (lines 4–6). Allowing (excluding) $v = w$ would enable (exclude) the presence of self-loops in the *p53* subnetwork.

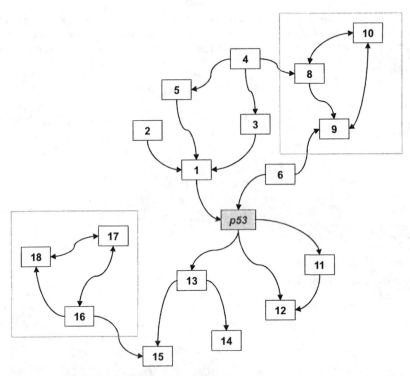

FIGURE 8.6 A subnetwork centered at a particular vertex v includes vertex v, all those vertices from which vertex v can be reached and all those vertices that are reachable from vertex v. Here, v is *p53*. The "*p53*" subnetwork is the subgraph $G_{p53} = (V_{p53}, E_{p53})$ with $V_{p53} = \{p53, 1, 2, 4, 5, 6, 11, 12, 13, 14, 15\}$ and $E_{p53} = \{\{1, p53\}, \{6, p53\}, \{2, 1\}, \{5, 1\}, \{3, 1\}, \{4, 3\}, \{4, 5\}, \{p53, 11\}, \{p53, 12\}, \{p53, 13\}, \{13, 14\}, \{13, 15\}\}$. The vertices $\{8, 9, 10\}$ and $\{16, 17, 18\}$ are not part of the *p53* subnetwork.

The "complete" TF-gene networks and its *p53* subnetwork are rather compact (Table 8.1). The average shortest path length, that represents the statistical diameter of a network, is found to be 2. The maximal shortest path length, that represents the actual diameter of a network, is 5 for both networks. That does not exclude the presence of longer paths because the shortest ones were only considered. Such small shortest paths signify the possibility of faster propagation of the regulatory communication between TF-genes. The average clustering coefficient (Chapter 3) of the networks clearly exceeds that of classical random networks with the same number of vertices and edges (Table 8.1), thus indicating the small-world properties of the both networks. The networks are inhomogeneous. Their degree distribution follows a power law and is best described by the scale-free model. In the double-logarithmic plots, the distribution is well approximated by a straight line (Fig. 8.7) that fits the power law $P(k) \sim k^{-\gamma}$. The relatively small value of γ_{in} (0.56 ± 0.26) found for *p53* subnetwork indicates a star-like organization of this subnetwork that fits to the way of how this network was defined and extracted.

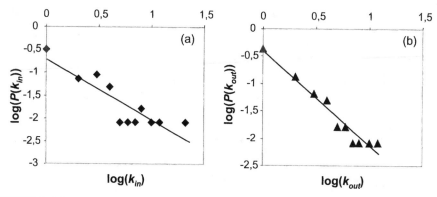

FIGURE 8.7 The degree distribution $P(k)$ in the mammalian network of genes coding for transcription factors shows its scale-free topology. (a) incoming degrees, k_{in}; (b) outgoing degrees, k_{out}. In all these cases, the distribution is well approximated by a straight line in the double-logarithmic plot, thereby fitting the power law. The straight lines are made by linear regression. (Reproduced from Ref. [52] with permission from JSBi.)

Modular organization is the hallmark of biological systems [25]. Molecules and genes within a module are thought to cooperate in order to provide a particular function. Such modules can be presented in a network as relatively independent parts or they can be organized in a hierarchical fashion. The hierarchical modularity is compatible with the scale-free topology but does not automatically arise from this topology. It relates to many small, highly interconnected groups of vertices that form larger but less cohesive topological modules [53]. This hierarchy can be characterized in a quantitative manner by analyzing the scaling of the clustering coefficient $C(k)$ that is the average value of the clustering coefficients of all vertices with degree k (Chapter 3). For random networks and for simple scale-free networks, $C(k)$ is independent of k. However, models that lead to a perfect hierarchical modularity predict that $C(k)$ depends on k and is proportional to a reverse k, $C(k) \sim k^{-1}$ [10,17,53]. The dependence can be weaker if not all modules follow a hierarchical type of organization. This scaling law offers a straightforward method to identify the presence of hierarchy in a network and to quantify this hierarchy.

The both mammalian transcription networks — "complete" network of TF-genes and $p53$ subnetwork — show the dependence of the clustering coefficient on the degree of a vertex (Fig. 8.8 a). The scaling of the clustering coefficient follows a power law $C(k) \sim k^{-\omega}$, as a straight line of the slope ω on a log–log plot (Fig. 8.8 a). The scaling exponent ω (0.7 ± 0.2 and 1.1 ± 0.2 for the reference and $p53$ transcription networks, respectively) reasonably approaches the value 1, which is the signature of hierarchical modularity. This means that the topology of both mammalian networks of TF-genes tends to have a hierarchical modular architecture, which might help to more efficiently reorganize the transcription networks according to the current needs of the cell. Such a hierarchical modularity have been observed in the transcription networks of bacteria and yeast as well [62a] and, thereby, appears to be a common property of transcription systems in different organisms.

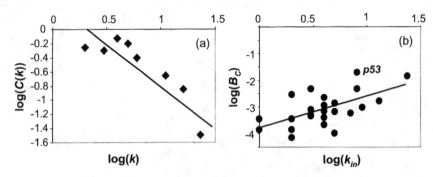

FIGURE 8.8 Scaling of the clustering coefficient (*a*) and betweenness centrality (*b*) in the mammalian network of transcription factor genes. (*a*) The mean clustering coefficient of all vertices with *k* links, $C(k)$, depends on *k* and this dependence can be well approximated with a power law $C(k) \sim k^{-\omega}$. (*b*) Positive correlation between the values of the betweenness centrality and the degree of a vertex. High-degree TF-genes tend to display a larger value of betweenness. The straight lines are computed by linear regression. (Reproduced from Ref. [52] with permission from JSBi.)

The betweenness centrality, B_c is a measure of the intermediary role of an individual element in the communication between all other elements (Chapter 4). It allows quantify how influential and important a given TF-gene in a whole transcription network is. There is a clear positive correlation between the B_c values and the degree of vertices in mammalian networks of TF-genes (Fig. 8.8 b). That evidences the interrelation between local (degree) and integral (B_c) topologies of the transcription networks and supports the view [10] that hubs, that is elements with highly enriched local topology, represent the most influential elements of a network and tend to be essential for sustaining the integrity of a network. Note that this correlation is not absolute: There are several low-degree TF-genes with B_c values that are comparable with and even exceed that of several hubs (Fig. 8.8b).

To identify the mammalian transcription factor genes of high impact, different topological features of individual genes can be taken into consideration. Table 8.2 represents the results of such an analysis applied to the TF-genes in the "complete" network. The data in the table are arranged according to the corresponding B_c values. The top part of the list presented there is significantly enriched with TF-genes *p53*, c-*fos*, c-*jun*, *SRF*, and c-*myc*, which are known to be involved in controlling the cell cycle and display features of tumor-suppressors or proto-oncogenes [65,66]. Many of top ranking TF-genes, but not all of them, are attributed with relatively high values of degrees and as a rule with high clustering coefficient (Chapter 3). Among several topological characteristics of the mammalian transcription networks, B_c, appears to be the most representative in regard to the biological significance of individual elements.

8.8 INTERCELLULAR MOLECULAR REGULATORY NETWORKS

In contrast to relatively simple unicellular organisms, such as bacteria, yeast, multicellular organisms are made by specialized cell types, tissues, and organs,

TABLE 8.2 Topological Features of Some Individual TF-Genes in the Mammalian Network of Transcription Factor Genes.

Name	k_{in}	k_{out}	k	c	B_c
p53	8	10	18	0.142	0.0188
c-fos	24	3	27	0.032	0.0139
Egr1	3	6	9	0.333	0.0047
c-jun	8	5	13	0.197	0.0046
WT1	2	1	3	0	0.0029
SRF	4	8	12	0.333	0.0022
c-myc	13	3	16	0.143	0.0016
HNF4A	5	5	10	0.333	0.0014
HOXA1	4	1	5	0.5	0.0011
RAR-β	9	4	13	0.393	0.0009
ONECUT1	3	2	5	0	0.0007
HOXB1	4	2	6	0.5	0.0007
Egr2	5	1	6	0.1	0.0006
IRF-1	7	1	8	0	0.0005
TCF1	3	6	9	0.5	0.0004
ELK1	3	4	7	0.7	0.0004
CRE-BP1	3	5	8	0.333	0.0004
JUND	4	1	5	0.4	0.0003
STAT3	2	4	6	0.5	0.0003
RARA	1	6	7	0.5	0.0003
C/EBPα	4	2	6	0.4	0.0002
NFKB1	1	1	2	0	0.0001
Pax-6	1	2	3	0	0.0001

Note: The top part of the complete list is presented. k—degree, c—clustering coefficient, B_c—betweenness centrality of a given TF-gene. The Table is arranged according to B_c Values.
Source: Reproduced from [71] with Permission from JSBi.

various functions of which must be properly coordinated. In mammals, the endocrine system strongly contributes to this coordination. Cells constantly sense their environment in an organism. On the basis the obtained information, they constantly correct the internal processes and prepare themselves for the necessity to proliferate, move or die. A human organism consists of about 10^{14} cells that comprised more than 200 different cell types. There is a great need for understanding how multiple intercellular communications are organized and function within a whole integral system of an organism.

Recently, a new database, EndoNet, on the endocrine cell–cell signaling in human has been developed [51], which enables the analysis of intercellular regulatory pathways. The data model includes two classes of components and can be viewed as a bipartite directed graph. One class represents the signaling molecules (hormones, cytokines, growth factors, or other messengers), which are secreted by defined donor cells. The other class represents the acceptor or target cells expressing the corresponding receptors. The identity and anatomical environment of cell types, tissues, and organs are provided through references to the CYTOMER ontology [48].

CYTOMER is a relational database on organs/tissues, cell types, physiological systems, and developmental stages in a human organism. In EndoNet, all entries for signaling molecules and receptors are provided with links to external information resources, that including TRANSPATH database on intracellular signal transduction [36,37]. That promises to greatly increase the scale of molecular regulatory circuits available for analyses. That might help to bridge the gap between known genotypes and their molecular and clinical phenotypes in the area of medical research and its applications.

8.9   SUMMARY

Regulatory networks play a decisive role in the evolution and existence of organisms. The number of regulators in evolutionary different organisms increases with the size of a genome faster than the number of genes. It is the evolution of the gene regulatory networks and not the genes themselves that play the critical role in making organisms different from one another.

Signal transduction and gene regulation networks are central players of intracellular regulation. Various extracellular ligands, such as hormones, growth factors, cytokines, chemokines, and so on, approach the cell from outside. The vast majority of them do not enter the cell. Typically, such a ligand binds to the corresponding receptor embedded in the cell membrane. Each such binding triggers a cascade of downstream signal transduction reactions. The targets of signal transduction pathways are metabolic enzymes and transcription factors.

Transcription factors can be in two states and transit between active and inactive states. After getting a proper signal and becoming active, transcription factors specifically bind to the regulatory regions of genes and change the level of the expression of these genes. In higher eukaryotes, transcription factors often function in a cooperative manner: a particular set of different transcription factors is necessary to initiate the transcription of a particular gene. The presence of all such factors in a set is required. One gene can be switched on by means of several sets of transcription factors, thereby enabling the transcription of the gene under different circumstances. Each transcription factor can take part in regulation of several genes.

Most signaling molecules themselves are the products of gene expression and therefore are part of multiple regulatory circuits. Gene regulatory networks refer to a wide range of systems dealing with various aspects of complex interrelationships between genes and their products in a cell. The topology of signal transduction and gene regulation networks abstracts the inherent connectivity of many objects within these systems, while ignoring their detailed form. It might be considered as a kind of skeleton that provides a qualitative framework, on which quantitative data can further be superimposed for reasons of quantitative modeling and simulation.

The generalized topology of a signal transduction path significantly differs from that of a metabolic path. Signal transduction paths consist of a rather limited mass flow and mainly provide an "information transmission" along a sequence of reactions.

Regulatory networks appear to be highly inhomogeneous. Signal transduction and transcription networks display small-world properties and scale-free topology. They are sparse, compact, and demonstrate increased clustering. Most components participate in a small number of interactions, but few participate in dozens and function as hubs. At least some regulatory networks display the negative assortative mixing by vertex degree: direct links between high-degree proteins are systematically suppressed, whereas those between a high-degree and low-degree pairs of proteins are favored. As it follows from the scaling of the clustering coefficient, networks of transcription factor genes show the presence of hierarchical modularity. The composition of regulatory networks may significantly change with time or environmental conditions, thereby rewiring the network. Thus, while responding to diverse stimuli, few yeast transcription factors serve as permanent hubs, but most act transiently and appear under certain conditions only.

Intercellular molecular regulatory networks can be in the focus of topological analysis in the nearest future.

Scale-free and modular topology of regulatory networks is the result of biological evolution. The network topology is optimized in regard to multiple cellular functions and necessity to provide living cells and organisms with robustness, homeostasis, flexibility, and capacity for development.

8.10 EXERCISES

1. Given a network of transcription factor genes. You know from the literature that three genes A, B, and C in this network are connected by directed edges $\{A, B\}$ and $\{A, C\}$, which indicate that gene A stimulates the expression of gene B and gene C. You are making an experiment and find that gene A is expressed: that is, the product of this gene, transcription factor A is synthesized and the corresponding protein is detected by a western blot. However, your experiment shows that gene B and gene C are not expressed. Does it mean that the network taken from the literature obviously contains mistakes in regard to the edges $\{A, B\}$ and $\{A, C\}$? If not, what might be the reason for their inactivity? Give your explanation for this apparent contradiction.

2. Consider the same network as in Exercise 1. Your other experiment indicates that after stimulation of gene A, gene C is expressed as well, but gene B is not expressed. Therefore, the causal link $A \rightarrow B$ works, while the link $A \rightarrow B$ is still inactive. Give your explanation for this particular case.

3. Given a gene network, consisting of vertices (i.e., genes) and directed edges (causal links between genes). Let us consider vertex v and edges around this vertex. What would provide a stronger impact on the network topology: (a) deleting the vertex v or (b) deleting one of its incoming or outgoing edges? Explain your conclusion.

4. When working on Exercises 2 and 3, you may refer to Fig. 8.4. Keep in mind that each edge represents a multiple conditional event that includes more than

one functional step. Therefore, an edge indicates the possibility of action, which can be realized under a set of conditions. Note that a complete set of several transcription factors, which cooperatively act on a given gene, is necessary. Note that each of these transcription factors must be activated via its upstream signaling.

5. Given a gene network $G = (V, E)$ and given its subnetwork $G' = (V', E')$ centered at gene X. That is, G' includes gene X, all those genes from which gene X is reachable and all those genes that are reachable from gene X.

   (a) What kind of changes in this subnetwork G' might be expected if all incoming edges of gene X would be deleted?
   (b) Which changes in the subnetwork G' might be expected if all outgoing edges of gene X would be deleted?
   (c) Which changes in the subnetwork G' might be expected if all incoming and all outgoing edges of gene X would be deleted? Would that be comparable with the effect of deleting vertex X?

REFERENCES

1. R. Albert. Scale-free networks in cell biology. *Journal of Cell Science*, 118:4947–4957, 2005.

2. R. Albert, H. Jeong, and A.-L. Barabási. Lethality and centrality in protein networks. *Nature*, 401:130–131, 1999.

3. R. Albert, H. Jeong, and A.-L. Barabási. Error and attack tolerance of complex networks. *Nature*, 406:378–382, 2000.

4. E. Almaas, B. Kovacs, T. Vicsek, Z. N. Oltvai, and A.-L. Barabási. Global organization of metabolic fluxes in the bacterium *Escherichia coli*. *Nature*, 427:839–843, 2004.

5. U. Alon. *An Introduction to Systems Biology. Design Principles of Biological Circuits*. CRC Press, Boca Raton, FL, 2007.

6. E. A. Ananko, N. L. Podkolodny, I. L. Stepanenko, O. A. Podkolodnaya, D. A. Rasskazov, D. S. Miginsky, V. A. Likhoshvai, A. V. Ratushny, N. N. Podkolodnaya, and N. A. Kolchanov. Genenet in 2005. *Nucleic Acids Research*, 33:D425–D427, 2005.

7. M. M. Babu, S. A. Teichmann, and L. Aravind. Evolutionary dynamics of prokaryotic transcriptional regulatory networks. *Journal of Molecular Biology*, 358:614–633, 2006.

8. G. Balazsi, A.-L. Barabási, and Z. N. Oltvai. Topological units of environmental signal processing in the transcriptional regulatory network of *Escherichia coli*. *Proceedings of the National Academy of Sciences USA*, 102:7841–7846, 2005.

9. A.-L. Barabási. *Linked. The New Science of Networks*. Perseus Books, New York, 2002.

10. A.-L. Barabási and Z. N. Oltvai. Network biology: Understanding the cell's functional organization. *Nature Reviews in Genetics*, 5:101–113, 2004.

11. S. Bergmann, J. Ihmels, and N. Barkai. Similarities and differences in genome-wide expression data of six organisms. *PLoS Biology*, 2:85–93, 2004.

12. D. Bray. Molecular networks: The top–down view. *Science*, 301:1864–1865, 2003.

13. J. M. Carlson and J. Doyle. Complexity and robustness. *Proceedings of the National Academy of Sciences USA*, 99:2538–2545, 2002.

14. S. B. Carroll. Evolution at two levels: On genes and form. *PLoS Biology*, 3:e245, 2005.

15. R. W. Carthew. Gene regulation by micrornas. *Current Opinion in Genetics and Development*, 16:203–208, 2006.

16. C. Choi, M Krull, O. Kel-Margoulis, S. Pistor, A. Potapov, N. Voss, and E. Wingender. TRANSPATH — a high quality database on signal transduction. *Comparative and Functional Genomics*, 5:163–168, 2004.

17. S. N. Dorogovtsev, A. V. Goltsev, and J. F. Mendes. Pseudofractal scale-free web. *Physical Reviews E*, 65:e066122, 2002.

18. S. N. Dorogovtsev and J. F. Mendes. Evolution of networks. *Advances in Physics*, 51:1079–1187, 2002.

19. J. Doyle and J.M. Carlson. Highly optimized tolerance: Robustness and design in complex systems. *Physical Review Letters*, 84:2529–2532, 2000.

20. J. C. Doyle, D. L. Alderson, L. Li, S. Low, M. Roughan, S. Shalunov, R. Tanaka, and W. Willinger. The "robust yet fragile" nature of the internet. *Proceedings of the National Academy of Sciences USA*, 102:14479–14502, 2005.

21. L. Giot, J. S. Bader, C. Brouwer, A. Chaudhuri, B. Kuang, Y. Li, Y. L. Hao, C. E. Ooi, B. Godwin, E. Vitols, G. Vijayadamodar, P. Pochart, H. Machineni, M. Welsh, Y. Kong, B. Zerhusen, R. Malcolm, Z. Varrone, A. Collis, M. Minto, S. Burgess, L. McDaniel, E. Stimpson, F. Spriggs, J. Williams, K. Neurath, N. Ioime, M. Agee, E. Voss, K. Furtak, R. Renzulli, N. Aanensen, S. Carrolla, E. Bickelhaupt, Y. Lazovatsky, A. DaSilva, J. Zhong, C. A. Stanyon, Jr. R. L. Finley, K. P. White, M. Braverman, T. Jarvie, S. Gold, M. Leach, J. Knight, R. A. Shimkets, M. P. McKenna, J. Chant, and J. M. Rothberg. A protein interaction map of *Drosophila melanogaster*. *Science*, 302:1727–1736, 2003.

22. N. Guelzim, S. Bottani, P. Bourgine, and F. Kepes. Topological and causal structure of the yeast transcriptional regulatory network. *Nature Genetics*, 31:60–63, 2002.

23. D. Hanahan and R. A. Weinberg. The hallmarks of cancer. *Cell*, 100:57–70, 2000.

24. C. T. Harbison, D. B.Gordon, T. I. Lee, N. J. Rinaldi, K. D. Macisaac, T. W. Danford, N. M. Hannett, J. B. Tagne, D. B. Reynolds, J. Yoo, E.G. Jennings, J. Zeitlinger, D. K. Pokholok, M. Kellis, P. A. Rolfe, K. T. Takusagawa, E. S. Lander, D. K. Gifford, E. Fraenkel, and R. A. Young. Transcriptional regulatory code of a eukaryotic genome. *Nature*, 431:99–104, 2004.

25. L. H. Hartwell, J. J. Hopfield, S. Leibler, and A. W. Murray. From molecular to modular cell biology. *Nature*, 402:C47–C52, 1999.

26. H. Jeong, S. P. Mason, A.-L. Barabási, and Z. N. Oltvai. Lethality and centrality in protein networks. *Nature*, 411:41–42, 2001.

27. H. Jeong, B. Tombor, R. Albert, Z. N. Oltvai, and A.-L. Barabási. The large-scale organization of metabolic networks. *Nature*, 107:651–654, 2000.

28. G. Joshi-Tope, M. Gillespie, I. Vastrik, P. D'Eustachio, E. Schmidt, B. de Bono, B. Jassal, G. R. Gopinath, G. R. Wu, L. Matthews, S. Lewis, E. Birney, and L. Stein. Reactome: A knowledgebase of biological pathways. *Nucleic Acids Research*, 33:D428–D432, 2005.

29. A. R. Joyce and B. O. Palsson. The model organism as a system: integrating "omics" data sets. *Nature Reviews Molecular Cell Biology*, 7:198–210, 2006.

30. M. Kanehisa, S. Goto, M. Hattori, K. F. Aoki-Kinoshita, M. Itoh, S. Kawashima, T. Katayama, M. Araki, and M. Hirakawa. From genomics to chemical genomics: New developments in KEGG. *Nucleic Acids Research*, 34:D354–D357, 2006.

31. M. Kanehisa, S. Goto, S. Kawashima, Y. Okuno, and M. Hattori. The KEGG resource for deciphering the genome. *Nucleic Acids Research*, 32:D277–D280, 2004.

32. O. Kel-Margoulis, V. Matys, C. Choi, I. Reuter, M. Krull, A. P. Potapov, N. Voss, I. Liebich, A. Kel, and E. Wingender. Databases on gene regulation. V. B. Bajic and T. T. We, editors, *Information Processing and Living Systems*, pp. 709–727, Imperial College Press, London, 2005.

33. I. M. Keseler, J. Collado-Vides, S. Gama-Castro, J. Ingraham, S. Paley, I. T. Paulsen, M. Peralta-Gil, and P. D. Karp. EcoCyc: A comprehensive database resource for *Escherichia coli*. *Nucleic Acids Research*, 33:D334–D337, 2005.

34. H. Kitano. Systems biology: A brief overview. *Science*, 295:1662–1664, 2002.

35. H. Kitano. Biological robustness. *Nature Reviews in Genetics*, 5:826–837, 2004.

36. M. Krull, S. Pistor, N. Voss, A. Kel, I. Reuter, D. Kronenberg, H. Michael, K. Schwarzer, A. Potapov, C. Choi, O. Kel-Margoulis, and E. Wingender. TRANSPATH: An information resource for storing and visualizing signaling pathways and their pathological aberrations. *Nucleic Acids Research*, 34:D546–D551, 2006.

37. M. Krull, N. Voss, C. Choi, S. Pistor, A. Potapov, and E. Wingender. TRANSPATH: An integrated database on signal transduction and a tool for array analysis. *Nucleic Acids Research*, 31:97–100, 2003.

38. C. Lemer, E. Antezana, F. Couche, F. Fays, X. Santolaria, R. Janky, Y. Deville, J. Richelle, and S. J. Wodak. The aMAZE lightbench: A web interface to a relational database of cellular processes. *Nucleic Acids Research*, 32:D443–D448, 2004.

39. L. Lozada-Chavez, S. C. Janga, J. Collado-Vides. Bacterial regulatory networks are extremely flexible in evolution. *Nucleic Acids Research*, 34:3434–3445, 2006.

40. N. M. Luscombe, M. M. Babu, H. Yu, Snyder, S. A. Teichmann, and M. Gerstein. Genomic analysis of regulatory network dynamics reveals large topological changes. *Nature*, 431:308–312, 2004.

41. H. Ma and A. P. Zeng. Reconstruction of metabolic networks from genome data and analysis of their global structure for various organisms. *Bioinformatics*, 19:270–277, 2003.

42. H. W. Ma, J. Buer, and A. P. Zeng. Hierarchical structure and modules in the *Escherichia coli* transcriptional regulatory network revealed by a new top-down approach. *BMC Bioinformatics*, 5:e199, 2004.

43. A. Ma'ayan, R. D. Blitzer, and R. Iyengar. Toward predictive models of mammalian cells. *Annual Review of Biophysics and Biomolecular Structure*, 34:319–349, 2005.

44. S. Maslov and K. Sneppen. Specificity and stability in topology of protein networks. *Science*, 296:910–913, 2002.

45. S. Maslov and K. Sneppen. Computational architecture of the yeast regulatory network. *Physical Biology*, 2:S94–S100, 2005.

46. V. Matys, E. Fricke, R. Geffers, E. Gösling, M. Haubrock, R. Hehl, K. Hornischer, D. Karas, A. E. Kel, O. V. Kel-Margoulis, D. U. Kloos, S. Land, B. Lewicki-Potapov, H. Michael, R. Münch, I. Reuter, S. Rotert, H. Saxel, M. Scheer, S. Thiele, and E. Wingender. TRANSFAC: Transcriptional regulation, from patterns to profiles. *Nucleic Acids Research*, 31:374–378, 2003.

47. V. Matys, O. V. Kel-Margoulis, E. Fricke, I. Liebich, S. Land, A. Barre-Dirrie, I. Reuter, D. Chekmenev, M. Krull, K. Hornischer, N. Voss, P. Stegmaier, B. Lewicki-Potapov, H. Saxel, A. E. Kel, and E. Wingender. TRANSFAC and its module TRNASCompel: Transcriptional gene regulation in eukaryotes. *Nucleic Acids Research*, 34:D108–D110, 2006.

48. H. Michael, X. Chen, E. Fricke, M. Haubrock, R. Ricanek, and E. Wingender. Deriving an ontology for human gene expression sources from the CYTOMER database on human organs and cell types. *In Silico Biology*, 5:e7, 2004.

49. M. E. J. Newman. Assortative mixing in networks. *Physical Review Letters*, 89:e208701, 2002.

50. M. E. J. Newman. The structure and function of complex networks. *SIAM Review*, 45:167–256, 2003.

51. A. Potapov, I. Liebich, J. Dönitz, K. Schwarzer, N. Sasse, T. Schoeps, T. Crass, and E. Wingender. EndoNet: An information resource about endocrine networks. *Nucleic Acids Research*, 34:D540–D545, 2006.

52. A. Potapov, N. Voss, N. Sasse, and E. Wingender. Topology of mammalian transcription networks. *Genome Informatics*, 14:270–278, 2005.

53. E. Ravasz and A.-L. Barabási. Hierarchical organization in complex networks. *Physical Reviews E*, 67:e026112, 2003.

54. E. Ravasz, A. L. Somera, D. A. Mongru, Z. N. Oltvai, and A.-L. Barabási. Hierarchical organization of modularity in metabolic networks. *Science*, 297:1551–1555, 2002.

55. H. Salgado, S. Gama-Castro, A. Martinez-Antonio, E. Diaz-Peredo, F. Sanchez-Solano, M. Peralta-Gil, D. Garcia-Alonso, V. Jimenez-Jacinto, A. Santos-Zavaleta, C. Bonavides-Martinez, and J. Collado-Vides. RegulonDB (version 4.0): transcriptional regulation, operon organization and growth conditions in *Escherichia coli* k-12. *Nucleic Acids Research*, 32:D303–D306, 2004.

56. H. Salgado, S. Gama-Castro, M. Peralta-Gil, E. Diaz-Peredo, F. Sanchez-Solano, A. Santos-Zavaleta, I. Martinez-Flores, V. Jimenez-Jacinto, C. Bonavides-Martinez, J. Segura-Salazar, A. Martinez-Antonio, and J. Collado-Vides. RegulonDB (version 5.0): *Escherichia coli* K-12 transcriptional regulatory network, operon organization, and growth conditions. *Nucleic Acids Research*, 34:D394–D397, 2006.

57. T. Schlitt and A. Brazma. Modelling gene networks at different organizational levels. *FEBS Letters*, 579:1859–1866, 2005.

58. R. Sedgewick. *Algorithms in C^{++}*. Pearson Education Inc., Addison-Wesley, Addison Reading, MA, 2002.

59. K. Sneppen and G. Zocchi. *Physics in Molecular Biology*. Cambridge University Press, Cambridge, UK, 2005.

60. J. Stelling, U. Sauer, Z. Szallasi, F. J. Doyle 3rd, and J. Doyle. Robustness of cellular functions. *Cell*, 118:675–685, 2004.

61. D. Thieffry, A. M. Huerta, E. Perez-Rueda, and J. Collado-Vides. From specific gene regulation to genomic networks: a global analysis of transcriptional regulation in *Escherichia coli*. *Bioessays*, 20:433–440, 1998.

62. A. H. Tong, G. Lesage, G. D. Bader, H. Ding, H. Xu, X. Xin, J. Young, G. F. Berriz, R. L. Brost, M. Chang, Y. Chen, X. Cheng, G. Chua, H. Friesen, D. S. Goldberg, J. Haynes, C. Humphries, G. He, S. Hussein, L. Ke, N. Krogan, Z. Li, J. N. Levinson, H. Lu, P. Menard, C. Munyana, A. B. Parsons, O. Ryan, R. Tonikian, T. Roberts, A. M. Sdicu, J. Shapiro,

B. Sheikh, B. Suter, S. L. Wong, L. V. Zhang, H. Zhu, C. G. Burd, S. Munro, C. Sander, J. Rine, J. Greenblatt, M. Peter, A. Bretscher, G. Bell, F. P. Roth, G. W. Brown, B. Andrews, H. Bussey, and C. Boone. Global mapping of the yeast genetic interaction network. *Science*, 303:808–813, 2004.

62a. A. Vazquez, R. Dobrin, D. Sergi, J.-P. Eckmann, Z. N. Oltvai, A. L. Barabasi. The topological relationship between the large-scale attributes and local interaction patterns of complex networks. *Proceedings of the National Academy of Sciences USA*, 101:17940–17945, 2004.

63. E. van Nimwegen. Scaling laws in the functional content of genomes. *Trends in Genetics*, 19:479–484, 2003.

64. V. van Noort, B. Snel, and M. A. Huynen. The yeast coexpression network has a small-world, scale-free architecture and can be explained by a simple model. *EMBO Reports*, 5:280–284, 2004.

65. B. Vogelstein and K. W. Kinzler. Cancer genes and the pathways they control. *Nature Medicine*, 10:789–799, 2004.

66. B. Vogelstein, D. Lane, and A. J. Levine. Surfing the p53 network. *Nature*, 408:307–310, 2000.

67. A. Wagner. The yeast protein interaction network evolves rapidly and contains few redundant duplicate genes. *Molecular Biology and Evolution*, 18:1283–1292, 2001.

68. A. Wagner and D. A. Fell. The small world inside large metabolic networks. *Proceedings of the Royal Society of London B: Biological Sciences*, 268:1803–1810, 2001.

69. D. J. Watts and S. H. Strogatz. Collective dynamics of "small-world" networks. *Nature*, 393:440–442, 1998.

70. H. V. Westerhoff and L. Alberghina. Systems biology: did we know it all along? *Systems Biology: Definitions and Perspectives*, pp. 3–9. Springer, New York, 2005.

71. S. H. Yook, Z. N. Oltvai, and A.-L. Barabási. Functional and topological characterization of protein interaction networks. *Proteomics*, 4:928–942, 2004.

9

PROTEIN INTERACTION NETWORKS

Frederik Börnke

9.1 INTRODUCTION

Proteins control and mediate the vast majority of biological processes in a living cell. They act as catalysts, transport or store other molecules, provide mechanical strength, confer immunity, transmit signals, and control growth and development. Proteins are polymers of amino acids, covalently linked through peptide bonds into a chain. The function of a protein is determined by its three dimensional structure, which in turn is defined by its amino acid sequence. Monitoring the alterations in expression of specific proteins in response to changing environments or across different developmental stages of a given organism has provided substantial insight into the complex regulatory networks controlling life. Proteins operate entirely on the basis of interactions with other molecules such as low molecular weight compounds, lipids, nucleic acids, or other proteins. The functionally active form of a protein is rarely its monomer, that is, the single molecule. Rather close association with partner proteins, or assembly into larger protein complexes is necessary for biological activity. Besides the obvious role of protein–protein interactions in the assembly of the cell's structural components, such as the cytoskeleton, they are also crucial for processes ranging from transcription, splicing, and translation to cell cycle control, secretion, and the assembly of enzymatic complexes. Prominent examples for the latter are the organization of enzymes catalyzing sequential steps in a metabolic pathway, such as glycolysis or fatty acid biosynthesis, into multienzyme complexes (for an overview see Ref. [61]). A major advantage of such spatial organization is the transfer of biosynthetic intermediates between catalytic sites without diffusion into the enzyme's surrounding (Fig. 9.1a).

Analysis of Biological Networks, Edited by Björn H. Junker and Falk Schreiber

This so-called *metabolic channeling* can be envisioned as a means of attaining high local substrate concentration, regulate distribution of intermediates shared by competing pathways, and sequester reactive or toxic intermediates. Besides this rather static protein–protein interactions, there are also a large number of transient interactions, that is, interactions occurring for a limited time only. For example, all protein modifications necessarily involve such transient protein–protein interactions (Fig. 9.1b). These include the interactions of all protein kinases, protein phosphatases, glycosyl transferases, proteases, etc. with their target proteins. The transmission of regulatory signals from the external environment to relevant locations in the cell depends heavily on protein–protein interactions. This type of interaction is much more difficult to study, because the physiological condition under which the binding occurs has to be determined in advance. Forces that mediate protein–protein interactions include electrostatic interactions, hydrogen bonds, the van der Waals attraction, and hydrophobic effects. It has been proposed that hydrophobic forces drive protein–protein interactions and hydrogen bonds and salt bridges confer specificity. On the protein level, interactions can be mediated at one extreme by a small region of a protein fitting into

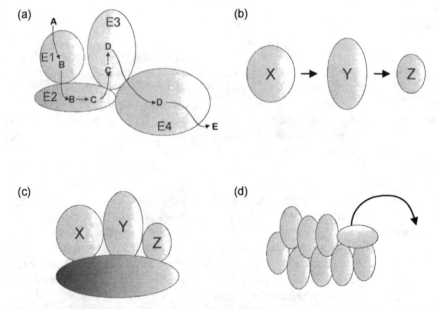

FIGURE 9.1 Examples for protein–protein interactions in various cellular processes. (*a*) Multienzyme complex. Compound A is converted to compound E by four sequentially acting enzymes (E1 – E4). Organization of the pathway into a multienzyme complex enables direct transfer of metabolic intermediates between enzymes and thus enhances, for example, catalytic efficiency. (*b*) Signal transduction chain. Transmission of the signal relies on the transient interaction between the participating proteins. (*c*) Protein scaffolds. Signal transduction pathways can be tethered to a particular cellular location by scaffolding proteins. (*d*) Protein interactions form the basis for the function of larger protein complex, for example, ribosomes.

a cleft of a partner protein and at another extreme by two surfaces interacting over a large area. From the above mentioned examples, it becomes obvious that protein–protein interactions are much more widespread than once suspected and the degree of regulation they confer is large. Thus, to gain a thorough understanding of biological function, it is important to look at a protein in the context of other interacting proteins. Moreover, disease is often related to alterations in certain protein–protein interactions. Hence, the manipulation of protein–protein interactions that contribute to certain diseases provides a potential therapeutic strategy. The complete sequence of genomes of many bacteria, viruses, and small and large eukaryotes has provided a vast new resource to define gene function at the morphological, biochemical, and physiological level. Sequence information by itself, however, does not lead to a clear insight into the underlying principles of cellular systems. This is mainly because the biological function of the plethora of predicted genes remains experimentally uncharacterized. Thus, an understanding of biological mechanisms and disease processes demands a "systems" approach that goes beyond the one-at-a-time studies of single components to more global analyses of the structure, function, and dynamics of the networks in which proteins function. Recent technological advancements using highly parallelized and automated approaches opened the possibility to assess protein–protein interactions on a genome wide scale. This led to the establishment of complete protein-linkage maps, called the *interactome*, which can be regarded as "framework" information. Protein networks not only aid functional annotation of unknown proteins by opening the possibility to group unknowns into a known biological context ("guilt by association" principle), moreover, increasingly detailed and reliable biological models can be generated by integrating other functional genomic and proteomic data sets into interaction maps [23]. In this chapter, several experimental approaches that are currently used to generate protein–protein interaction data are introduced. In Section 9.3.1, it is briefly reviewed how these data are used to generate protein interaction networks. As all experimental techniques to determine protein–protein interactions suffer from noise and systematic biases; in Section 9.3.2, computational methods for improving confidence and quality of interaction data are introduced. In Section 9.4, approaches for characterizing the topology of networks, such as finding hubs and analyzing subnetworks in terms of common motifs are discussed. Finally, evolution and conservation of protein networks are briefly reviewed.

9.2 DETECTING PROTEIN INTERACTIONS

Classically, protein–protein interactions have been studied using so-called physical methods such as affinity chromatography or coimmunoprecipitation (Fig. 9.2) [48,67]. Although these approaches have been proven to be extremely useful, they are not without problems. The quantity and quality of the starting material are of major importance for all of these approaches. Coimmunoprecipitation in principle can proceed from crude protein extracts as starting material; however, usually only minimal amounts of interacting protein can be recovered limiting downstream analysis. On the

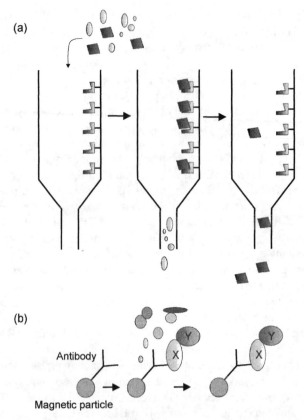

FIGURE 9.2   Two fundamental principles to experimentally analyze protein–protein interactions. (*a*) Protein affinity chromatography. Extract proteins are passed over a column containing immobilized protein. Proteins that do not bind flow through the column, ligand proteins that bind are retained. Subsequently, bound proteins can be eluted and further analyzed. (*b*) Coimmunoprecipitation. An antibody directed against a protein of interest is coupled to a magnetic particle. The antibody binds its target protein with all its associated proteins in a cell extract. A magnet immobilizes the protein complex bound to the antibody and contaminating nonbinding proteins can be washed away.

contrary, affinity chromatography requiring relatively large amounts of pure protein to be coupled to the column matrix can yield sufficient amounts of interacting proteins for subsequent biochemical analysis. Identification of the corresponding protein and identification of the gene used to be a cumbersome process. Only the recent development of powerful mass spectrometric methods for protein identification has led to a renaissance of biochemical methods for protein complex identification because of their superior sensitivity, speed, and versatility compared with traditional protein sequencing methods (for details see Section 9.2.2). Currently, two experimental strategies are used to generate proteome-wide interaction maps at high-throughput. They are the yeast two-hybrid system and analysis of protein complexes by affinity purification coupled to mass spectrometry (AP-MS). The yeast two-hybrid system, as a binary assay captures direct protein–protein interactions, whereas AP-MS identifies components of stable complexes. Both assays individually can provide useful information on protein function employing the "guilt by association" principle. The basic principles of the two approaches are discussed below.

9.2.1   The Yeast Two-Hybrid System

Since its advent some 15 years ago [17], the yeast two-hybrid system has marked a cornerstone in the identification and characterization of protein–protein interactions. The system takes advantage of the modular domain structure of most eukaryotic transcription factors. For instance, the yeast GAL4 protein, a global transcriptional activator of the galactose metabolic pathway, consists of a DNA-binding domain (DB) that binds to specific upstream sequences of GAL4 responsive genes and an activation domain (AD), which subsequently binds the proteins necessary to activate transcription. Both domains constitute only relatively small proportions of the entire GAL4 protein and they are functionally independent and structurally separable, respectively. When both domains of the GAL4 protein are expressed as individual polypeptides within the same cell, they fail to activate transcription of GAL4 responsive genes. However, the observation that transcription factor activity could be restored if these two proteins do physically interact provided the inspiration for the two-hybrid system [17]. To achieve reconstitution of transcription factor activity a "bait" protein is constructed by fusing a protein X to the DB and a "prey" is constructed by fusing a protein Y to the AD of the GAL4 protein, respectively. The bait and prey fusions are coexpressed in a yeast reporter strain, where interaction between X and Y brings the DB and AD in close enough proximity to reconstitute an active transcription factor (Fig. 9.3). Reconstitution of transcription factor activity is measured by assaying the activity of reporter genes that are placed downstream of GAL4 responsive elements. Commonly, activation of the reporter genes leads to growth on specific media and results in a color reaction after application of a particular coloring reagent. The yeast two-hybrid system not only enables to demonstrate the interaction between two known proteins but also allows for the detection of novel interactions by screening a given bait protein against a DNA library, representing the total of proteins an organism can

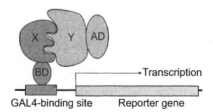

FIGURE 9.3   Principle of the yeast two-hybrid system. Two test proteins, X fused to the DNA binding domain and Y fused to the activation domain, are coexpressed in a yeast reporter strain. Upon interaction of X and Y, an active transcription factor is reconstituted and activates specifically transcription of the reporter genes.

express, fused to an AD. A typical two-hybrid screening experiment begins with a strain that contains a set of reporter genes, and that expresses a bait protein fused to the DB of GAL4. The strain that contains the bait is then transformed with a DNA library that expresses proteins fused to the AD. Each yeast cell will ideally take up a single protein out of the entire library and thus allows for the pair-wise analysis of all proteins encoded by the library against the bait protein. The resulting yeast cells are generally selected on specific medium where only yeast cells expressing the respective reporter gene grow. These are then tested for activation of the second reporter gene by placing them on plates containing the coloring reagent. From colonies showing activity of both reporter genes, the library DNA is isolated and the nucleotide sequence of the respective cDNA insert can subsequently be determined. The power of the yeast two-hybrid system lies in its superior speed and robustness. Since it is a genetic system, it can detect all types of protein–protein interactions over a wide range of physiological conditions. Its great sensitivity enables to detect even very weak or transient interactions. Moreover, detection of interaction between two proteins is directly linked to the genetic information encoding them. Although the yeast two-hybrid system offers a number of obvious advantages over other methods to detect protein–protein interactions that render it currently the most versatile system for large-scale interaction studies, it still has some associated problems. Since read-out of the interaction relies on the reconstitution of an active transcription factor, it is necessary for the interaction to occur within the nucleus of the yeast cell. Although the DNA constructs used to express the hybrid proteins usually are engineered to direct the proteins into the nucleus of the yeast cell, problems may arise with proteins having strong localization signals to other compartments of the cell, or are, for example, due to their size, not imported into the nucleus. One class of proteins that cannot be readily investigated in the yeast two-hybrid system are transcription factors because they carry an AD themselves and thus, when fused to a DB, lead to autoactivation of the reporter genes. Also proteins that have hydrophobic domains, such as membrane proteins, are unlikely to be functional in a two-hybrid assay. A major drawback of testing protein–protein interactions in a heterologous system, such as the yeast cell, is that interactions may depend on certain posttranslational modifications of one or the other partner, for example, phosphorylation, acetylation, or glycosylation. These modifications may not occur properly in the yeast system and thus an interaction is not detected. Furthermore, although the two-hybrid assay selects for binary interactions, it cannot be excluded that indirect interactions bridged by endogenous yeast proteins occur. Finally, interactions of no biological relevance can be detected if both partners naturally reside in different compartments of the cell and are only brought together artificially in the yeast nucleus. Consequently, interactions detected with the yeast two-hybrid system should be regarded as hypothesis until they are validated by independent experimental procedures [37]. In order to circumvent some of the above-mentioned limitations of the classical GAL4-based yeast two-hybrid system, alternative approaches have been developed recently. In general, in these systems, interaction of two test proteins leads to the functional reconstitution of a protein whose activity results in a particular phenotype allowing for the visual identification of cells expressing interacting proteins. For an overview concerning some of these alternative systems, the reader is referred to recent reviews [9,12]. Because of the above-mentioned advantages, the yeast two-hybrid system offers over other approaches and because of its amenability for automation using robotics, it has become the method of choice to generate genome-wide protein interaction maps. Typically, large-scale two-hybrid studies rely on interaction mating. In this method, the bait-protein and the prey-protein are transferred to yeast reporter strains of opposite mating types (MATa and MATα), and the strains are mated to determine if the two proteins interact. Mating occurs when cells of opposite mating types come into close contact, and results in the fusion of the two haploids to form a diploid yeast strain in which the two interacting proteins are produced. Thus, an interaction can be determined by measuring activation of the reporter genes in the diploid strain (Fig. 9.4). The main advantage of this approach is that it considerably reduces the number of transformations necessary to investigate a large number of interactions. This is because each bait strain can be mated to an arbitrary number of prey proteins using relatively simple mating steps. In cases where the same AD-tagged DNA library needs to be screened with multiple baits, only a single large-scale transformation is required. Two systematic approaches have been employed in large-scale yeast two-hybrid screens—the matrix approach and the library approach. In the matrix approach

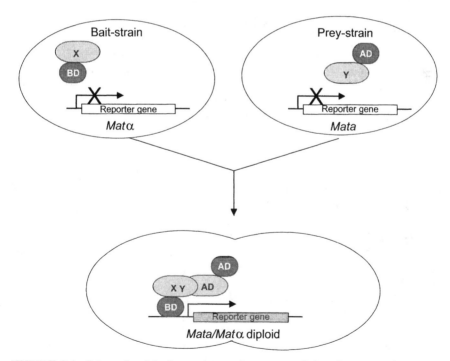

FIGURE 9.4 Schematic of the interaction mating strategy. Bait and prey proteins are expressed in MATa and MATα yeast reporter strains, respectively. After mating, both proteins are expressed in the same yeast cell and, if X and Y interact, reporter gene activity is detected.

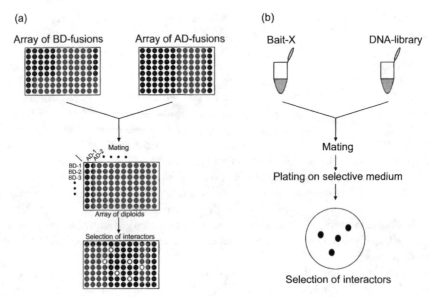

FIGURE 9.5   High-throughput yeast two-hybrid assays. (*a*), Matrix approach. An array of prey clones is created in one multititre plate and, using robotics, transferred to a second plate containing an array of BD-clones. Both clones are allowed to mate, and diploids expressing interacting proteins are detected based on reporter gene expression. (*b*), Library screening approach. Yeast cells containing a particular BD-X fusion are mated to with cells pretransformed with a library of random DNA fragments. Diploids are plated onto selective medium, and interaction is scored by virtue of reporter gene expression. In contrast to the matrix approach where the identity of each fusion protein is known, diploids that passed the assay need to be picked and library DNA has to be isolated for sequencing.

(Fig. 9.5a), a set of full-length open reading frames (ORFs; i.e., the portion of the gene encoding the corresponding protein) is amplified using the polymerase chain reaction and cloned into both BD- and AD-fusion DNA constructs. The resulting constructs are separately transformed in MATa and MATα reporter strains. Each specific pairing of DB- and AD-fusions generated via mating is assayed for interaction. This approach was initially used to identify protein–protein interactions between drosophila proteins involved in cell cycle regulation [50]. Later it was applied to analyze the specific pairs of protein–protein interactions among all of the 266 ORFs of the vaccinia virus. Of the ca. 70,000 combinations assayed, 37 protein–protein interactions were found, including 28 that were previously unknown [45]. The matrix approach has also been used to tackle larger genomes. Recently, two groups reported on large-scale approaches to detect genome-wide protein–protein interactions in *S. cerevisiae* [34,66]. In the approach followed by Ito et al. [34], all of the approximately 6000 yeast ORFs were cloned individually as BD fusion baits in one yeast strain and as AD fusion preys in another strain of the opposite mating type. These collections of AD and BD fusion clones were divided into 65 pools each containing 96 clones. Of the possible

4225 (65 × 65) mating reactions, the authors performed 430 thus covering approximately 10% of all possible combinations. As a result, 183 independent two-hybrid interactions were detected, of which more than half were entirely novel [34]. This systematic approach was later completed and eventually identified a total of 841 interactions [33]. In a complementary study, Uetz et al. [66] cloned 192 yeast ORFs as BD fusions and then with the 6000 yeast ORF AD fusions. This resulted in the identification of 287 interactions. Surprisingly, when comparing the data sets of Ito et al. [33,34] to the one from Uetz et al. [66], the overlap was only 20%; despite the fact that both groups used the same 6000 ORFs [33]. The reason for this small overlap is difficult to explain, but it is most likely that the differences are due to the use of different experimental components. For instance, the different plasmids used by the two groups may give rise to different expression levels or might affect folding of particular proteins, or the use PCR to amplify the yeast ORFs may have introduced mutations that abolish interactions. Thus, although the results demonstrate the power of the matrix approach to conduct large-scale interaction studies, care should be taken when judging the significance of newly identified interactions. As with all two-hybrid experiments, the matrix approach is also prone to detect a high level of false positive and negatives. Since biological validation of interaction data generated in high-throughput experiments is not readily possible, the various interaction data have be to tested for accuracy on confident sets of interactions using computational methods ([68]; see also discussion below). Given the importance of protein–protein interactions for the understanding of disease mechanisms and signaling cascades, a comprehensive interaction map of human proteins is highly desirable. Toward this end, Stelz et al. [62] used a yeast two-hybrid matrix approach to screen more than 5500 human proteins for potential interactions, building an interaction dataset that connects 1705 human proteins via 3186 interactions. Matrix experiments provide the advantage of knowing the identity of each DB and AD pair, thus circumventing the need for sequencing. However, this strategy provides no information about the protein domain that confers a given interaction. Moreover, full-length proteins, as they are used in matrix approaches, might contain domains hindering the interaction with other proteins. Proteins that contain hydrophobic transmembrane domains or strong targeting signals will be unable to reach the nucleus and thus can give rise to false-negative results. The use of libraries consisting of random DNA fragments fused to the AD to screen individual BD-fusions or pools of bait proteins partially circumvents this limitation (Fig. 9.5b). This type of library approach was first applied to determine the protein interaction network of the *Escherichia coli* phage T7 proteome, comprising 55 proteins, by screening libraries of random DNA fragments fused to either the BD or the AD against each other. This yielded 22 interactions between phage proteins, of which only four had been described previously [7]. Fromont-Racine and coworkers [21] used 15 yeast ORFs involved in mRNA splicing to screen a yeast AD-tagged library. They were able to identify 170 interactions, corresponding to 145 different yeast ORFs [21]. Later, the first partial protein interaction map for a multicellular organism, the nematode *Cenorhabditis elegans*, was created by using 27 proteins known to be involved in vulval development to carry out exhaustive library screens [70]. As a result, 148 interaction were identified, including 15 known interactions and 109 interactions that had previously been predicted based on the *C. elegans* genome sequence [70]. More recently, this study was extended using 1873 ORFs related to multicellular function as baits to screen two different AD-tagged cDNA libraries, and with the interactions previously described, the *C. elegans* protein interaction data set currently contains about 5500 interactions [41]. Up to now, the most comprehensive large-scale two-hybrid study on a multicellular organism was performed on the proteome of the fly *Drosophila melanogaster* [24]. A total of 10, 623 ORFs were cloned as DB-fusions and screened against DNA libraries to produce a draft protein interaction map of 7058 proteins 20,405 interactions. In summary, the large-scale screenings carried out so far clearly demonstrate the power of the yeast two-hybrid system in the identification of protein–protein interactions on a genome wide scale in various organisms. Although, without independent validation data from these high-throughput approaches should be treated with caution protein interaction maps from these data sets will definitely advance our understanding how cellular processes are linked by the physical interactions of the proteins involved.

9.2.2 Affinity Capture of Protein Complexes

Recently, mass spectrometry (MS) began to play an important role in the identification of protein–protein interactions as it allows for the detection and identification of peptides with unprecedented sensitivity. For the identification of protein–protein interactions, MS based technologies are generally used in conjunction with protein-complex affinity purification methods (affinity purification coupled to MS; AP-MS, see Fig. 9.6). Traditional affinity methods use either antibodies or affinity ligands to purify protein complexes (see also Section 9.2 and Fig. 9.2); however, these approaches heavily depend on the availability of specific antibodies or other capturing ligands. A major step toward a generic approach for protein-complex purification was the use of standard affinity tags attached to a protein of interest. A wide variety of affinity tags is currently available that are either short peptides (e.g., the His6-tag consisting of six consecutive histidine residues) or small proteins (e.g., protein A from *Staphylococcus aureus*) [64]. A common feature of these tags is that they allow purification of fusion proteins with relatively high purity and yield using generalized protocols. In a typical AP-MS experiment, fusion of the target protein sequence and the affinity tag is done using standard DNA cloning techniques. The recombinant gene thus created is then introduced and expressed in the host cell for which a transformation procedure must be available. The tagged protein, if biologically functional, is assembled into its native protein complex *in vivo*. Protein complexes are captured from the total proteins extracted from the cell by affinity purification on the respective affinity matrix. The purified material is then fractionated using gel electrophoresis that separates proteins according to their mobility in an electric field, and copurified proteins are subsequently excised from the gel and identified by MS (Fig. 9.4). Successful purification requires a method that is stringent enough to differentiate the complex of interest from all other proteins present in the cell extract but in turn is gentle enough not to compromise the integrity of the complex during the purification process. Major progress in protein

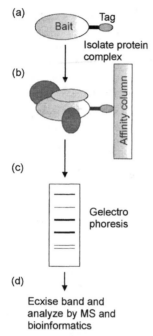

FIGURE 9.6   Analyzing protein complexes by affinity purification coupled to mass spectrometry (AP-MS). (*a*) An affinity-tagged bait protein is expressed in a host cell. (*b*) The bait protein is affinity-purified along with associated proteins. (*c*) Purified protein complexes are by gelelectrophoresis, so that proteins become separated according to molecular mass. (*d*) Proteins are excised from the gel and analyzed by mass spectrometry.

complex purification was made by the development of the tandem affinity purification (TAP) method [49,51]. This method is based on two successive affinity chromatography steps. The introduction of two consecutive affinity purification steps greatly enhances the specificity of the purification procedure. Although originally designed for protein complex purification from yeast, the TAP-tag and derivatives of the original tag have successfully been used to isolate protein complexes from mammalian cells, plants, and bacteria [10,11,52,53,55]. One of the major advantages of AP-MS as compared with the yeast two-hybrid system is that it identifies interactions that occur in the native cellular environment, provided that temporal and spatial expression of the target protein is normal, although purification of complexes can lead to both loss of real interactions and gain of spurious ones. Moreover, if interactions depend on posttranslational modifications of one or more components of the complex, they can be identified by AP-MS but usually not in the yeast two-hybrid system. While the yeast two-hybrid system detects binary interactions, AP-MS detects higher order interactions, and data from these experiments generally do not provide information about how components of the complex are connected to each other. Two pioneering studies on the large-scale AP-MS analysis of purified protein complexes from *S. cerevisiae* have been performed [22,30]. These two approaches varied significantly in their experimental design. Ho et al. [30] used transient overexpression of flag-tagged ORFs, which were subjected to one-step affinity purification followed by gel electrophoresis (SDS-PAGE). Complex constituents were identified by tandem MS. Gavin et al. [22] integrated a TAP-tag cassette by homologous recombination at the $3'$ of each ORF. The advantages of this tagging strategy are that there is no competition *in vivo* for the untagged endogenous protein and that expression tagged protein is under control of its own promoter. After affinity purification and SDS-PAGE, matrix-assisted laser desorption ionization (MALDI)-MS was used for protein identification [22]. From the 1617 yeast strains generated by Gavin et al. [22], 589 protein complexes were purified yielding 4111 interactions between 1440 individual proteins. Ho et al. [30] used 725 protein baits and detected 3618 interactions that involved 1578 proteins. Recently, an attempt was made to map more completely the yeast interactome using AP-MS. Toward this end, Krogan Ho et al. [39] generated *S. cerevisiae* strains with in-frame insertions of TAP-tags at the $3'$ end of every predicted yeast ORF. This enabled the authors to purify 4562 different yeast proteins by tandem affinity purification. To increase coverage and accuracy of protein identification following the affinity purification each preparation was analyzed by both MALDI-MS and liquid chromatography tandem MS (LC-MS/MS). Machine learning was used to integrate the MS scores and assign probabilities to the protein–protein interactions. Among the 4087 different proteins identified with high confidence by MS from 2357 successful purifications, the core data set comprises 7123 protein–protein interactions involving 2708 proteins [39].

9.2.3   Computational Methods to Predict Protein Interactions

An alternative to experimental determination of protein interactions is prediction by various computational genomics approaches. Generally, these approaches take into account the genomic "context" of a given gene or protein to retrieve functional information. Examples of genomic context include conservation of gene identity and position in genomes of different species, genome-wide analysis of gene fusion, metabolic reconstruction, gene coregulation, and expression (reviewed in Refs. [15,16]). The phylogenetic profiling approach is based on the assumption that protein pairs are likely to functionally or even physically interact when their corresponding genes are located in close proximity to each other in the genome and when this arrangement is conserved across multiple genomes [31]. Using a similar method, functional interaction of proteins is predicted on the basis of patterns of domain fusions [42]. Sometimes two protein domains exist as separate proteins in one genome, but they are fused together into a single bifunctional protein in another genome. In such a case, the domains are likely to be functionally related. Two methods exploit the hypothesis that interacting proteins tend to coevolve. In the first method, the coevolution of interacting protein families is measured by the similarity of phylogenetic trees constructed from multiple sequence alignments of the two protein families [25]. When applied on a genomic-scale, phylogenetic trees for all proteins can be constructed and similarity between those trees indicates interaction (see also Chapter 11). In the second method, the coevolutionary signal in multiple sequence alignments is further analyzed in terms correlated mutations, which means that a protein pair is likely to interact if there is accumulation of correlated mutations between the two partners [46]. It is also possible to predict protein–protein interactions from sequence information using machine-learning techniques. For example, using a database of known interactions, a support vector machine learning system can be trained to predict interactions based on sequence information and associated physicochemical properties, like charge, hydrophobicity, and surface tensions [8]. A number of methods have been developed to extract protein interactions from literature. These methods can be grouped into two categories. Methods in the first category use machine learning techniques to screen the literature for articles containing information about protein interactions [43]; selected articles are then curated by hand. Methods in the second category automatically extract protein interaction events from biomedical articles using natural language processing technologies [20] or statistical analysis of cooccurrence of names of biomolecules [32].

9.2.4   Other Ways to Identify Protein Interactions

Protein Microarrays   The microarray concept, initially developed to monitor mRNA expression in a multiparallel manner, has also been adapted for protein–protein interaction analyses. In a typical protein microarray experiment for the detection of protein–protein interactions, a set of recombinant proteins is immobilized onto a solid surface and subsequently incubated with a labeled protein probe in the liquid phase. Interaction of the labeled protein with a protein on the solid matrix generates a signal on the array, which can subsequently be detected. The fabrication of protein microarrays requires the availability of expression clones, ideally for every protein encoded by the respective genome, and the technology for protein expression and purification. It was a major breakthrough in protein microarray technology when Snyders' group reported the yeast proteome array in 2001 [76], where 5800 yeast ORFs were expressed and presented on a single glass-slide. To test for protein–protein interactions, the yeast proteome array was probed with biotinylated calmodulin in the presence of calcium. Calmodulin is a protein of major importance in the regulation of cellular processes dependent on calcium. It acts by binding to target proteins and thereby altering their biological activity. In addition to known partners, the calmodulin probe identified 33 novel potential calmodulin binding proteins on the protein array [76]. Since then, protein microarrays have been used to study many kinds of protein interactions and biochemical activities on a global scale (for a recent overview see Ref. [40]).

Synthetic Lethality   In yeast, genetic interaction between two genes can be studied by the so-called synthetic lethality. That is when two nonessential genes cause lethality when mutated at the same time in a cell. Such genes are often functionally associated and their encoded proteins may also interact physically. This type of interaction is currently being studied in all-versus-all approach in yeast [65]. To this end, a query yeast strain carrying a mutation in a nonessential gene is crossed to of 4700 deletion mutants in the so-called synthetic genetic array (SGA).

Inviable double-mutant meiotic progeny identify functional relationships between genes. SGA analysis of genes with roles in cytoskeletal organization, DNA synthesis and repair, or uncharacterized functions so far generated a network of 291 interactions among 204 genes [65]. The advantage of this approach is that it is an *in vivo* technique, although an indirect one; and it is amenable to unbiased genome-wide screens. However, although synthetic lethality indicates a functional association between two given genes it does not necessarily mean that the corresponding proteins do physically interact and thus a network of genetic interaction data has different information content than a protein interaction network.

9.3   ESTABLISHING PROTEIN INTERACTION NETWORKS

9.3.1   Data Storage and Network Generation

To effectively exploit the large amounts of protein–protein interaction data generated by current experimental and computational methods for biological research, it is necessary for these data to be stored in a consistent and reliable way. To this end, a number of protein–protein interaction databases have been developed recently and are now publicly available. These databases greatly differ by their coverage and contents and only a selection of databases is described here (Table 9.1). For a more comprehensive description of protein interaction databases, including more specialized ones, the reader is referred to some excellent recent reviews [18,54] and the listings at http://www.pathguide.org. The databases listed in Table 9.1 combine protein interaction data from a number of sources such as from high- and low-throughput yeast two-hybrid and AP-MS analyses, respectively, as well from data mining of the literature. Therefore, these databases can be considered metaservers of protein interaction data. The Database of Interacting Proteins (DIP) was developed at the University of California Los Angeles is a relational database that contains experimentally determined protein–protein, protein–nucleotide, and protein–ligand interactions [57,73]. Interactions in DIP are curated both manually (by expert curators) and automatically (by text-mining approaches). This database is useful in identifying interacting partners of a protein of interest and visualizing the interactions between proteins. The

TABLE 9.1   Comprehensive Interaction Databases

Name	Abbr.	Entries	URL
Database of interacting proteins	DIP	55,732	http://dip.doe-mbi.ucla.edu/
Biomolecular Interaction Network Database	BIND	201,732	http://binddb.org
Molecular Interactions Database	MINT	61,330	http://mint.bio.uniroma2.it

interaction diagrams also provide researchers with a confidence level of every inter-action, as indicated by the width of edges connecting interacting proteins. Access to DIP is free to members of the academic community upon registration. The top-level search page of DIP allows different search strings to be applied on the data set. Proteins can be found by protein name (vertex), sequence motif, BLAST search, or article. Once a protein has been found, a new window can be opened showing the interactions, and information concerning the interaction partners can be browsed and displayed in several ways. The information in the DIP is available for download. The Biomolecular Interaction Network Database (BIND) contains by far the most com-prehensive interaction dataset and is designed to store full descriptions of interactions, molecular complexes, and pathways [1,2]. All interactions contained in BIND have been experimentally verified and published in at least one peer-reviewed publication and are collected by literature mining or investigator submission, with review by cu-rators before incorporation. Incorporated visualization tools help visualizing complex multiprotein interactions. The basic search mode uses text entries to query BIND. The results page from a search shows interaction between what was queried and what is in the database in a binary set. Querying the database with a protein of interest, either using the text search or the BLAST search option using sequence information, gener-ates a results page with interaction listed in sets of two—the protein of interest with its potential partners. The results page also lists comprehensive additional information concerning the interaction partners such as a general description, molecular function, cellular component, biological process, experiments, and links. The Molecular Inter-action Database (MINT), developed at the University of Rome Tor Vergata, focuses primarily on experimentally verified protein–protein interactions from mammalian genomics [75]. The MINT data set was generated by mining scientific literature us-ing a respective algorithm and then reviewed by expert curators. MINT combines basic protein and gene information with data on binary interactions. The database can be searched for individual proteins by name, keyword, structure, or accession number. Such a search results in a list of entries matching the search criteria, and clicking on an individual result leads to a page describing a particular entry, containing a variety of annotation information, as well as a link to all binary interactions involving this protein. Additionally, interactions from MINT can be viewed graphically as an inter-action network using the interactive JAVA-based MINT viewer. Each protein in such an interaction network is linked back to the corresponding information page, and in-teracting proteins are linked to the corresponding binary interaction page. A long list of interacting proteins or a table of protein pairs falls short of capturing what actually happens in a cell, which is a dynamic process that occurs in at least four dimensions (including time). Instead, the use of graphics suits the human preference of visual perception over every other sensory system. Commonly, protein interaction networks are build from experimental data sets using network visualization tools, which in the simplest case represent protein interaction networks as a graph composed of vertices (proteins) connected by edges (interactions) (see also Chapter 2). A number of purely automatic and general algorithms have been developed for visualizing biological net-works. These tools rely on a layout algorithm to organize a graph of vertices and edges into an easily navigable layout that is usually featured by minimizing the number of edges that cross each other, and grouping groups of vertices that are highly connected to each other. Two commonly used open source visualization tools for protein networks are BioLayout [14] and Cytoscape [58], which both are based on JAVA and thus readily portable between a variety of computer environments. They also allow the interactive editing of graphs, including movement of vertices, vertex labeling, and changing graph appearance. BioLayout utilizes the weighted Fruchterman–Rheingold algorithm and has a number of options for graph customization, data overlay, export, and graph analysis. Cytoscape provides a number of different layout algorithms for producing useful visualizations such as circular, hierarchical, organic, embedded, and random layouts. Circular and hierarchical algorithms try to lay out networks as their name suggests. Organic and embedded are two versions of a force-directed layout algorithm.

9.3.2 Benchmarking High-Throughput Interaction Data

An essential problem inherent to large-scale protein interaction data sets is that only a relative small fraction of interactions in networks are known with any certainty, which leads to difficulties in estimating the rate of both false positive and false negative rates. This has far reaching consequences on the extraction of biological meaningful information from theses data sets. Therefore, several attempts have been made to minimize the number of false positives in high-throughput studies. On the experimental level, interactions can be scored based on their reproducibility [66] or their frequency of occurrence [33]. Recently, statistics and bioinformatics approaches have been used to benchmark the quality of high-throughput interaction data sets. In a pioneering study von Mering et al. [68] compared yeast high-throughput interaction data sets from several sources and used interactions found in more than one data set to estimate the fraction of true positives. The overlap found was surprisingly small; however, the authors concluded that this was due to the inherent biases in different experimental methods. Database annotations have been used by Sprinzak et al. [60] in the form of colocalization data of interacting proteins and their proposed cellular role as a measure to estimate the reliability of interaction data. However, this approach is limited to model organisms with well-annotated genomes. In addition to these evaluations of the overall quality of the interaction datasets, attempts have been made to identify the most reliable subsets of high-throughput data. These attempts usually involve combining multiple sources of experimental information. However, since there is only marginal overlap between data sets [56,68], the number of interactions validated this way is very small. This number can be increased if one also takes into account known interactions between paralogs of the putative interacting pairs. This approach, as demonstrated by Deane et al. [13], allows one to identify roughly half of the true interactions within a typical high-throughput data set. Another method of quality evaluation has been proposed by Bader and colleagues [4]. Basically, this method exploits the observation that interacting proteins tend to form highly connected clusters within interaction networks; it is therefore possible to asses the quality of a prospective interaction by examining the length of the shortest path that connects the potential interactors.

9.4   ANALYZING PROTEIN INTERACTION NETWORKS

9.4.1   Network Topology and Functional Implications

The topological properties of protein interaction networks have been intensively studied since the first large-scale data sets were published. Interaction networks, as other biological networks, have been shown to be the so-called "small-world" networks meaning that any two vertices can be connected with a path of a few links only (see Chapter 3) [6]. This short path length indicates that local perturbations (e.g., regulation of the biological activity of a given protein) could reach the whole network very quickly. Another topological term frequently attributed to interaction networks is "scale-free" [5]. Scale-free networks are characterized by a few highly connected vertices (hubs) and many less connected peripheral vertices. The distribution of the vertex degree k follows a power law $\left(P(k) \sim k^{-\gamma} \right)$. The scale-free nature explains several properties of protein interaction networks. For example, hubs have a high probability of representing essential and evolutionarily conserved proteins playing a central role in the dynamic organization of systems-level cellular properties. In turn, the apparent scale-free topology is linked to the robustness of interaction networks, being largely insensitive to random removal of single vertices but particularly sensitive to targeted removal of hubs [27,35]. Indeed, Jeong and coworkers [35] could show that random mutations in the yeast genome do not appear to affect the overall topology of the network. By contrast, when the most connected proteins are computationally eliminated, the network diameter increases rapidly (i.e., the average path length). Although proteins with five or fewer links constitute approximately 93% of the total number of proteins in the data set of Jeong et al., they found that only approximately 21% of them are essential. By contrast, only some 0.7% of the yeast proteins with known phenotypic profiles have more than 15 links (i.e., they have a high degree centrality; see also Chapter 4), but single deletion of 62% of these prove lethal. This implies that highly connected proteins with a central role in the network's architecture are three times more likely to be essential than proteins with only a small number of links to other proteins. This phenomenon has been observed in protein interaction networks of yeast, *C. elegans* and *D. melanogaster* [26] and is commonly referred to as the *centrality–lethality rule*. This rule is widely believed to reflect the significance of network architecture in determining network function. However, this view has recently been challenged by He and Zhang [29]. In this work, the authors proposed a small fraction of randomly distributed essential protein interactions, each of which is lethal to an organism when disrupted. Under this scenario, a hub is more likely to be essential than a nonhub simply because the hub has more interactions and thus a higher chance to engage in an essential interaction [29]. Hence, the centrality–lethality rule would be explained without involvement of the network architecture.

9.4.2   Functional Modules in Protein Interaction Networks

As interaction networks become increasingly large and complex, there is a growing need to break them down into more manageable subnetworks or "modules" [28].

The module, which is loosely defined as a cohort of proteins that perform the same cellular task, should be dedicated largely by the topology of the network itself and thus spatially separable from the whole system. Functional modules are useful for annotating uncharacterized proteins, for studying the evolution of interacting systems, and for getting a general overview of the immediate functional partners of a protein. The existence of such modules has been proven experimentally, and several modules like the ribosome, the replication complex, glycolysis, or the mitotic spindle apparatus have been successfully reconstituted *in vitro*, while others have been confirmed *in vivo* [28]. As many of these modules cannot be predicted merely on their constituent protein's annotation, several groups have developed algorithms to identify functional clusters in protein interaction networks. For example, Spirin and Mirny [59] developed an algorithm that was able to recover many previously known protein complexes (e.g., the anaphase-promoting complex) and functional modules (e.g., the yeast pheromone-response pathway). In addition, new complexes and new members of known complexes were identified and thus these methods can provide information about single proteins and their biological context. Another approach to predict functional modules in complex interaction networks relies on mathematical tools for network decomposition. For example, k-core decomposition, a method that is based on the recursive removal of the least connected vertices from a given network, was used to detect a novel nucleolar network in yeast [3]. In protein interaction networks, fully connected subgraphs, that is, motifs (see also Chapter 5) with every vertex linked to every other vertex, the so-called *cliques*, have been found to have a high functional significance [59,74]. The simplest higher order structural element in protein networks is the triangle-that is statistically overrepresented in interaction networks and thus might represent higher order hubs [74]. Given the high biological significance of triangular structures in protein interaction networks, an even stronger functional link of the members of a completely connected clique can be assumed. Indeed, most of the cliques found in the genome-wide yeast interaction network constitute central subunits of known protein complexes. For example, proteins found in fully connected pentagons, a motif highly overrepresented in the yeast protein interaction network, contained components of the yeast proteasome complexes RPN, PSA, PSB, and PRS [72]. However, in order not to remain merely hypothetical, it is important to combine conclusions from graph-theoretical analyses of molecular networks with experimental validation of the predictions by focused or systematic perturbation analysis.

9.4.3 Evolution of Protein Interaction Networks

Gene duplication and subsequent functional divergence are the fundamental mechanisms of the evolution of genomes and complexity in general. One copy of the gene can maintain its essential function, while the other copy is free to mutate and evolve (reviewed in Ref. [63]). It has been proposed that the evolution of complex molecular networks is based on a similar mechanism [69]. A two-step model involving gene duplication and rewiring of interactions has been proposed to explain the evolution of protein interaction networks and their specific topology. Duplication and divergence shape the network architecture on two different levels, on the level of single vertices and edges as well as on the higher order level of network motifs and modules [47]. On the contrary, it has been suggested that proteins involved in multiple interactions are more conserved than proteins with a smaller number of interaction partners [19]. However, another study showed that only a small fraction of proteins with the largest number of interactions, that is, hubs, tend to evolve slower than the bulk of the proteins [36]. Thus, the correlation found by Fraser and coworkers maybe an artifact caused by a small subset of proteins rather than a general phenomenon [19].

9.4.4 Comparative Interactomics

The accumulation of comprehensive high-throughput protein interaction data from yeast, and now fly and worm, combined with the paucity of information from other organisms, has led to the question of how protein interaction networks compare between organisms and to what extent interactions and subnetworks in one organism are conserved in another. In order to detect such homologous interactions and pathways, Kelley and coworkers [38] developed the program PathBlast (http://www.pathblast.org), which aligns two protein interaction networks combining interaction topology and sequence similarity. Following this approach, it was possible to show that the protein interaction networks of yeast and *Helicobacter pylori* harbor a significant number of evolutionarily conserved pathways. One spectacular example among the conserved subnetworks is a group of proteins involved in membrane transport in bacteria and in nuclear-cytoplasmic shuttling in yeast [38]. This finding indicates that nuclear-cytoplasmic transport in eukaryotes may have originated from a homologous system in bacterial plasma membranes. Following a similar approach Wojcik and colleagues predicted a protein interaction map for *E. coli* from *H. pylori* interaction data [71]. Matthews and coworkers [44] performed systematic sequence similarity searches for pairs of potential of known interacting proteins in yeast to identify potentially conserved interactions, dubbed "interlogs," in *C. elegans*. Starting from a large number of published yeast two-hybrid interactions between yeast proteins, searches for candidate interlogs identified networks of potential physical interaction among *C. elegans* proteins. At least 16% of the predicted protein interactions in these networks could be confirmed by yeast two-hybrid analysis, suggesting that these interactions are indeed confirmed [44]. The significant conservation of interactions confirms that a feasible strategy for network reconstruction is to transfer interactions from organisms in which they have been measured.

9.5 SUMMARY

One of the major challenges in the postgenome era is to determine how the complement of expressed cellular proteins—the proteome—is organized into functional, higher order networks. Interaction proteomics, the systematic identification of protein interactions within an organism, promises to facilitate systems-level studies of biological processes. Two powerful approaches have become popular in the study of protein–protein interactions. The first is the yeast two-hybrid system, a yeast-based assay that detects binary protein–protein interactions on a genetic basis, and the second is the purification of protein complexes coupled to mass spectrometry. The latter approach allows investigating higher order protein complexes in their physiological environment. Both methods have been used to generate comprehensive high-throughput protein interaction data sets from a number of model organisms. These large data sets harbor information that is not immediately obvious without integrative analysis. Although integration and analysis have traditionally been carried out by humans, the sheer amount of data now calls for bioinformatics approaches. Graph layout is now extensively used for protein interaction network visualization and analysis. Bioinformatics analysis implicates that protein interaction networks adopt a scale-free topology that explains their error tolerance or vulnerability, depending on whether hubs of peripheral proteins are attacked. It is becoming increasingly apparent that the intermediate and local levels of network organization—the modules, motifs and cliques—play important roles as the operational units of biological function. Networks also allow the prediction of protein function form their interaction partners and therefore the formulation of analytical hypotheses. Finally, comparative network analysis predicts interactions for distantly related species based on conserved interactions, even if their genome sequences are only weakly conserved.

9.6   EXERCISES

1. Name three limitations of the yeast two-hybrid system that are circumvented by using complementary approaches to identify protein–protein interactions such as AP-MS.

2. In this exercise you will be comparing protein interaction data for a small family of proteins called glutaredoxins.

   (a) Go to the Saccharomyces Genome Database (SGD) at http://www. yeastgenome.org and identify the five glutaredoxin genes found in this organism. List the gene names, all aliases, the systematic names, and a brief description for each gene.

   (b) Identify the interactions for each glutaredoxin identified by Ito and coworkers. Go to their Web site at http://itolab.cb.k.u-tokyo.ac.jp/Y2H/ and download the full data file. From this file, draw an interaction map that illustrates all interactions identified for each glutaredoxin in this experiment. Identify name or function of each gene product.

   (c) Identify the interactions for each glutaredoxin identified by Uetz and coworkers. Go to their Web site at http://portal.curagen.com/cgi-bin/com.curagen.portal.servlet.PortalYeastList. Try to identify the interactions and draw an interaction map for each glutaredoxin in this experiment. Identify name or function for each gene product.

(d) Compare the interactions that you identified from the original data sets. Where are overlaps or differences? Discuss the reason for possible discrepancies.

(e) For this exercise you will take preliminary look at the Cytoscape network visualization tool. To download Cytoscape on your computer go to http://www.cytoscape.org and follow the instructions for installation. To download the Cytoscape-formatted interaction data for Grx2, go to SGD and click the "Batch download" link on the left side of the page. Enter Grx2 under "Option 1," select "physical interactions," and then submit. Follow the instructions for downloading the file. Start Cytoscape, in the "File-load-Network" menu, enter the file name of the interaction file that you downloaded from SGD. Under the "Layout" Menu, select "Apply Spring Embedded Layout-all nodes." Find Grx2 in this network. To which nodes is it connected? What is this protein and what is its function. Compare what you see in the Cytoscape network to what you have identified from the Ito and Uetz data, respectively. What is the most highly connected node in the Cytoscape network? Using the functional description for the protein from SGD, provide a biological explanation for this highly connected node.

REFERENCES

1. G. D. Bader, D. Betel, and C. W. V. Hogue. Bind: The biomolecular interaction network database. *Nucleic Acids Research*, 31:248–250, 2003.
2. G. D. Bader, I. Donaldson, C. Wolting, B. F. F. Ouellette, T. Pawson, and C. W. V. Hogue. Bind—the biomolecular interaction network database. *Nucleic Acids Research*, 29:242–245, 2001.
3. G. D. Bader and C. W. Hogue. Analyzing yeast protein–protein interaction data obtained from different sources. *Nature Biotechnology*, 20:991–997, 2002.
4. J. S. Bader, A. Chaudhuri, J. M. Rothberg, and J. Chant. Gaining confidence in high-throughput protein interaction networks. *Nature Biotechnology*, 22:78–85, 2004.
5. A.-L. Barabási and R. Albert. Emergence of scaling in random networks. *Science*, 286:509–512, 1999.
6. A.-L. Barabási and Z. N. Oltvai. Network biology: Understanding the cell's functional organization. *Nature Reviews in Genetics*, 5:101–113, 2004.
7. P. L. Bartel, J. A. Roecklein, D. SenGupta, and S. Fields. A protein linkage map of *Escherichia coli* bacteriophage t7. *Nature Genetics*, 12:72–77, 1996.
8. J. R. Bock and D. A. Gough. Predicting protein–protein interactions from primary structure. *Bioinformatics*, 17:455–460, 2001.
9. F. Börnke. *Biosensors and Bioassays Based on Microorganisms*, chapter Microbial systems to assay protein–protein interactions in vivo: The yeast two-hybrid system and beyond, pages 229–253. Research Signpost, Kerala, 2006.

10. T. Bürckstümmer, K. L. Bennett, A. Preradovic, G. Schutze, O. Hantschel, G. Superti-Furga, and A. Bauch. An efficient tandem affinity purification procedure for interaction proteomics in mammalian cells. *Nature Methods*, 3:1013–1019, 2006.

11. G. Butland, J. M. Peregrin-Alvarez, J. Li, W. Yang, X. Yang, V. Canadien, A. Starostine, D. Richards, B. Beattie, N. Krogan et al. Interaction network containing conserved and essential protein complexes in *Escherichia coli*. *Nature*, 433:531–537, 2005.

12. B. Causier and B. Davies. Analysing protein–protein interactions with the yeast two-hybrid system. *Plant Molecular Biology*, 50:855–870, 2002.

13. C. M. Deane, L. Salwinski, I. Xenarios, and D. Eisenberg. Protein interactions: Two methods for assessment of the reliability of high throughput observations. *Molecular and Cellular Proteomics*, 1:349–356, 2002.

14. A. J. Enright and C. A. Ouzounis. Biolayout - an automatic graph layout algorithm for similarity visualization. *Bioinformatics*, 17:853–854, 2001.

15. A. J. Enright and C. A. Ouzounis. *Protein–Protein Interactions: A Laboratory Manual*, Analysis of genome-wide protein interactions using computational approaches, pp. 595–621. Cold Spring Harbor Laboratory Press, Cold Spring Harbor, New York, 2002.

16. A. J. Enright, L. Skrabanek, and G. D. Bader. Computational prediction of protein–protein interactions. *The Proteomics Protocols Handbook*, pp. 629–652. Humana Press, Totowa, NJ, 2005.

17. S. Fields and O. K. Song. A novel genetic system to detect protein–protein interactions. *Nature*, 340:245–246, 1989.

18. T. B. Fischer, M. Paczkowski, M. F. Zettel, and J. Tsai. A guide to protein interaction databases. *The Proteomics Protocols Handbook*, pp. 753–799. Humana Press, Totowa, NJ, 2005.

19. H. B. Fraser, A. E. Hirsh, L. M. Steinmetz, C. Scharfe, and M. W. Feldman. Evolutionary rate in the protein interaction network. *Science*, 296:750–752, 2002.

20. C. Friedman, P. Kra, H. Yu, M. Krauthammer, and A. Rzhetsky. Genies: A natural-language processing system for the extraction of molecular pathways from journal articles. *Bioinformatics*, 17:S74–82, 2001.

21. M. Fromont-Racine, J. C. Rain, and P. Legrain. Toward a functional analysis of the yeast genome through exhaustive two-hybrid screens. *Nature Genetics*, 16:277–282, 1997.

22. A. C. Gavin, M. Bosche, R. Krause, P. Grandi, M. Marzioch, A. Bauer, J. Schultz, J. M. Rick, A. M. Michon, C. M. Cruciat, M. Remor, C. Hofert, M. Schelder, M. Brajenovic, H. Ruffner, A. Merino, K. Klein, M. Hudak, D. Dickson, T. Rudi, V. Gnau, A. Bauch, S. Bastuck, B. Huhse, C. Leutwein, M. A. Heurtier, R. R. Copley, A. Edelmann, E. Querfurth, V. Rybin, G. Drewes, M. Raida, T. Bouwmeester, P. Bork, B. Seraphin, B. Kuster, G. Neubauer, and G Superti-Furga. Functional organization of the yeast proteome by systematic analysis of protein complexes. *Nature*, 415:141–147, 2002.

23. H. Ge, A. J. Walhout, and M. Vidal. Integrating ómicinformation: A bridge between genomics and systems biology. *Trends in Genetics*, 19:551–560, 2003.

24. L. Giot, J. S. Bader, C. Brouwer, A. Chaudhuri, B. Kuang, Y. Li, Y. L. Hao, C. E. Ooi, B. Godwin, E. Vitols, G. Vijayadamodar, P. Pochart, H. Machineni, M. Welsh, Y. Kong, B. Zerhusen, R. Malcolm, Z. Varrone, A. Collis, M. Minto, S. Burgess, L. McDaniel, E. Stimpson, F. Spriggs, J. Williams, K. Neurath, N. Ioime, M. Agee, E. Voss, K. Furtak, R. Renzulli, N. Aanensen, S. Carrolla, E. Bickelhaupt, Y. Lazovatsky, A. DaSilva, J. Zhong, C. A. Stanyon, Jr. R. L. Finley, K. P. White, M. Braverman, T. Jarvie, S. Gold,

M. Leach, J. Knight, R. A. Shimkets, M. P. McKenna, J. Chant, and J. M. Rothberg. A protein interaction map of *Drosophila melanogaster*. *Science*, 302:1727–1736, 2003.

25. C. S. Goh and F. E. Cohen. Co-evolutionary analysis reveals insights into protein–protein interactions. *Journal of Molecular Biology*, 324:177–192, 2002.

26. M. W. Hahn and A. D. Kern. Comparative genomics of centrality and essentiality in three eukaryotic protein-interaction networks. *Molecular Biology and Evolution*, 22:803–806, 2005.

27. J. D. Han, N. Bertin, T. Hao, D. S. Goldberg, G. F. Berriz, L. V. Zhang, D. Dupuy, A. J. Walhout, M. E. Cusick, F. P. Roth, and M. Vidal. Evidence for dynamically organized modularity in the yeast protein–protein interaction network. *Nature*, 430:88–93, 2004.

28. L. H. Hartwell, J. J. Hopfield, S. Leibler, and A. W. Murray. From molecular to modular cell biology. *Nature*, 402:C47–52, 1999.

29. X. He and J. Zhang. Why do hubs tend to be essential in protein networks? *PLoS Genetics*, 2:e88, 2006.

30. Y. Ho, A. Gruhler, A. Heilbut, G. D. Bader, L. Moore, S. L. Adams, A. Millar, P. Taylor, K. Bennett, K. Boutilier, L. Yang, C. Wolting, I. Donaldson, S. Schandorff, J. Shew-narane, M. Vo, J. Taggart, M. Goudreault, B. Muskat, C. Alfarano, D. Dewar, Z. Lin, K. Michalickova, A. R. Willems, H. Sassi, P. A. Nielsen, K. J. Rasmussen, J. R. Ander-sen, L. E. Johansen, L. H. Hansen, H. Jespersen, A. Podtelejnikov, E. Nielsen, J. Crawford, V. Poulsen, B. D. Sorensen, J. Matthiesen, R. C. Hendrickson, F. Gleeson, T. Pawson, M. F. Moran, D. Durocher, M. Mann, C. W. Hogue, D. Figeys, and M. Tyers. Systematic identification of protein complexes in *Saccharomyces cerevisiae* by mass spectrometry. *Nature*, 415:180–183, 2002.

31. J. C. Mellor, I. Yanai, and C. DeLisi. Identifying functional links between genes using conserved chromosomal proximity. *Trends in Genetics*, 18:176–179, 2002.

32. I. Iliopoulos, A. J. Enright, and C. A. Ouzounis. Textquest: Document clustering of med-line abstracts for concept discovery in molecular biology. *Pacific Symposium on Biocom-puting*, pp. 384–395, 2001.

33. T. Ito, T. Chiba, R. Ozawa, M. Yoshida, M. Hattori, and Y. Sakaki. A comprehensive two-hybrid analysis to explore the yeast protein interactome. *Proceedings of the National Academy of Sciences USA*, 98:4569–4574, 2001.

34. T. Ito, K. Tashiro, S. Muta, R. Ozawa, T. Chiba, M. Nishizawa, K. Yamamoto, S. Kuhara, and Y. Sakaki. Toward a protein–protein interaction map of the budding yeast: A compre-hensive system to examine two-hybrid interactions in all possible combinations between the yeast proteins. *Proceedings of the National Academy of Sciences USA*, 97:1143–1147, 2000.

35. H. Jeong, S. P. Mason, A.-L. Barabási, and Z. N. Oltvai. Lethality and centrality in protein networks. *Nature*, 411:41–42, 2001.

36. I. K. Jordan, Y. Wolf, and E. Koonin. No simple dependence between protein evolution rate and the number of protein–protein interactions: Only the most prolific interactors tend to evolve slowly. *BMC Evolutionary Biology*, 3:e1, 2003.

37. W. G. Kaelin, D. C. Pallas, J. A. Decaprio, F. J. Kaye, and D. M. Livingston. Identification of cellular proteins that can interact specifically with the t/e1a-binding region of the retinoblastoma gene-product. *Cell*, 64:521–532, 1991.

38. B. P. Kelley, R. Sharan, R. M. Karp, T. Sittler, D. E. Root, B. R. Stockwell, and T. Ideker. Conserved pathways within bacteria and yeast as revealed by global protein network

alignment. *Proceedings of the National Academy of Sciences USA*, 100:11394–11399, 2003.

39. N. J. Krogan, G. Cagney, H. Yu, G. Zhong, X. Guo, A. Ignatchenko, J. Li, S. Pu, N. Datta, A. P. Tikuisis, T. Punna, J. M. Peregrin-Alvarez, M. Shales, X. Zhang, M. Davey, M. D. Robinson, A. Paccanaro, J. E. Bray, A. Sheung, B. Beattie, D. P. Richards, V. Canadien, A. Lalev, F. Mena, P. Wong, A. Starostine, M. M. Canete, J. Vlasblom, S. Wu, C. Orsi, S. R. Collins, S. Chandran, R. Haw, J. J. Rilstone, K. Gandi, N. J. Thompson, G. Musso, P. St Onge, S. Ghanny, M. H. Lam, G. Butland, A. M. Altaf-Ul, S. Kanaya, A. Shilatifard, E. O'Shea, J. S. Weissman, C. J. Ingles, T. R. Hughes, J. Parkinson, M. Gerstein, S. J. Wodak, A. Emili, and J. F. Greenblatt. Global landscape of protein complexes in the yeast *Saccharomyces cerevisiae*. *Nature*, 440:637–643, 2006.

40. L. A. Kung and M. Snyder. Proteome chips for whole-organism assays. *Nature Reviews Molecular Cell Biology*, 7:617–622, 2006.

41. S. M. Li, C. M. Armstrong, N. Bertin, H. Ge, S. Milstein, M. Boxem, P. O. Vidalain, J. D. J. Han, A. Chesneau, T. Hao, D. S. Goldberg, N. Li, M. Martinez, J. F. Rual, P. Lamesch, L. Xu, M. Tewari, S. L. Wong, L. V. Zhang, G. F. Berriz, L. Jacotot, P. Vaglio, J. Reboul, T. Hirozane-Kishikawa, Q. Li, H. W. Gabel, A. Elewa, B. Baumgartner, D. J. Rose, H. Yu, S. Bosak, R. Sequerra, A. Fraser, S. E. Mango, W. M. Saxton, S. Strome, S. van den Heuvel, F. Piano, J. Vandenhaute, C. Sardet, M. Gerstein, L. Doucette-Stamm, K. C. Gunsalus, J. W. Harper, M. E. Cusick, F. P. Roth, D. E. Hill, and M. Vidal. A map of the interactome network of the metazoan *C. elegans*. *Science*, 303:540–543, 2004.

42. E. M. Marcotte, M. Pellegrini, H. L. Ng, D. W. Rice, T. O. Yeates, and D. Eisenberg. Detecting protein function and protein–protein interactions from genome sequences. *Science*, 285:751–753, 1999.

43. E. M. Marcotte, I. Xenarios, and D. Eisenberg. Mining literature for protein–protein interactions. *Bioinformatics*, 17:359–363, 2001.

44. L. R. Matthews, P. Vaglio, J. Reboul, H. Ge, B. P. Davis, J. Garrels, S. Vincent, and M. Vidal. Identification of potential interaction networks using sequence-based searches for conserved protein–protein interactions or "Interologs." *Genome Research*, 11:2120–2126, 2001.

45. S. McCraith, T. Holtzman, B. Moss, and S. Fields. Genome-wide analysis of vaccinia virus protein–protein interactions. *Proceedings of the National Academy of Sciences USA*, 97:4879–4884, 2000.

46. F. Pazos and A. Valencia. *In silico* two-hybrid system for the selection of physically interacting protein pairs. *Proteins*, 47:219–227, 2002.

47. J. B. Pereira-Leal and S. A. Teichmann. Novel specificities emerge by stepwise duplication of functional modules. *Genome Research*, 15:552–559, 2005.

48. E. M. Phizicky and S. Fields. Protein–protein interactions—methods for detection and analysis. *Microbiological Reviews*, 59:94–123, 1995.

49. O. Puig, F. Caspary, G. Rigaut, B. Rutz, E. Bouveret, E. Bragado-Nilsson, M. Wilm, and B. Seraphin. The tandem affinity purification (tap) method: A general procedure of protein complex purification. *Methods*, 24:218–229, 2001.

50. Jr. R. L. Finley and R. Bren. Interaction mating reveals binary and ternary connections between drosophila cell cycle regulators. *Proceedings of the National Academy of Sciences USA*, 91:12980–12984, 1994.

51. G. Rigaut, A. Shevchenko, B. Rutz, M. Wilm, M. Mann, and B. Seraphin. A generic protein purification method for protein complex characterization and proteome exploration. *Nature Biotechnology*, 17:1030–1032, 1999.

52. J. S. Rohila, M. Chen, R. Cerny, and M. E. Fromm. Improved tandem affinity purification tag and methods for isolation of protein heterocomplexes from plants. *Plant Journal*, 38:172–181, 2004.

53. J. S. Rohila, M. Chen, S. Chen, J. Chen, R. Cerny, C. Dardick, P. Canlas, X. Xu, M. Gribskov, S. Kanrar, J. K. Zhu, P. Ronald, and M. E. Fromm. Protein–protein interactions of tandem affinity purification-tagged protein kinases in rice. *Plant Journal*, 46:1–13, 2006.

54. C. Rohl, Y. Price, T. B. Fischer, M. Paczkowski, M. F. Zettel, and J. Tsai. Cataloging the relationships between proteins: A review of interaction databases. *Molecular Biotechnology*, 34:69–93, 2006.

55. V. Rubio, Y. Shen, Y. Saijo, Y. Liu, G. Gusmaroli, S. P. Dinesh-Kumar, and X. W. Deng. An alternative tandem affinity purification strategy applied to *Arabidopsis* protein complex isolation. *Plant Journal*, 41:767–778, 2005.

56. L. Salwinski and D. Eisenberg. Computational methods of analysis of protein–protein interactions. *Current Opinion in Structural Biology*, 13:377–382, 2003.

57. L. Salwinski, C. S. Miller, A. J. Smith, F. K. Pettit, J. U. Bowie, and D. Eisenberg. The database of interacting proteins: 2004 update. *Nucleic Acids Research*, 32:D449–451, 2004.

58. P. Shannon, A. Markiel, O. Ozier, N. S. Baliga, J. T. Wang, D. Ramage, N. Amin, B. Schwikowski, and T. Ideker. Cytoscape: A software environment for integrated models of biomolecular interaction networks. *Genome Research*, 13:2498–2504, 2003.

59. V. Spirin and L. A. Mirny. Protein complexes and functional modules in molecular networks. *Proceedings of the National Academy of Sciences USA*, 100:12123–12128, 2003.

60. E. Sprinzak, S. Sattath, and H. Margalit. How reliable are experimental protein–protein interaction data? *Journal of Molecular Biology*, 327:919–923, 2003.

61. P. A. Srere. Complexes of sequential metabolic enzymes. *Annual Review of Biochemistry*, 56:89–124, 1987.

62. U. Stelzl, U. Worm, M. Lalowski, C. Haenig, F. H. Brembeck, H. Goehler, M. Stroedicke, M. Zenkner, A. Schoenherr, S. Koeppen, J. Timm, S. Mintzlaff, C. Abraham, N. Bock, S. Kietzmann, A. Goedde, E. Toksoz, A. Droege, S. Krobitsch, B. Korn, W. Birchmeier, H. Lehrach, and E. E. Wanker. A human protein–protein interaction network: A resource for annotating the proteome. *Cell*, 122:957–968, 2005.

63. J. S. Taylor and J. Raes. Duplication and divergence: The evolution of new genes and old ideas. *Annual Review of Genetics*, 38:615–643, 2004.

64. K. Terpe. Overview of tag protein fusions: From molecular and biochemical fundamentals to commercial systems. *Applied Microbiology and Biotechnology*, 60:523–533, 2003.

65. A. H. Tong, M. Evangelista, A. B. Parsons, H. Xu, G. D. Bader, N. Page, M. Robinson, S. Raghibizadeh, C. W. Hogue, H. Bussey, B. Andrews, M. Tyers, and C. Boone. Systematic genetic analysis with ordered arrays of yeast deletion mutants. *Science*, 294:2364–2368, 2001.

66. P. Uetz, L. Giot, G. Cagney, T. A. Mansfield, R. S. Judson, J. R. Knight, D. Lockshon, V. Narayan, M. Srinivasan, P. Pochart, A. Qureshi-Emili, Y. Li, B. Godwin, D. Conover, T. Kalbfleisch, G. Vijayadamodar, M. Yang, M. Johnston, S. Fields, and J. M. Rothberg.

A comprehensive analysis of protein–protein interactions in *Saccharomyces cerevisiae*. *Nature*, 403:623–627, 2000.

67. A. Vitale. Physical methos. *Plant Molecular Biology*, 50:825–836, 2002.

68. C. von Mering, R. Krause, B. Snel, M. Cornell, S. G. Oliver, S. Fields, and P. Bork. Comparative assessment of large-scale data sets of protein–protein interactions. *Nature*, 417:309–403, 2002.

69. A. Wagner. How the global structure of protein interaction networks evolves. *Proceedings of the Biological Sciences of the Royal Society*, 270:457–466, 2003.

70. A. J. M. Walhout, R. Sordella, X. W. Lu, J. L. Hartley, G. F. Temple, M. A. Brasch, N. Thierry-Mieg, and M. Vidal. Protein interaction mapping in *C. elegans* using proteins involved in vulval development. *Science*, 287:116–122, 2000.

71. J. Wojcik, I. G. Boneca, and P. Legrain. Prediction, assessment and validation of protein interaction maps in bacteria. *Journal of Molecular Biology*, 323:763–770, 2002.

72. S. Wuchty, Z. N. Oltvai, and A.-L. Barabási. Evolutionary conservation of motif constituents in the yeast protein interaction network. *Nature Genetics*, 35:176–179, 2003.

73. I. Xenarios, L. Salwinski, X. J. Duan, P. Higney, S.-M. Kim, and D. Eisenberg. Dip, the database of interacting proteins: A research tool for studying cellular networks of protein interactions. *Nucleic Acids Research*, 30:303–305, 2002.

74. E. Yeger-Lotem, S. Sattath, N. Kashtan, S. Itzkovitz, R. Milo, R. Y. Pinter, U. Alon, and H. Margalit. Network motifs in integrated cellular networks of transcription-regulation and protein–protein interaction. *Proceedings of the National Academy of Sciences USA*, 101:5934–5939, 2004.

75. A. Zanzoni, L. Montecchi-Palazzi, M. Quondam, G. Ausiello, M. Helmer-Citterich, and G. Cesareni. Mint: A molecular interaction database. *FEBS Letters*, 513:135–140, 2002.

76. H. Zhu, M. Bilgin, R. Bangham, D. Hall, A. Casamayor, P. Bertone, N. Lan, R. Jansen, S. Bidlingmaier, T. Houfek, T. Mitchell, P. Miller, R. A. Dean, M. Gerstein, and M. Snyder. Global analysis of protein activities using proteome chips. *Science*, 293:2101–2105, 2001.

10

METABOLIC NETWORKS

Márcio Rosa da Silva, Jibin Sun, Hongwu Ma,
Feng He, and An-Ping Zeng

10.1 INTRODUCTION

Metabolic network refers to the network composed of metabolites and their interconversions (biochemical reactions) in an organism. Metabolites are usually small molecules such as glucose and amino acids but can also be macromolecules such as polysaccharides and glycan. The interconversion is usually catalyzed by enzymes (proteins). Only a few reactions in the cell are spontaneous and thus nonenzymatic. Metabolic pathway, an important concept in biochemistry, is a series of successive or tightly associated biochemical reactions for a specific metabolic function, for instance, glycolysis, lysine degradation, and penicillin biosynthesis. A metabolic pathway can be considered as a small local area of a metabolic network, whereas a metabolic network gives a better and more complete view of the cellular metabolism.

A complete metabolic network should show all the possible modes of material flows in the cell, therefore indicating all the metabolic potential and capacity of the cell. In other words, metabolic network is the material processing center for a functioning cell. The cell relies on this network to uptake and digest substrates from the environment, to generate energy (e.g., in form of ATP) and to synthesize components that are necessary for its growth and survival.

It is of great interest to study metabolic network for its fundamental importance in biology in general and in particular because many applications are directly built on the use of cellular metabolism. Biotechnologists modify the cells and use them as

Analysis of Biological Networks, Edited by Björn H. Junker and Falk Schreiber
Copyright © 2008 John Wiley & Sons, Inc.

cellular factories to produce bulky or fine chemicals, antibiotics, industrial enzymes, antibodies, and so on. In biomedicine, people cure metabolic diseases of human beings through better understanding the metabolic mechanism, and control infections of pathogens by making use of the metabolic differences between human beings and pathogens.

The sequencing of genome and Bioinformatics and functional genomic studies of sequences make it now possible to reconstruct the metabolic network of a specific organism at genome level and thus open up new horizons in many areas of biotechnology and life science. In this chapter, we briefly illustrate some methods and concepts used for the reconstruction, structural, and functional analysis of genome-based metabolic network.

10.2 VISUALIZATION AND GRAPH REPRESENTATION

A metabolic network consists of metabolites that are converted into others through biochemical reactions catalyzed by enzymes. Thus, in a general approach of visualization and analysis, it can be represented as a graph with two different kinds of vertices (metabolite and enzyme) in an alternating order, which is typical for bipartite graphs (as explained in Chapter 2). For simplicity and for a more straightforward analysis, metabolic networks can be represented as a simple direct graph with vertices representing metabolites and edges corresponding to reactions converting one metabolite into the other [18,24]. As a complement, a reaction graph can also be used [25]. In the reaction graph, the vertices represent reactions and there will be an edge between two vertices if the product of one reaction is the substrate of the other.

This simplification is helpful for structural analysis as many graph algorithms do not consider different types of vertices as in a bipartite graph. The drawback of this representation is that if the currency metabolites [24] are not removed, biologically meaningless shortcuts may be introduced in the network as explained in the following section. Figure 10.1 shows (using Cytoscape [43]) the metabolic network of *E. coli* represented as a simple graph. In this network, vertices represent metabolites and edges between two metabolites exist if one metabolite can be converted into the other. For a more complete overview of graph representation of networks see Chapter 2.

10.3 RECONSTRUCTION OF GENOME-SCALE METABOLIC NETWORKS

Piece by piece, information on biochemical reactions and metabolic pathways were accumulated during the long history of biochemistry study. The human knowledge on metabolism was condensed on the famous Boehringer Mannheim wall chart of metabolic pathways [3]. However, until the emergence of the first complete bacterial genome in 1995, it is impossible to access the complete and species-specific

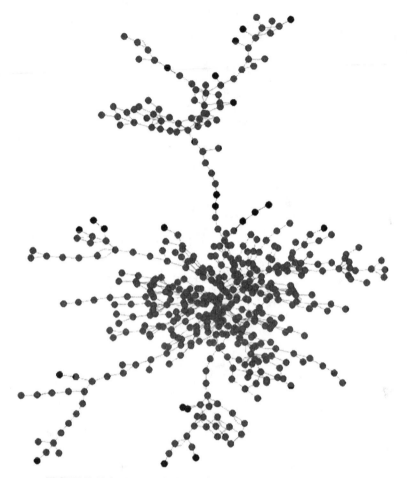

FIGURE 10.1   Metabolic network from *E. coli* shown as a graph.

metabolic network. Since then, many genome were or are being sequenced in a fascinating speed. According to NCBI (Status April, 2006), 348 genomes are completed, including human-being itself and many clinically or industrially important organisms, while another 844 genome sequencing projects are ongoing. An era to understand the organisms at a systems-level comes. Many of these networks are available online: Kyoto Encyclopedia of Genes and Genomes [4], EcoCyc and BioCyc [1]. EcoCyc is an extensively human-curated database specialized for Escherichia coli K-12 whereas KEGG and BioCyc maintain databases for many organisms but with little human curation effort.

In this section, a simple algorithm for metabolic network reconstruction will be introduced. The reader are encouraged to read literature [11,24,46] to access more detailed information.

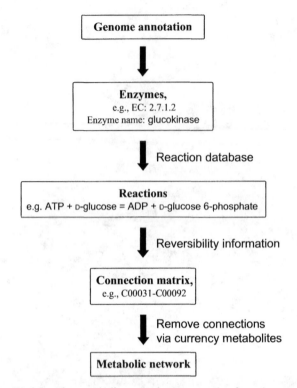

FIGURE 10.2   Workflow for metabolic network reconstruction from the genome annotation.

Figure 10.2 shows the workflow to reconstruct the metabolic network from genome annotation. For a given genome with annotations, the EC numbers (Enzyme Commission numbers, a number in the form x.x.x.x representing the biochemical reaction catalyzed by the corresponding enzyme) was extracted. If the EC numbers were not assigned in the original annotation, the enzyme name can also be used for the following workflow. The EC numbers or enzyme names were searched in the reaction database such as KEGG LIGAND [20]. If only enzyme names instead EC numbers are available, it is sometimes difficult to find the reactions catalyzed by the enzyme unambiguous since one enzyme often has variant names. In this case, human curation is necessary.

Once the reaction list is created, the reactions are further converted to connection matrix: substrate–product pairs are recoded in a format that can be more easily handled by the computer. For example, in Fig. 10.2, a connection C00031-C00092 was generated from the reaction: ATP + D-glucose = ADP + D-glucose 6-phosphate. Here C00031 and C00092 represent D-glucose and D-glucose 6-phosphate, respectively. The connection matrix file can be directly interpreted by many network visualization tools such as *Cytoscape* [2] for visualization and analysis.

The reversibility information of the reactions should be considered when the connection matrix was built. Unfortunately, the reversibility information is not included in the famous reaction databases like LIGAND. We made efforts to curate the LIGAND database and achieved an improved reaction database in EXCEL format [5,24].

The connection matrix should be further treated by removing connections via currency metabolites such as H_2O, CO_2, ATP, etc [24]. Currency metabolites are the metabolites that are mainly used as carriers for transferring electrons and certain functional groups (hydrogen, phosphate, amino group, one carbon unit, methyl group, etc.). For example, in the reaction Glucose + ATP = G6P + ADP. ADP and ATP are currency metabolites for transferring phosphate to glucose. The currency metabolites are often not shown in the metabolic pathway maps in KEGG.

Currency metabolites are normally used as carriers for transferring electrons and certain functional groups (phosphate group, amino group, one carbon unit, methyl group, etc.). When considering the connections through currency metabolites, structure analysis often produces biologically meaningless results. For example, in the glycolysis pathway, the path length (number of reaction steps in the pathway) from glucose to pyruvate should be nine in terms of biochemistry (see Fig. 10.3). However, if ATP and ADP are considered as vertices in the network there would be only two steps from glucose to pyruvate (the first reaction uses glucose and produces ADP, while the last reaction consumes ADP and produces pyruvate). This is obviously biologically not meaningful. Different approaches have been proposed

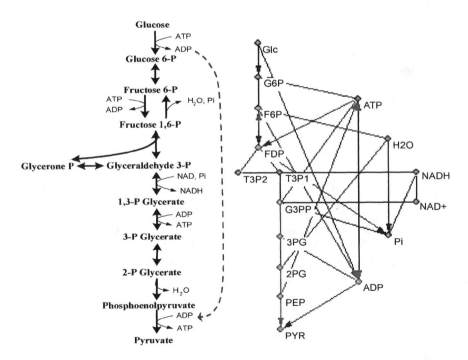

(a) Conventional way of presentation in biochemistry

(b) Connection structure in a graphic representation with current metabolites also as connections

FIGURE 10.3 The glycolysis pathway as a part of metabolic network [24].

to address this problem. A simple way is to exclude the top-ranked metabolites based on their connection degree (number of edges connected with a metabolite). The problem is that certain primary metabolites such as pyruvate may also have high connection degrees. Moreover, currency metabolites cannot be defined *per se* by compounds but should be defined according to the reaction. For example, glutamate (GLU) and 2-oxoglutarate (AKG) are currency metabolites for transferring amino groups in many reactions, but they are primary metabolites in the following reaction:

$$AKG + NH_3 + NADPH = GLU + NADP^+ + H_2O$$

The connections through them should be considered. The same situations are for NADH, NAD$^+$, ATP, etc. Another problem is for the kind of reactions like

$$AcORN + GLU = ORN + AcGLU$$

The acetyl group (Ac) is transferred between GLU and ORN (ornithine) in this reaction. Only the connections AcORN-ORN, GLU-AcGLU should be included, but AcORN-AcGLU and GLU-ORN should be excluded. Otherwise the path length from GLU to ORN will be 1, and this is not in accordance with the real biochemical pathway. Therefore the reactions were manually checked to remove the biologically meaningless connections [24]. As an example, Fig. 10.4 depicts the two graphs (with and without connections through currency metabolites) for the reconstructed metabolic network of *Streptococcus pneumonia*. It can be seen that the one without currency metabolites is more realistic and more amendable for analysis. In contrast, the true network structure in the graph with currency metabolites is masked by the large number of edges through currency metabolites. Therefore, the removal of

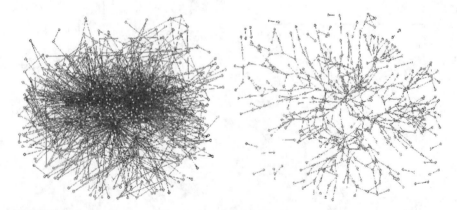

FIGURE 10.4   Metabolite graph representation of metabolic network of *Streptoccus pneumoniae*. The left network includes the connections through currency metabolites and the right one does not. Edges with arrow represent irreversible reactions and those without arrow represent reversible reactions.

connections through currency metabolites is an essential step to draw biologically meaningful conclusions from graph analysis of metabolic networks.

10.4   CONNECTIVITY AND CENTRALITY IN METABOLIC NETWORKS

The study of genome-based metabolic networks has given remarkable new insights into the fundamental aspects of cellular metabolism. One of the important findings is that like many nonbiological complex systems, metabolic networks exhibit typical characteristics of small-world network, namely a power law connection degree distribution [45], high cluster coefficients and a short network diameter [10,18,47]. This small world structure is regarded as one of the design principles of many robust and error-tolerant networks such as the computer network, neural network, and certain social and economic networks [6,45].

Jeong et al. [18] showed that metabolic networks were scale free networks when considering the connections through currency metabolites. Ma and Zeng [24] verified that the structure of metabolic networks still have the characteristics of a scale-free network after deleting the connections through currency metabolites. One important feature of scale-free network is the power law distribution of connection degree among the vertices [45]. In contrast to the scale-free network, a random newtwork has a Poisson distribution of the connection degree. The connection degree is defined as the number of connections linked with each metabolite (vertex). Considering the direction, the number of connections starting from the metabolite is called output degree, and the number of connections ending at the metabolite is called input degree. The output degree distributions of four typical organisms (*Homo sapiens* (eucaryote), *Escherichia coli* (gram negative bacteria), *Bacillus subtilis* (gram positive bacteria), and *Aeropyrum pernix* (archaea) for eukaryotes, proteobacteria, gram positive bacteria and archaea, respectively) are shown in Fig. 10.5. In this figure, $P(k)$ is the fraction of vertices that have a k-degree of outputs. It was calculated by dividing the number of metabolites, which had k output connections with the total number of metabolites in the organism. Except for the first point with $k = 1$, a clear power law distribution (linear relations in the logarithmic scale coordinates) can be ascertained (Fig. 10.5):

$$P(k) = \alpha k^{-\gamma} \tag{10.1}$$

or for the log–log plot:

$$\log P(k) = \log \alpha - \gamma \log k \tag{10.2}$$

where $-\gamma$ is the slope of the linear approximation of the curve.

The input degree distributions for these organisms also have similar power law relations (data not shown). The power law degree distribution exists in networks of all the organisms studied in the work of Ma and Zeng [24]. This indicates that the

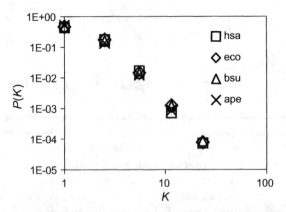

FIGURE 10.5 Output degree distribution in four typical organisms. $P(k)$ is the fraction of vertices that have a k-degree of output connections. The original data have been logarithmically binned according to Huynen and Nimwegen [17]. hsa: *Homo sapiens* (eukaryote), eco: *Escherichia coli* (gram negative bacteria), bsu: *Bacillus subtilis* (gram positive bacteria), ape: *Aeropyrum pernix* (archaea).

metabolic network without connections through currency metabolites is still a scale-free network.

Centrality measurements (see also Chapter 4) can help to identify important vertices in the network. In social network analysis, network activity for a vertex is measured using the concept of degree—the number of direct connections a vertex has.

Degree centrality can show the vertices with highest number of connections in the network. This helps to identify the "hubs" in the network. In a metabolic network, this means the vertices that can be converted into more metabolites.

Betweenness centrality may help to find some important vertices in the network. Vertices with the highest betweenness centrality are the ones with the highest number of shortest pathways going through them. For a metabolic network, this may mean vertices that participate in more metabolites conversions. This is not always true because in many cases the shortest path is not the one used in the reality. One example in Fig. 10.6 shows the biosynthesis of arginine from glutamate. The shortest path from glutamate to arginine is the path **B**. But this pathway is not used to synthesize arginine. The initial part of this path is actually used for the proline synthesis (path **C**). Path **A** is longer than **B**, but it is the metabolic pathway actually used for the biosynthesis of arginine.

With the parameter *closeness centrality*, one can identify vertices in the central part of a network and vertices in the periphery part. Vertices in the central part have a shorter distance to other vertices in the network (they are closer to other vertices, therefore the name "closeness"). In a metabolic network, this means that these metabolites can be converted to others in fewer steps.

Tables 10.1 to 10.3 show the top 10 vertices in terms of different definitions of centrality for the metabolic network from *Bacillus subtilis*.

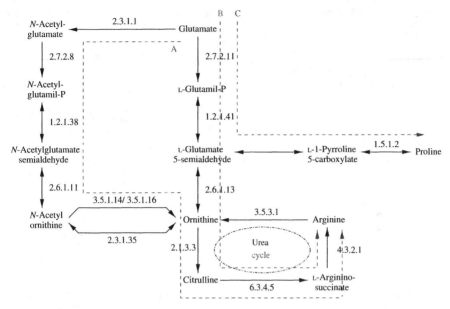

FIGURE 10.6   Conversion from glutamate to arginine. Numbers in reactions are EC numbers.

TABLE 10.1   Ten Metabolites with Highest Betweenness Centrality for _B. Subtilis_ Network

Rank	Betweenness centrality	KEGG[a]	Name
1	0.432132	C00022	Pyruvate
2	0.310035	C00117	D-Ribose 5-phosphate
3	0.297295	C00119	5-Phospho-alpha-D-ribose 1-diphosphate
4	0.282609	C00118	D-Glyceraldehyde 3-phosphate
5	0.278441	C03090	5-Phosphoribosylamine
6	0.275247	C03838	5'-Phosphoribosylglycinamide
7	0.272031	C04376	5'-Phosphoribosyl-N-formylglycinamide
8	0.268792	C04640	2-(Formamido)-N1-(5'-phosphoribosyl) acetamidine
9	0.265530	C03373	Aminoimidazole ribotide
10	0.262246	C04751	1-(5-Phospho-D-ribosyl)-5-amino-4-imidazolecarboxylate

[a] Compound index number from KEGG [19].

TABLE 10.2   Ten Metabolites with Highest Closeness Centrality for *B. Subtilis* Network

Rank	Closeness centrality	KEGG[a]	Name
1	0.124260	C00022	Pyruvate
2	0.122129	C04442	2-Dehydro-3-deoxy-6-phospho-D-gluconate
3	0.122058	C11437	1-Deoxy-D-xylulose 5-phosphate
4	0.122022	C00074	Phosphoenolpyruvate
5	0.121247	C00118	D-Glyceraldehyde 3-phosphate
6	0.120309	C04691	3-Deoxy-arabino-heptulonate 7-phosphate
7	0.119590	C00036	Oxaloacetate
8	0.118243	C00279	D-Erythrose 4-phosphate
9	0.115543	C00111	Dihydroxyacetone phosphate
10	0.115195	C00085	D-Fructose 6-phosphate

[a] Compound index number from KEGG [19]

TABLE 10.3   Ten Metabolites with Highest Degree Centrality for *B. Subtilis* Network

Rank	Degree centrality	KEGG[a]	Name
1	0.038095	C00022	Pyruvate
2	0.026190	C00024	Acetyl-CoA
3	0.023810	C00025	Glutamate
4	0.021429	C00092	D-Glucose 6-phosphate
5	0.019048	C00085	D-Fructose 6-phosphate
5	0.019048	C00124	D-Galactose
5	0.019048	C00111	Dihydroxyacetone phosphate
5	0.019048	C00118	D-Glyceraldehyde 3-phosphate
9	0.016667	C00031	D-Glucose
9	0.016667	C00248	Lipoamide
9	0.016667	C00049	Aspartate

[a] Compound index number from KEGG [19]

10.5   MODULARITY AND DECOMPOSITION OF METABOLIC NETWORKS

In biochemistry, it is well established that modules consisting of several interacting bioreactions or metabolic pathways build discrete functional units of metabolism [14,26]. These modules are further nested to form a complex metabolic network. However, the structural analysis of metabolic networks indicates a small-world structure [18,24] where all the vertices in the whole network are linked through a short path. The modular organization seems missing in this small-world structure. To resolve the apparent contradiction between the small-world structure and modularity organization Ravasz et al. [35] proposed a hierarchical modularity model for metabolic networks. According to this model, metabolic networks of organisms are organized as many small, but highly connected modules that combine in a hierarchical manner to larger, less cohesive units. Several studies using concepts such as the reaction betweenness centrality distribution and the dependency of metabolites have further verified that metabolic networks are organized in a hierarchical way [12,16]. These results indicate that hierarchical modularity is also an important feature of metabolic networks. Modularity has been shown to be common in the organization of robust and sustainable complex systems [14]. Therefore, identifying the modular organization of metabolic network by certain network decomposition methods can help us in better understanding the organization principle of complex systems.

A possible large-scale design principle is that one part (module) of the network constitutes a densely connected core that is also central in terms of network distance, and the rest of the network forms a periphery [15]. For example, in a network of airline connections, one would most certainly pass such a core-airport on any many-flight itinerary. In metabolic networks, the core part is the central metabolism where many metabolites will be converted to supply the necessary metabolites to other, less central, pathways (modules in the periphery part).

Methods for a rational decomposition of metabolic network into relatively independent functional subsets are essential to better understand the modularity and organization principle of large-scale, genome-wide networks.

Several methods such as elementary flux mode analysis and extreme pathway analysis [37,38,40,41] have been developed for analyzing the pathway structure of metabolic networks (see Section 10.6). These methods have been shown to be useful tools for investigating the metabolic capacity and pathway structure of the metabolic networks [30,31,34,41,44]. These methods are, however, hampered by the combinatorial explosion problem when applied to large-scale networks such as those reconstructed from genomic data. For large-scale networks reconstructed from genome information, decomposition methods should be first used to divide the whole network into small subsystems. The pathway structure of these subsystems may then be properly analyzed by these methods [38,42].

There are many methods that can be used for the decomposition of metabolic networks. The objective is to optimize modularity over all possible divisions to find the best one. This may be infeasible for systems larger than 20 or 30 vertices because the optimization of modularity is very costly [27]. Various approximate optimization methods are available: simulated annealing [21], genetic algorithms, and so forth.

Other methods make use of a hierarchical clustering. These methods use different criteria for the clustering such as distance [25], connection degree [12,42], and modularity coefficient [27] (explained in Section 10.5.2).

For example, to decompose a network based on distance [25], the first step is to give a proper distance definition to show the dissimilarity between the vertices. With the distances calculated, one can then build a hierarchical tree for finding clusters of

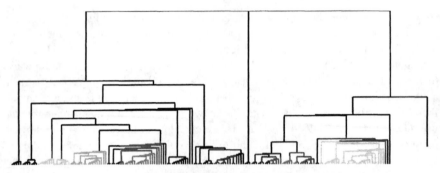

FIGURE 10.7   Tree showing the decomposition of *E. coli* metabolic network based on distance of vertices.

functionally closely related reactions. An example of such a tree is shown in Fig. 10.7 for the decomposition of *E. coli* network using distance as a measure.

Trees of the type of Fig.10.7 are sometimes called *dendrograms*. Cuts through this dendrogram at different levels give divisions of the network into larger or smaller numbers of modules (communities). To select the best number of modules the *modularity coefficient* can be used.

10.5.1   Modularity Coefficient

A good partition of a network into modules must comprise many within-module edges and as few as possible between-module edges. However, if we just try to minimize the number of between-module edges (or, equivalently, maximize the number of within-module edges), the optimal partition consists of a single module and no between-module edges. To avoid such a problem, one needs to find better ways to measure the quality of the decomposition. Newman [28] proposed a measure called modularity for this purpose. Modularity is defined as [13,28]

$$M = \sum_{s=1}^{N_M} \left[\frac{l_s}{L} - \left(\frac{d_s}{2L} \right)^2 \right] \tag{10.3}$$

where N_M is the number of modules, L is the number of edges in the network, l_s is the number of edges between vertices in module s, and d_s is the sum of the degrees (number of edges) of the vertices in module s.

If the number of within-communities is no better then random, we will get $M = 0$. Values approaching $M = 1$, which is the maximum, indicate a strong community structure. Values for social networks studied by Newman fall in the range from about 0.3 to 0.7 if there is a community structure present [27]. For metabolic networks, this value is around 0.8 (Table 10.4), which shows the strong community structure in this kind of networks [8].

TABLE 10.4   Modularity Coefficient for the Results of Decomposition for 5 Organisms [8]

Organism	No. modules	Modularity
A. pernix	12	0.792013476
B. subtilis	17	0.846591184
E. coli	16	0.846683825
S. cerevisiae	15	0.838768307
H. sapiens	17	0.866001288

10.5.2   Modularity-Based Decomposition

The modularity coefficient can be used not only as a measure of the quality of network decomposition, but also as a parameter for the clustering [27], therefore applied for the modularity-based decomposition algorithm. In this algorithm, each vertex in the network is first considered to be a module itself. So at the beginning, the number of modules is the same as the number of vertices. Then, the modules are joined in pairs to form new modules. The criterion to choose the modules to join is based on changes in modularity. The change in the modularity caused by the union of modules i and j is calculated [8] as

$$\Delta M = \frac{l_{i,j}}{L} - \left(\frac{d_{i,j}}{2L}\right)^2 \qquad (10.4)$$

where $l_{i,j}$ is the number of edges between the two modules, L is the total number of edges in the network, and $d_{i,j}$ is the total degree of the vertices in the two modules.

New modules that results in the greatest increase (or smallest decrease) in modularity will be created first. The progress of the algorithm can be represented as a dendrogram like in the distance-based method. The modularity coefficient is used to decide where to cut the dendrogram. The cut is done at the point that will generate a module distribution with the highest value of modularity.

Table 10.4   shows the modularity coefficient for the decomposition of the metabolic networks of *Aeropyrum pernix*, *Bacillus subtilis*, *Escherichia coli*, *Saccharomyces cerevisiae*, and *Homo sapiens* using this method. The networks used are adapted from the work of Ma and Zeng [24]. Figure 10.8 shows the modules generated for the decomposition of *E. coli* network using the modularity method illustrated above. In this figure, vertices represent the modules while vertex size represents the number of metabolites included in the module. Edge labels show the number of connections between modules. Table 10.5 shows the pathways present in each module after decomposition of the same network.

From Table 10.5, we can see that the automatic decomposition algorithm using the modularity method results in a reasonable separation of the metabolites. Functionally

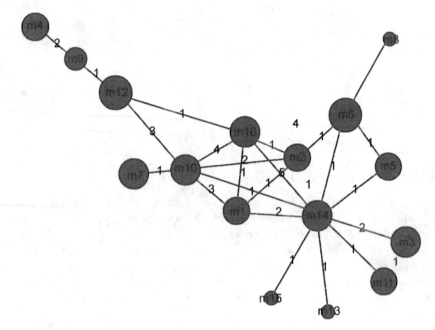

FIGURE 10.8 Module decomposition for *E. coli* based on modularity algorithm.

associated metabolites (traditionally recognized as a metabolic pathway) are now
successfully classified into the same modules in most cases. And tightly related pathways are also decomposed into the same modules, e.g., 1, 3, 4, 6, 7, 10, 12, 13, and
14. Some pathways are broken into different modules. The reason could be because
certain metabolite is involved in many pathways but presented as a single vertex in
the network. The algorithm will choose to put such a vertex in a module and other
vertices of the same pathways to another module(s) to generate the best community
structure. One way to minimize this problem is the use of reaction graph, where
vertices represent reactions instead of the normal representation with vertices representing metabolites [25].

10.6 ELEMENTARY FLUX MODES AND EXTREME PATHWAYS

Another important issue of metabolic network analysis is how to identify and analyze
metabolic pathways at a genome scale. To address this question, two related concepts,
elementary flux modes (EFMs) [32,40,41] and extreme pathways (EPs) [37,39], were
proposed by the groups of Schuster and Palsson. EFM is defined as a minimal set of
enzymes that can operate at a steady state with all irreversible reactions proceeding
in the appropriate direction. "Minimal" means that if only the enzymes belonging to
this set were operating, complete inhibition of one of these would lead to cessation
of any steady-state flux in the system. Before the calculation of EFMs, two types of

TABLE 10.5   Pathways in the Modules Generated by Modularity-Based Method

Module ID	Pathways
1	Glycerophospholipid metabolism
	Glycerolipid metabolism
2	Phenylalanine, tyrosine, and tryptophan biosynthesis
	Glycolysis/gluconeogenesis
3	Arginine and proline metabolism
	Urea cycle and metabolism of amino groups
4	Folate biosynthesis
	Riboflavin metabolism
	Purine metabolism
5	Lysine biosynthesis
	Peptidoglycan biosynthesis
6	Glycine, serine, and threonine metabolism
	Methionine metabolism
7	Galactose metabolism
	Nucleotide sugars metabolism
8	Pantothenate and CoA biosynthesis
9	Purine metabolism
10	Fructose and mannose metabolism
	Pentose and glucuronate interconversions
11	Pyrimidine metabolism
12	Pentose and glucuronate interconversions
	Purine metabolism
	Pentose phosphate pathway
13	Benzoate degradation via CoA ligation
	Butanoate metabolism
14	Citrate cycle (TCA cycle)
	Reductive carboxylate cycle (CO_2 fixation)
15	Valine, leucine, and isoleucine degradation
16	Valine, leucine, and isoleucine biosynthesis

metabolites , external metabolites and internal metabolites, must be defined first. If the formation of the metabolite can be balanced by its consumption (steady-state assumption) in the studied system, this metabolite can be defined as an internal metabolite (e.g., **A**, **B**, **C**, **D**, and **E** in Fig. 10.9). Otherwise, if the metabolites are the sources or sinks (nutrients and waste products, stored or excreted products or precursors for further transformations), for example, **Aex**, **Cex**, and **Eex** in Fig. 10.9, these metabolites can be defined as external metabolites. The currency metabolites (e.g., ATP and ADP in Fig. 10.9) can also be defined as external metabolites. Actually, the definition of external and internal metabolites is very complex, depending on biological systems and also research aims. The concept of EPs is related to EFMs. However, for EFMs reversible reactions are unnecessary to be split into two opposite irreversible reactions. Their similarities and differences are discussed in details by Klamt and Stelling [23].

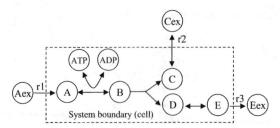

FIGURE 10.9   Illustration of an elementary flux mode.

Up to now a broad of significant applications are demonstrated based on the two concepts. The following are just a few examples.

1. The minimal medium requirement for *Haemophilus influenzae* and *Helicobacter pylori* was analyzed with EPs [36,38]. The predicted minimal medium composition is highly consistent with experimental results.

2. Reactions, which are not connected to the metabolic network, are identified by employing EFMs [9] and EPs [36,38]. These network dead ends or gaps correspond to reactions with reactants or products that are not produced or consumed by other parts of the metabolic network model, suggesting poorly characterized sections of the studied network. Some of these gaps are currently being reconciled with genome annotations.

3. Enzyme subsets of genome-scale metabolic network were identified by using EFMs [32] and EPs [31]. Enzyme subsets are also called systemically-correlated reactions of genome-scale metabolic network. These correlated sets could have significant implications in understanding the regulatory structure of metabolic network.

4. Pathway redundancy was analyzed by utilizing EFMs [44] and EPs [33], respectively. Pathway redundancy is a measure of how many systemically independent pathways have equivalent input and output fluxes, a quantitative description of network flexibility.

5. Klamt and Gilles introduced a concept of minimal cut sets for metabolic networks in virtue of the concept of the EFMs [22]. A minimal cut set (MCS) is a minimal (irreducible) set of reactions in the network whose inactivation will definitely lead to a failure in certain network functions. With the method of MCSs, a number of potential applications can be achieved, including network verification, phenotype prediction, assessing structural robustness and fragility, metabolic flux analysis, and target identification in drug discovery.

6. Cellular functions including growth and regulation were also studied by means of EFMs [44] and EPs [7]. Theoretical transcript ratios for growth on two alternative substrates were calculated by using EFMs. The results are in agreement with experimental observation [29]. Covert and

Palsson [29] examined the reduction of the solution space due to regulatory constraints by utilizing EPs. The imposition of environmental conditions and regulatory mechanisms greatly reduces the number of active extreme pathways. This approach was demonstrated for a skeleton system of core metabolism, which has 80 extreme pathways. As regulatory constraints were applied to the system, the number of feasible extreme pathways was reduced to between 2 and 26 extreme pathways, a reduction of between 67.5% and 97.5%. This method provides a way to interpret how regulatory mechanisms are used to constrain network functions and produce a small range of physiologically meaningful behaviors from all allowable network functions.

In short, some structural and functional characteristics are discovered by analysis of EFMs and EPs. But more applications are expected based on these two concepts. For example, one could apply MCS approach to find more robust drug targets.

10.7 SUMMARY

Metabolic network refers to the network composed of metabolites and their interconversions (biochemical reactions) in an organism. The sequencing of genomes and development of functional genomics make it now possible to reconstruct and understand the structure and function of metabolic networks at large scale. New computational tools and biological concepts are being developed to this end. This chapter briefly describes several useful tools and concepts recently developed. The issues covered range from reconstruction, visualization, and graph representation of genome-scale metabolic networks for structural analysis (e.g., connectivity and centrality analyses) over modularity and decomposition of the networks to rather fundamental and detailed studies such as elementary flux modes and extreme pathways. It should be emphasized that present methods and concepts primarily address static properties and functions of metabolic networks. New tools and concepts are desperately needed to understand the dynamic structure and function of metabolic networks.

10.8 EXERCISES

1. Download Pajek from: http://vlado.fmf.uni-lj.si/pub/networks/pajek/; Download a list of reaction IDs in *E. coli* glycolysis pathway from ftp://ftp.genome. jp/pub/kegg/pathway/organisms/eco/eco00010.rn and the reaction equations from: ftp://ftp.genome.jp/pub/kegg/ligand/reaction/reaction.lst, ftp://ftp. genome.jp/pub/kegg/ligand/reaction/reaction_name.lst. Based on these files, please:

    (a) Write a program to extract a list of reactions in *E. coli* glycolysis pathway including reaction equations;

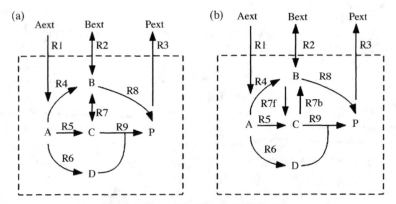

FIGURE 10.10 Example networks for calculation of EFMs and EPs. The stoichiometric coefficients are 1 or -1. The dash lines are the boundary of the systems. The metabolites inside the systems are internal metabolites. Otherwise, they are external metabolites (Aext, Bext, and Pext).

(b) Convert this list of reactions to a graph using metabolites as nodes and visualize it in Pajek;

(c) Remove the currency metabolites such as ATP, H_2O in the graph and see how different the network looked like in Pajek;

(d) Calculate the path length from glucose (C00267) to pyruvate (C00022) and the average path length in the two graphs;

(e) Calculate the degree distribution for the two graphs;

(f) Download other pathways and reconstruct the whole network for *E. coli* and analyze the degree distribution and path length for the whole network;

(g) Try to use other functions in Pajek to analyze the centrality and connectivity of the networks.

2. Using the networks created in the last exercise:

(a) Convert them to be visualized with cytoscape (http://cytoscape.org). To convert these networks, pyNetConv can be used. pyNetConv can be downloaded from http://pynetconv.sourceforge.net. After converting the networks, try to use cytoscape analysis tools.

(b) Use the cluster tool from http://pynetconv.sourceforge.net/cluster to decompose the network into smaller modules.

3. Calculate the EFMs for the example network **(a)** (Fig. 10.10a) and the EPs for the network **(b)** (Fig. 10.10b) given the system boundary shown as dash lines. How many EFMs or EPs exist and what are they in terms of reactions? Redo the calculation considering that the system boundary is redefined and only Aext and Bext are considered as external metabolites. Discuss the effect of selection of external metabolites on EFMs and EPs.

The solution of this exercise can be found with the help of the review [23].

REFERENCES

1. Biocyc. http://biocyc.org.

2. Cytoscape. http://cytoscape.org.

3. Expasy—biochemical pathways. http://www.expasy.org/cgi-bin/show_thumbnails.pl.

4. Kegg pathway database. http://www.genome.ad.jp/kegg/metabolism.html.

5. Systems biology. http://www.gbf.de/SystemsBiology.

6. R. Albert and A. L. Barabasi. Statistical mechanics of complex networks. *Reviews of Modern Physics*, 74(1):47–97, 2002.

7. M. W. Covert and B. O. Palsson. Constraints-based models: regulation of gene expression reduces the steady-state solution space. *Journal Theoretical Biology*, 221(3):309–325, 2003.

8. M. Rosa da Silva. Bioinformatics tools for the visualization and structural analysis of metabolic networks. Ph.D. thesis, Technische Universität Carolo-Wilhelmina zu Braunschweig, Germany, 2006.

9. T. Dandekar, S. Schuster, B. Snel, M. Huynen, and P. Bork. Pathway alignment: Application to the comparative analysis of glycolytic enzymes. *Biochemical Journal*, 343 Pt 1:115–124, 1999.

10. D. A. Fell and A. Wagner. The small world of metabolism. *Nature Biotechnology*, 18(11):1121–1122, 2000.

11. C. Francke, R. J. Siezen, and B. Teusink. Reconstructing the metabolic network of a bacterium from its genome. *Trends in Microbiology*, 13(11):550–558, 2005.

12. J. Gagneur, D. B. Jackson, and G. Casari. Hierarchical analysis of dependency in metabolic networks. *Bioinformatics*, 19(8):1027–1034, 2003.

13. R. Guimera and L. A. Nunes Amaral. Functional cartography of complex metabolic networks. *Nature*, 433(7028):895–900, 2005.

14. L. H. Hartwell, J. J. Hopfield, S. Leibler, and A. W. Murray. From molecular to modular cell biology. *Nature*, 402(6761 Suppl):C47–C52, 1999.

15. P. Holme. Core-periphery organization of complex networks. *Physics. Review. E: Statistical. Nonlinear, Soft Matter Physic.*, 72(4 Pt 2):046111, 2005.

16. P. Holme, M. Huss, and H. Jeong. Subnetwork hierarchies of biochemical pathways. *Bioinformatics*, 19(4):532–538, 2003.

17. M. A. Huynen and E. van Nimwegen. The frequency distribution of gene family sizes in complete genomes. *Molecular Biology and Evolution*, 15(5):583–589, 1998.

18. H. Jeong, B. Tombor, R. Albert, Z. N. Oltvai, and A. L. Barabasi. The large-scale organization of metabolic networks. *Nature*, 407(6804):651–654, 2000.

19. M. Kanehisa and S. Goto. Kegg: kyoto encyclopedia of genes and genomes. *Nucleic Acids Research*, 28(1):27–30, 2000.

20. M. Kanehisa, S. Goto, M. Hattori, K. F. Aoki-Kinoshita, M. Itoh, S. Kawashima, T. Katayama, M. Araki, and M. Hirakawa. From genomics to chemical genomics: New developments in KEGG. *Nucleic Acids Research*, 34(Database issue):354–357, 2006.

21. S. Kirkpatrick, C. D. Gelatt, and M. P. Vecchi. Optimization by simulated annealing. *Science*, (220):671–680, 1983.

22. S. Klamt and E. D. Gilles. Minimal cut sets in biochemical reaction networks. *Bioinformatics*, 20(2):226–234, 2004.

23. S. Klamt and J. Stelling. Two approaches for metabolic pathway analysis? *Trends in Biotechnology*, 21(2):64–69, 2003.

24. H. Ma and A. P. Zeng. Reconstruction of metabolic networks from genome data and analysis of their global structure for various organisms. *Bioinformatics*, 19(2):270–277, 2003.

25. H. W. Ma, X. M. Zhao, Y. J. Yuan, and A. P. Zeng. Decomposition of metabolic network into functional modules based on the global connectivity structure of reaction graph. *Bioinformatics*, 20(12):1870–1876, 2004.

26. F. C. Neidhardt, J. L. Ingraham, and M. Schaechter. *Physiology of the Bacterial Cell: A Molecular Approach*. Sinauer Associates, Inc. Publishers, 1990.

27. M. E. Newman. Fast algorithm for detecting community structure in networks. *Physics Review E: Statistical, Nonlinear and Soft Matter Physics*, 69(6 Pt 2):066133, 2004.

28. M. E. Newman and M. Girvan. Finding and evaluating community structure in networks. *Physics Review E: Statistical, Nonlinear, and Soft Matter Physics*, 69(2 Pt 2):026113, 2004.

29. M. K. Oh, L. Rohlin, K. C. Kao, and J. C. Liao. Global expression profiling of acetate-grown *Escherichia coli*. *Journal of Biological Chemistry*, 277(15):13175–13183, 2002.

30. B. O. Palsson, N. D. Price, and J. A. Papin. Development of network-based pathway definitions: the need to analyze real metabolic networks. *Trends in Biotechnology*, 21(5):195–198, 2003.

31. J. A. Papin, N. D. Price, and B. O. Palsson. Extreme pathway lengths and reaction participation in genome-scale metabolic networks. *Genome Research*, 12(12):1889–1900, 2002.

32. T. Pfeiffer, I. Sanchez-Valdenebro, J. C. Nuno, F. Montero, and S. Schuster. Metatool: For studying metabolic networks. *Bioinformatics*, 15(3):251–257, 1999.

33. N. D. Price, J. A. Papin, and B. O. Palsson. Determination of redundancy and systems properties of the metabolic network of helicobacter pylori using genome-scale extreme pathway analysis. *Genome Research*, 12(5):760–769, 2002.

34. N. D. Price, J. L. Reed, J. A. Papin, I. Famili, and B. O. Palsson. Analysis of metabolic capabilities using singular value decomposition of extreme pathway matrices. *Biophysical Journal*, 84(2 Pt 1):794–804, 2003.

35. E. Ravasz, A. L. Somera, D. A. Mongru, Z. N. Oltvai, and A. L. Barabasi. Hierarchical organization of modularity in metabolic networks. *Science*, 297(5586):1551–1555, 2002.

36. C. H. Schilling, M. W. Covert, I. Famili, G. M. Church, J. S. Edwards, and B. O. Palsson. Genome-scale metabolic model of helicobacter pylori 26695. *Journal of Bacteriology*, 184(16):4582–4593, 2002.

37. C. H. Schilling, D. Letscher, and B. O. Palsson. Theory for the systemic definition of metabolic pathways and their use in interpreting metabolic function from a pathway-oriented perspective. *Journal of Theoretical Biology*, 203(3):229–248, 2000.

38. C. H. Schilling and B. O. Palsson. Assessment of the metabolic capabilities of haemophilus influenzae rd through a genome-scale pathway analysis. *Journal of Theoretical Biology*, 203(3):249–283, 2000.

39. C. H. Schilling, S. Schuster, B. O. Palsson, and R. Heinrich. Metabolic pathway analysis: basic concepts and scientific applications in the post-genomic era. *Biotechnology Progress*, 15(3):296–303, 1999.

40. S. Schuster, T. Dandekar, and D. A. Fell. Detection of elementary flux modes in biochemical networks: a promising tool for pathway analysis and metabolic engineering. *Trends in Biotechnology*, 17(2):53–60, 1999.

41. S. Schuster, D. A. Fell, and T. Dandekar. A general definition of metabolic pathways useful for systematic organization and analysis of complex metabolic networks. *Nature Biotechnology*, 18(3):326–332, 2000.

42. S. Schuster, T. Pfeiffer, F. Moldenhauer, I. Koch, and T. Dandekar. Exploring the pathway structure of metabolism: decomposition into subnetworks and application to mycoplasma pneumoniae. *Bioinformatics*, 18(2):351–361, 2002.

43. P. Shannon, A. Markiel, O. Ozier, N. S. Baliga, J. T. Wang, D. Ramage, N. Amin, B. Schwikowski, and T. Ideker. Cytoscape: a software environment for integrated models of biomolecular interaction networks. *Genome Research*, 13(11):2498–2504, 2003.

44. J. Stelling, S. Klamt, K. Bettenbrock, S. Schuster, and E. D. Gilles. Metabolic network structure determines key aspects of functionality and regulation. *Nature*, 420(6912):190–193, 2002.

45. S. H. Strogatz. Exploring complex networks. *Nature*, 410(6825):268–276, 2001.

46. J. Sun and A. P. Zeng. Identics–identification of coding sequence and in silico reconstruction of the metabolic network directly from unannotated low-coverage bacterial genome sequence. *BMC Bioinformatics*, 5:112, 2004.

47. A. Wagner and D. A. Fell. The small world inside large metabolic networks. *Proceeding of the Biological Sciences*, 268(1478):1803–1810, 2001.

11

PHYLOGENETIC NETWORKS

Birgit Gemeinholzer

11.1 INTRODUCTION

In the following sections, biologists will become acquainted with network reconstruction methodologies and different variants of reticulate networks. Computer scientists will be introduced to simple models of reticulate evolution. Phylogenetic network detection and reconstruction methods are still at an early stage of development. To date, no methods are available to differentiate between signals reflecting sequence noise (e.g., due to technical limitations in the lab or signals independent of evolutionary processes) and phylogenetic patterns reflecting true evolutionary reticulation processes. This introduction to state-of-the-art approaches to tackle problematic phylogenetic network reconstruction covers the most frequently used techniques without claiming to be exhaustive.

The sections are divided into several subsections. After a prefatory section covering the problematic of reconstructing evolutionary reticulation processes, a brief introduction into character selection, character coding, and matrix structures is presented in Section 11.2. Then, common tree reconstruction methodologies are briefly introduced as they serve as a basis for phylogenetic network algorithms; explanations, however, are restricted to a minimum. For further reading, please refer to specialized textbooks [36,40]. Readers with prior knowledge may skip this chapter. In Section 11.4, the most commonly used algorithm descriptions to reconstruct phylogenetic networks are presented and several software packages are described in each subsection; however, many more exist and others are under development. Further lists of phylogeny programs with short descriptions, links to the programs and other phylogeny Internet

Analysis of Biological Networks, Edited by Björn H. Junker and Falk Schreiber
Copyright © 2008 John Wiley & Sons, Inc.

platforms can be found at the following sites [1–4]. The final section summarizes network reconstruction algorithms, methods and programs, and their application; some exercises are also presented.

Over the last 30 years, phylogenetic reconstruction has developed into a major research goal for biologists as an indispensable interpretive framework for the analysis of evolutionary processes by representing the interrelationships among biological entities [26,38,42,50]. Reconstructed phylogenies exhibit frameworks that allow the organization and interpretation of the evolution of organismal characteristics—from structure and physiology to genomics—provide patterns and hypotheses about lineage divergence, and illustrate the dynamics of speciation processes. Phylogenies can depict the order and timing of speciation events and, to some extent, extinction when fossil data are available [17,29]. Thus, phylogenies play a vital role in studies of adaptation to, for example, ecological requirements and evolutionary constraints (mechanisms that limit the development of specific evolutionary patterns such as constrained body size) [26,49,51–54]. Furthermore, phylogenies can help to elucidate functional relationships within living organisms [30,32,70], which is of high importance, for example, for pharmaceutical industries to make functional predictions for product developments of vaccines, herbicides, and so on.

As phylogenies are important for biology, a wide variety of methods for phylogenetic reconstructions has been generated by different specialists: biologists, mathematicians, statisticians, or computer scientists. However, most of them assume that the underlying evolutionary history of a given data set can be represented in a tree-like structure. The trees are usually interpreted as putative evolutionary-history reconstructions representing ancestors and descendants along with their character-state changes [10]. However, a tree-like interpretation of data does not always reflect the true phylogeny, as reticulate evolution (union of different evolutionary lineages triggered by hybridization) is a common mechanism in biology and noise in sequence data (due to data errors, biased sampling methodologies, etc.) is frequent. Ford Doolittle [24] famously wrote: "Molecular phylogeneticists will have failed to find the 'true tree,' not because their methods are inadequate or because they have chosen the wrong genes, but because the history of life cannot properly be represented as a tree." Noise as well as reticulate evolution cannot be modeled by bifurcating trees without loosing information but may arouse in parallel in different organisms and then transform a tree into a network. If tree algorithms are applied, reticulating taxa (featuring characters of different evolutionary lineages due to preceding hybridization) are forced into a nonreticulating tree topology, in which they might either occupy positions intermediate between two parental taxa or be placed basal to the group that includes its most derived parent [21,22].

Recent advancements below the species level, for example, in population genetics theory, and the availability of large data sets of comparative genetic information at the population level gave rise to the development of powerful tools to investigate the unique characteristics of intraspecific data. Intraspecific relationships are not hierarchical, as they are the result of sexual reproduction and recombination, and mutations are less frequent—consequently the amount of informative phylogenetic characters (diagnostic for groups of organisms) are fewer as well. Furthermore, on population level, ancestral characteristics are expected to persist and often still exist next to their descendants—which cannot be depicted by bifurcating trees where tips represent recent organisms and nodes reflect divergence in ancestral relationships. Related to the persistence of ancestral characteristics is the fact that multiple descendants are often present in populations leading to a multifurcation in a tree-like structure.

11.2 CHARACTER SELECTION, CHARACTER CODING, AND MATRICES FOR PHYLOGENETIC RECONSTRUCTION

Phylogenetic reconstructions are based upon phylogenetic hypotheses that are created by numerical analyses (phylogenetic analyses) of characters. Thereby, characters can include virtually any organismal attribute that has a heritable basis and reflects evolutionary pathways. Characters must comprise different character states to be of phylogenetic importance and can include physical forms (morphology, anatomy, cytology), biochemical characters, behavioral characters, biogeographical characters and so on. Characters for which the genetic basis is not usually known are often referred to as "morphological," whereas phylogenetic characters derived from DNA or protein data are called "molecular" characters. Here, the genetic diversity and relationship between or within different taxa (plural of taxon: group or category, at any level, in a system for classifying organisms) is analyzed. This can be based on the determination and comparison of defined regions, called DNA sequences, the predominant source of phylogenetic characters used for analyses above the species level. Below species level, DNA-fingerprinting methods are more often applied to reconstruct phylogenies. Here, the presence or absence of mutations throughout complete genomes without prior knowledge of genome structure and sequence data is screened and compared. Different DNA-fingerprinting methods exist, RFLP (restriction fragment length polymorphism), RAPD (random amplified polymorphic DNA), ISSR-PCR (intersimple sequence repeat polymerase chain reaction), AFLP (amplified fragment length polymorphism), and many others, the advantages or disadvantages of which are revised elsewhere [68]. However, all DNA-fingerprinting methods result in the scoring of presence or absence of mutations, whereas DNA sequence data provides four character states (adenine (A)/guanine (G)/cytosine(C)/ thymine (T) or 20 character states for proteins. Numerous molecular characters are available and their discrete attributes allow explicit coding (A/G/C/T or present/absent (1/0)) compared to the often continuous morphological characters (e.g, leaf length 6–8 cm or 7–11 cm, flower color yellow to orange or orange to red), and can either be binary (0/1) or multistate (adenine (A)/guanine (G)/cytosine (C)/thymine (T), small/medium/large, etc.). Molecular characters are available in nearly all organisms allowing the comparison of groups with no morphological similarities, for example, (1) elephants and green plants both contain cytochrome genes, the proteins of which serve a vital function in the transfer of energy within cells, or (2) juvenile and adult stages (e.g., tadpoles and frogs), which cannot be compared reasonably on a morphological basis. Other advantages are heritability, abundance of characters, different mutation rates of certain genomic regions, accepted models of sequence evolution, and independent organelle evolution (e.g., chloroplasts in plants are predominantly inherited maternally and, unlike nuclear DNA, not liable to hybridization). Disadvantages include the limited number of character states for nucleotides (only A/G/C/T), potential problems with paralogous sequences (regions that have been duplicated in the genome during evolution and might accumulate different mutations, possibly reflecting contradictory signals), evolutionary constraints on certain regions within a genome (e.g., deleterious mutations in functionally important regions lead to extinctions) and the unresolved problems of aligning sequences in comparable relative positions for different individuals [39,55]. Furthermore, gene evolution does not need to be identical to organism evolution and rapid evolution restricts the use of sequence data to a few million years. Moreover, molecular characters can seldom be obtained from fossils, require expensive laboratory equipment, and expertise and cannot be used for field identification.

Careful character selection is crucial for retrieving informative phylogenies. Whether of molecular, morphological, or other origin, characters must be coded for computational calculations. To this effect, a matrix (database) containing all taxa and all characters with their respective character states must be established (Tables 11.1–11.3) and, in turn, converted into a software compatible data format (for phylogenetic analyses usually the NEXUS- or PHYLIP-format, Table 11.4). As phylogenies reflect

TABLE 11.1   Taxa Description

Taxon A	Perennial creeping shrubs with linear hairy leaves, yellow flowers, the black fruits are one-seeded berries, chromosome numbers $x=10$, Egypt.
Taxon B	Perennial creeping herbs with oblong hairy leaves, flowers red, fruits are one-seeded reddish drupe, chromosome number $x=8$, Tunisia.
Taxon C	Perennial creeping herbs with oblong hairy leaves, flowers yellow, fruits are one-seeded red drupes, chromosome number $x=8$, Algeria.
Taxon D	Perennial creeping shrub, leaves linear and hairy, flowers yellow, fruits are one-seeded black berries, chromosome numbers $x=8$, Algeria, Libya.
Taxon E	Annual erect hairless herbs, leaves linear, flowers yellow, fruits are many-seeded black berries, chromosome numbers $x=8$, Tunisia, Libya.
Taxon F	Annual erect hairless herb, leaves linear, flowers yellow, fruits are one-seeded black berries, chromosome numbers $x=8$, Germany, Poland.
Taxon G	Annual erect herb, leaves hairy, linear, flowers yellow, fruits are one-seeded black berries, chromosome number $x=8$, Italy, France, Spain.
Taxon H	Annual creeping hairy herb, leaves linear, flowers yellow, fruits are one-seeded black berries, chromosome numbers $x=8$, Italy, Austria, Germany.

TABLE 11.2   Characters and Character States for the Taxa Descriptions in Table 11.1

Character 1	life form: perennial (0); annual (1)
Character 2	stem: creeping (0); erect (1)
Character 3	hairs: present (0); absent (1)
Character 4	leaf form: linear (0); oblong (1)
Character 5	flower color: yellow (0); red (1)
Character 6	fruit type: berry (0); drupe (1)
Character 7	fruit color: black (0); red (1)
Character 8	seeds: one (0); many (1)
Character 9	chromosome numbers: x=10 (0); x=8 (1)
Character 10	natural distribution: Africa (0); Eurasia (1)

hypotheses regarding character and taxa evolution, it is important to define the reading order that determines the ancestral or derived character states: which state was acquired by an ancestor deeper in the phylogeny or which is the most recent common ancestors of the taxa under consideration. As it is assumed, that closely related taxa are more similar to each other, accumulation of mutations is an indication of divergence allowing for ancestral and derived character coding. If this information is lacking, the characters of one or more outgroup taxa (which, according to the hypothesis, are less closely related to each of the taxa under consideration than any are to each other) can be used to help resolve the polarity of characters. Choosing independent characters to reflect evolutionary relationships and scoring character states can be tricky. Equivalent character states that evolved independently in two or more taxa (convergent evolution) can be quite common and create complications since phylogenies are hypotheses and might not reflect the true picture. As character selection is the first crucial step in phylogenetic reconstructions, mistakes at this stage will further affect the complete analyses. For further reading please refer to phylogenetic textbooks, [62].

An example how data matrices are deduced from taxa descriptions is shown in the Tables 11.1–11.3.

TABLE 11.3   Data Matrix Based on the Taxa Descriptions in Table 11.1 and on the Characters and Character States of Table 11.2

	1	2	3	4	5	6	7	8	9	10
Taxon A	0	0	0	0	0	0	0	0	0	0
Taxon B	0	0	0	1	1	1	1	0	1	0
Taxon C	0	0	0	1	0	1	1	0	1	0
Taxon D	0	0	0	0	0	0	0	0	1	0
Taxon E	1	1	1	0	0	0	1	1	1	0
Taxon F	1	1	1	0	0	0	1	0	1	1
Taxon G	1	1	0	0	0	0	1	0	1	1
Taxon H	1	0	0	0	0	0	1	0	1	1

TABLE 11.4 Example, How the Data Matrix from Table 11.3 is Transformed into the Most Common Matrix Formats for Further Computational Processing: NEXUS- and PHYLIP Format (for 8 Taxa and 10 Characters)

NEXUS-format:	PHYLIP-format:
#NEXUS	8 10
BEGIN DATA;	MATRIX
DIMENSIONS NTAX=8 NCHAR=10;	Taxon A 0000000000
FORMAT Symbols="0 1";	Taxon B 0001111010
	Taxon C 0001011010
MATRIX	Taxon D 0000000010
Taxon_A 0000000000	Taxon E 1110001110
Taxon_B 0001111010	Taxon F 1110001011
Taxon_C 0001011010	Taxon G 1100001011
Taxon_D 0000000010	Taxon H 1000001011
Taxon_E 1110001110	
Taxon_F 1110001011	
Taxon_G 1100001011	
Taxon_H 1000001011	
;	
end;	

11.3 TREE RECONSTRUCTION METHODOLOGIES

For tree construction there are three main categories of approaches to postulate evolutionary scenarios: distance-, character-, and likelihood-based methodologies. The first converts the character matrix (sequence data, morphological data, etc.) into a distance matrix of pairwise differences (distances) between the taxa and assumes that the distances provide estimates for the evolutionary divergence among taxa (Table 11.5). The most common class of distance methods that search for the best tree that takes the observed distances into account are unweighted pair-group method with arithmetic means (UPGMA) and neighbor joining (NJ). For sequence data, NJ has almost completely replaced UPGMA in the current literature as, for optimization, evolutionary models have been implemented in the algorithm to reflect basic statistical quantities of molecular evolution such as nucleotide frequencies (the likelihood for a nucleotide to mutate is higher if the nucleotide is more frequent within a sequence), codon frequencies (mutations of a protein codon in relation to the total amount of this protein being present in the total sequence), and others [45]. For 0/1 matrices, however, UPGMA analysis is still the method of choice (Fig. 11.1). In both distance-based tree reconstruction methods, UPGMA and NJ, the two taxa with the highest similarity (closest distance) are grouped together, and then more distant taxa are added for tree reconstruction.

The major objection against distance methods is the loss of information through erecting a pairwise distance matrix for a dataset, which can never be retraced to the original dataset. Furthermore, the option to interpret estimated branch lengths as

TABLE 11.5   Distance Matrix: The Data Matrix from Table 11.3 Gets Transformed into a Distance Matrix (e.g., Taxon A and Taxon B Differ in 5 out of 10 Characters (50%)=0.5; Taxon A Differs from Taxon C in 4 Characters out of 10 (40%) = 0.4; and so on

	A	B	C	D	Taxon E	F	G	H
Taxon A	–	0.5	0.4	0.1	0.6	0.6	0.5	0.4
Taxon B		–	0.1	0.4	0.7	0.7	0.6	0.5
Taxon C			–	0.3	0.6	0.6	0.5	0.4
Taxon D				–	0.5	0.5	0.4	0.3
Taxon E					–	0.2	0.3	0.4
Taxon F						–	0.1	0.2
Taxon G							–	0.1
Taxon H								–

distances almost always underestimates the actual number of changes along lineages (e.g., if Taxon A and B vary on character 5 by 10 mutations, and Taxon B and C vary on character 4 by 10 mutations, A/B and B/C have the same evolutionary distance of 10%, independent of the site of character change). Distance methods are implemented in PAUP [61], PHYLIP [27,28], and MEGA [47], the most commonly used phylogeny reconstruction programs.

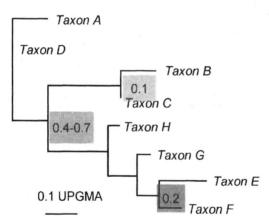

FIGURE 11.1   The tree is a UPGMA tree of the data/distance matrix in Tables 11.3 and 11.5 respectively. Horizontal branch lengths in the tree diagram reflect evolutionary distances; vertical branch lengths give cluster information only. The barrow in the lower left-hand corner depicts the length of 0.1 evolutionary distances. The shading within the distance matrix (Table 11.5) refers to the colors in the tree—presenting some groupings. If computer algorithms are used to calculate phylogenetic trees (e.g., distance trees or others), in general Newick–Formats are used as output, depicting taxa groupings in brackets with the branch lengths behind each taxon separated by colons. See Table 11.6 for the representation of the UPGMA tree in the Newick–Format. Computer programs which transform Newick–Formats into graphical outputs are, for example, TREE VIEW [5].

TABLE 11.6   **Newick–Format Representation of the UPGMA Tree Shown in Fig. 11.1.**

Newick-Format:

#NEXUS

Begin trees;
tree PAUP_1 = [&U]
(Taxon_A:1,((Taxon_B:1,Taxon_C:0):2,(((Taxon_E:2,Taxon_F:0):1,
Taxon_G:0):1,Taxon_H:0):2):1,Taxon_D:0);
End;

The second group of phylogenetic tree reconstruction methodologies in current use are character-based tree searching methods. They are, in general, slower and normally result in more than one equally good tree. They compare characters within each column (each site) directly and are called Parsimony methods. Parsimony is based on the assumption that the tree most likely to best reflect evolution is the one that requires the fewest numbers of mutations.

The basic premise of parsimony is that taxa sharing a common characteristic do so because they inherited that characteristic from a common ancestor. Exhaustive searching—which means the search for every possible tree—and branch-and-bound methods are used to optimize the search for the best tree, but due to computational limitations heuristic methods are necessary if the sample group exceeds 20. Parsimony methods are inconsistent if sequences evolve at different rates because they group fast evolving sequences together, which is called "long branches attraction," and for large data sets calculation time is a major drawback. The most commonly used parsimony programs are PAUP [61], PHYLIP [27,28], and MEGA [47]. To overcome the problem of speed for parsimony programs, Parsimony Ratchet algorithms have been invented by Nixon [68] and others. The ratchet programs are based on the assumption that frequently the "correct" tree topology can be derived from 1/10th of the characters of a normal input matrix and 9/10th of the data supports the topology, but does not lead to new implications. Search time is reduced by selection of a random set of characters from the actual data matrix with which tree topologies are reconstructed—for the tree with the fewest evolutionary steps of the reduced matrix, all characters are added and tree length is calculated. This may result in trees no longer representing a local optimum; however, the heuristic search continues until a new optimum is reached. The algorithm then reverts to the original weighting, and the search continues. This procedure is repeated *x*times until confidence limits propose the shortest tree discovered. Nixon [68] demonstrated the efficacy of the method on a 500-taxon data set, where the ratchet-based search found a tree two steps shorter than standard heuristic searches in less time. PAUPRAT by Sikes and Lewis [60] and TNT by Goloboff, Farris, and Nixon (together with NONA) [31] implement parsimony ratchet methods.

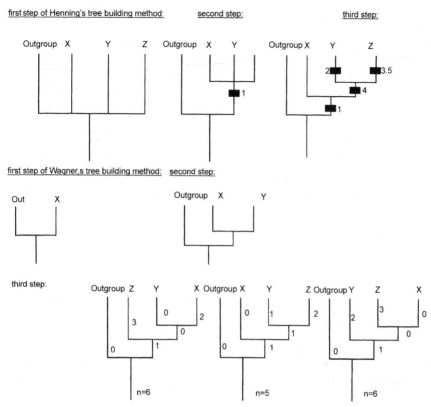

FIGURE 11.2 Parsimony tree building methods according to Henning and Wagner applied to the data of data matrix of Table 11.7. In Wagner's method, all possibilities of connecting one additional taxon to an existing tree are evaluated and the shortest tree is kept, as it might reflect the shortest evolutionary pathway, which is per definition most likely to represent evolutionary relationships. In this example (the tree in the middle of the bottom line) is the best and is also in concordance to Henning's best tree; the evolutionary relationships of the trees left and right in the bottom line (length $n=6$) have to be rejected.

The third main group of tree reconstruction methodologies are likelihood based and comprise maximum likelihood, and Bayesian tree reconstruction. Maximum likelihood [25] also uses each position in an alignment, evaluates all possible trees, and calculates the likelihood for each tree using an explicit model of

TABLE 11.7   Example of a Data Matrix

	1	2	3	4	5
Outgroup	0	0	0	0	0
Taxon X	1	0	0	0	0
Taxon Y	1	1	0	1	0
Taxon Z	1	0	1	1	1

evolution (while parsimony just looks for the fewest evolutionary changes). The likelihoods for each aligned position are then multiplied to provide a likelihood for the tree in total. The tree with the maximum likelihood is the most probable tree. Similar to maximum parsimony, maximum likelihood reconstructs ancestors at all nodes of each considered tree, but it also assigns branch lengths based on the probabilities of mutations. Likelihood functions for statistical inference are consistent and powerful. Various parameters of evolutionary processes, like relative probabilities, evolution rates across sites, and all possible mutational pathways that are compatible with the data are considered. However, the main drawback of this method, especially for large data sets, is computational time for tree reconstruction. Maximum likelihood is the slowest method of all, as the algorithm to find the maximum likelihood score must search through a multidimensional space of parameters. Maximum likelihood approaches have been incorporated in many different phylogenetic programs, like PAUP [61], PHYLIP [27,28], NONA [31], and PHYML [33], the fastest method for maximum likelihood reconstructions.

Bayesian analysis is closely related to maximum-likelihood tree reconstructions. The goal is to obtain the optimal hypothesis, which is the one that maximizes the posterior probability. To find this the so-called posterior probability distribution requires combining the likelihood and the prior probability distribution. Likelihood gives information about the data and the parameter within, while prior probability distribution comprises expectations of the data. With constant prior probabilities, the posterior distributions are simply proportional to the likelihood distributions, and the parameter value with the maximum likelihood has also the maximum posterior probability. The impact of the prior probabilities upon the posterior probabilities will diminish with increasing amounts of data. The advantages of the Bayesian methods are high computational speed and the possibility to incorporate complex models of sequence evolution. The most commonly used Bayesian phylogenetic software program is MR BAYES [43], based upon Markov Chain Monte Carlo (MCMC) simulations to search the tree space and infer the posterior distribution of topologies. Input data can be nucleic acid sequences protein sequences, or morphological characters.

11.4   PHYLOGENETIC NETWORKS

In contrast to bifurcating trees, phylogenetic networks can have multifurcations and, in principle, cycles (biologists might call them loops). The diversity of phylogenetic network reconstruction methods can be structured in different ways. Some authors use the classifications distance-based algorithms, maximum parsimony heuristics, and maximum likelihood heuristic, while others divide the methods according to data presentation techniques—either visualizing conflicts in data sets or representing more complex models of evolution, for example, reticulation graphs, which reflect hybridization graphs, or ancestor recombination graphs. However, most techniques have advantages and disadvantages. Therefore, a vast variety of different approaches have been proposed of which only a small selection can be presented here. All are still being improved and expanded, and several additional algorithms are still being implemented in one or the other technique to better reflect evolutionary mechanisms and strengthen their explanatory power.

Some network reconstruction methods are based on tree reconstruction and the addition of nontree edges using predefined criteria for optimization (e.g., minimize weighted sum of evolutionary events) until stopping rules terminate the network reconstruction. Nontree edges are typically chosen on the basis of incongruence between segments of sequences. According to parsimony criteria, no extra edges are added if the data can be represented by a tree. If the data implies a network, various segments within the network will require tree-like structures that can be optimized by tree-style criteria. The nodes in the network represent taxa, hypothetical ancestral taxa, or intermediary nodes. Disadvantageous is the dependence of the final network topology on the topology of the starting tree. The most widely used network methods in this class are statistical parsimony, galled trees and median networks (which are presented in further detail in Section 11.4.1–11.4.3), the variants of median networks, and the netting method. These methods perform best on the analysis of intraspecific data (e.g., for populations of one species). Problems may be encountered when the level of diversity increases, for example, when the networks become too complicated.

Another approach aims to find the best subtrees (e.g., in terms of parsimony scores) for various segments of the data and uses "the best trees" for combination into a network, while other models are based on the computation of all possible minimum spanning trees for a given data set and combine them into a minimum spanning network. The rationale underlying both hypotheses is based on the fact that different segments of the sequences evolved down different trees, that is, conflicts between the trees represent reticulation events. These approaches can be calculated with trees resulting from distance, parsimony, or likelihood analyses. The use of distance data alone implies that these phylogenetic network methods start with less information than those using the complete alignment. Nevertheless, there is evidence that much phylogenetic information is preserved in the distance matrix, even in the presence of reticulation [18,48,69]. Typical algorithms are median–joining networks [14] (explained in further detail here in Section 11.4.4), union of maximum parsimonious trees (UMP) [18], and molecular–variance parsimony [56], implemented in software programs like NETWORK [57] and ARLEQUIN [59].

Last but not least, there is a group of "compute splits algorithms" for which incompatibilities in the data set are analyzed first and then network edges are introduced to account for them. The outputs are the so-called splits graphs, which generalize the concept of a phylogenetic tree. Methods, which handle compatible and incompatible systems of splits include the pyramidal clustering model and slit decomposition (presented in Section 11.4.5 and 11.4.7), Neighbor-Net, and Consensus Networks. Software programs with implemented split decomposition algorithms are for example, SPLITS TREE [44] and PYRAMIDS [21].

11.4.1   Galled Trees

Gusfield [34,35] developed the idea of galled trees, which represent a small deviation from true trees based on the assumption that if a perfect phylogenetic tree cannot be derived from a data set, it should be deviated from a tree by as little as necessary. Rather than having a phylogenetic network with a complex interleaving of cycles (loops), it is preferable (if possible) to have a tree with a few extra edges (nodes), each creating a disjoint cycle—leading to a galled tree, which is two dimensional and contains loops. A galled-tree problem is defined as two graphs representing "incompatibilities" and "conflicts" between sites. Hereby "incompatibilities" are defined as four rows (taxa) in a data matrix where two columns (characters) contain all four of the ordered pairs 1:0; 0:1, 1:1, 0:0, which, for example, is the case in Fig. 11.3 for character 3 and 5 being incompatible due to taxa A, B, C, D that feature all four possible character combinations, which is called the "four-gametetest." In contrast, "conflicting" sequences are defined as those containing three of the above four character states (three-gamete-test) and also need networks for depiction. If conflicting or

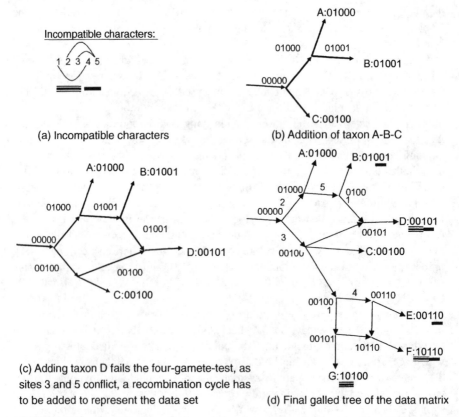

(a) Incompatible characters

(b) Addition of taxon A-B-C

(c) Adding taxon D fails the four-gamete-test, as sites 3 and 5 conflict, a recombination cycle has to be added to represent the data set

(d) Final galled tree of the data matrix

FIGURE 11.3   Example of a galled tree network reconstruction based on the data matrix of Table 11.8.

TABLE 11.8   Data Matrix

	1	2	3	4	5
AS (ancestral state)	0	0	0	0	0
Taxon A	0	1	0	0	0
Taxon B	0	1	0	0	1
Taxon C	0	0	1	0	0
Taxon D	0	0	1	0	1
Taxon E	0	0	1	1	0
Taxon F	1	0	1	1	0
Taxon G	1	0	1	0	0

incompatible trees are represented together, they result in a recombination cycle of a phylogenetic network (e.g., Fig. 11.3 from AS 00000 to Taxon D 10100). Galled Tree [6] is the program that constructs phylogenetic networks based upon a set of binary characters. In the resulting phylogenetic network, the recombination cycles are node disjoint (non-intersecting). The output is guaranteed to use the minimum number of recombinations over all possible phylogenetic networks and indicates how many matrices in the input file have a galled-tree, and how many have a perfect phylogeny with some ancestral sequences. As only single-crossover recombinations are allowed, galled-trees are restricted to input matrices with low or moderate recombination rates.

11.4.2   Statistical Parsimony

Statistical parsimony is a network approach especially created for the analyses of molecular characters from organisms of one or more populations that take population level phenomena into account. These analyses are called haplotype analyses, whereby the word haplotype refers to specific combinations of sequence variations within one organism that are likely to be inherited together. Individuals comprising the same haplotype (the same genetic constitution at a comparable site) are grouped together. Then, the relationships among different haplotypes are measured using genetic distances, that is, how many different mutations separate two haplotypes, as in minimum spanning trees. In contrast to bifurcating trees; haplotype, networks can have multifurcations and, in principle, cycles.

The statistical parsimony algorithm is based on the principles of minimum spanning tree reconstructions; however, first estimates the maximum number of differences among sequences caused by single substitutions and estimates a 95% statistical confidence limit, which is called parsimony limit or parsimony connection limit. Single substitutions that are below the 95% limit are ignored in favor of better network representations. Multiple substitutions at a single site are neglected. The algorithm then connects haplotypes that differ by one mutation. In the next step, haplotypes differing by two changes are surveyed, confidence limits are evaluated and taxa added to the network, then those differing by three mutations are screened, and so forth until all haplotypes have been added to the network or the parsimony connection limit is reached.

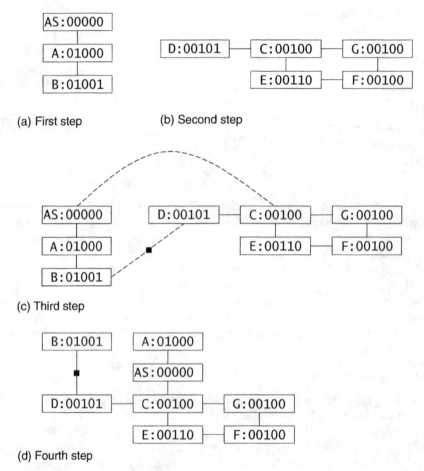

FIGURE 11.4 Example of a minimum spanning network reconstruction based on the data matrix of Table 11.8 (a 95% confidence limit is added for statistical parsimony, ignoring single substitutions in favor of improved data visualization; however, for the example here the data set than is too large for representation).

Larger data sets are needed to define 95% confidence limits of character states; therefore, the example shown in Fig. 11.4 is just representing the development of a minimum spanning tree—ignoring the enhancements of statistical parsimony.

Statistical parsimony is implemented in the TCS Java computer program [20], which analyses haplotypes within a population and considers multifurcations as well as reticulations (back crossings and hybridization events in evolution/retracing to one ancestral parent). TCS software opens nexus and phylip formats and has a graphic user interface to display the resulting networks. The program collapses sequences into haplotypes and calculates the frequencies of the haplotypes in the sample. The program outputs the sequences, the pair wise absolute distance matrix, probabilities of parsimony for mutational steps just beyond the 95% cut-off level, a test listing the connections and defining missing intermediates, and a graph output file containing the resulting network.

Templeton [63–67] used this statistical parsimony approach to formalize the correlation between number of haplotypes in a sample and its relationships (connections) to infer population histories, which resulted in the nested clade population analysis [NCPA]. This approach is a widely used technique for population analyses. It first consists of a network reconstruction procedure, based on the above-explained parsimony limit criteria. In a second step, Templeton's key, based on population genetic and biogeographic assumptions [64], is applied to group haplotypes in a biogeographic framework, based on population inheritance relationships (coalescence theory).

11.4.3   Median Network

In the median-network approach [13,15], constant sites are excluded from the data set (see example in Fig. 11.5). Sequences are converted to binary data, whereby each split is encoded as a binary character with states 0 and 1. Then, sites that support one type of split are grouped as one character; however, the amount of sites grouped together will influence the weighting of this split. Thus, this method represents haplotypes as binary 0–1 vectors. Median or consensus vectors are then calculated for each triplet of vectors until the median network is complete. Predictions from coalescent theories optimize and reduce the network, which otherwise results in high-dimensional hypercubes that cannot be displayed if the amount of haplotypes exceeds 30.

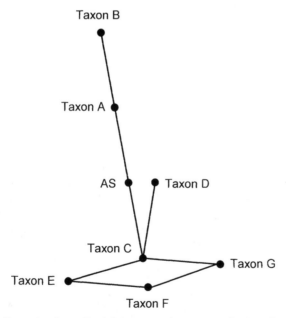

FIGURE 11.5   Example of a median joining network reconstruction based on the data matrix of Table 11.8 using NETWORK [15].

Several programs are based upon the median-network theory. SPECTRONET [41] is optimized for alignment in Nexus–Format and can handle large data sets up to 128 taxa and 50,000 nucleotides. The package works by computing a collection of weighted splits or bipartitions of the taxa and then allowing the user to interactively analyze the resulting collection. Other available programs are SPECTRUM [19] and NETWORK [57].

11.4.4   Median-Joining Networks

The median-joining network method was developed by Bandelt et al. [14] to depict reticulate evolution in phylogenies. The network reconstruction is first based upon the construction of minimum spanning trees (MSTs) using Kruskal's algorithm [46], which is an algorithm in graph theory that finds a minimum spanning tree for a connected weighted graph. It finds a subset of the edges that forms a tree that includes every vertex, where the total weight of all the edges in the tree is minimized. If the graph is not connected, then it finds a minimum spanning forest (a minimum spanning tree for each connected component).

Kruskal's algorithm [46] is an example of a greedy algorithm:

Kruskal's_algorithm (graph $G=(V, E)$)
    create a forest F (a set of trees), where each vertex in the graph G
    is a separate tree;
    create a set S containing all the edges in the graph G;
    while $S \neq \emptyset\{$
      remove an edge e with minimum weight from S;
      if e connects two different trees T_1 and T_2 {
        $F \longleftarrow F \cup e$;
        combine T_1 and T_2 into a single tree;
      } else
        discard e
    }
    return F;

At the termination of the algorithm, the forest has only one component and forms a minimum spanning tree of the graph.

Then, the trees are combined into a single network by sequential addition of new vertices called "median vectors," which are reconstructed under parsimony criteria (Farris' maximum parsimony algorithm). The median vectors represent missing intermediates, either due to incomplete data sampling, extinction of organisms, or reticulate evolution. Median-joining networks are established under the assumption of absent recombination, which limits the applicability to population analyses.

The algorithm is implemented in the program NETWORK [15], which is suitable for DNA or amino acid sequences, short tandem repeats, RFLP, AFLP, or microsatellite data as well as for language or linguistic matrices. The program is fast;

however, it does not resolve ties, and visualization of large datasets is problematic. It is possible to calculate age estimations for ancestral nodes or branching points. Two independent methods can be calculated: either reduced median networks [14], only for binary input data, or median-joining networks [14] for all types of data. The program is freely available [7] and based upon Windows and DOS executables.

11.4.5 Pyramids

The pyramidal clustering model was first introduced by Diday and Bertrand [23]. With this method, it is possible to detect groups of related elements within a data set. The model is an extension of the hierarchical clustering method (e.g., UP-GMA), as any element can belong simultaneously not only to one but also to two classes or clusters in a given data set and a pyramidal classification results in a set of compatible orders over the elements [11]. As input data, distance matrices (here called dissimilarity matrix) with data points between 0 and 1 are required. Several mathematical definitions constrain this clustering method, which can be reviewed in Aude [11]; here, however, a reconstruction example is presented in Section 11.4.6.

The program computing pyramidal clustering models is called PYRAMIDS and comprises several components to create and draw pyramidal representations of a set of sequences based upon their relative distances, resulting in planar graphs. The program includes a graphical interface to create figures in different formats, which can be transferred to a wide variety of environments. The Software runs on Sun, Linux, and Unix platforms; free online software is available [8].

11.4.6 Example of a Pyramidal Clustering Model

| | | Taxon | | |
	A	B	C	D	E
Taxon A	–	0.1	0.5	0.7	0.7
Taxon B		–	0.1	0.7	0.7
Taxon C			–	0.7	0.7
Taxon D				–	0.35
Taxon E					–

The distance matrix here is called dissimilarity matrix. A pair of taxa with the lowest distance is chosen, for example, A/B= 0.1 and grouped together in Class 1, which is added in an additional column to the dissimilarity matrix, using the mean aggregation index, for example,

Class 1

A   B

$$\mu(1, C) = 0.5(\mu(A, C) + \mu(B, C))$$

which for the above sample means:

$$\mu(1, C) = 0.5(0.5 + 0.1) = 0.3$$
$$\mu(1, D) = 0.5(0.7 + 0.7) = 0.7$$
$$\mu(1, E) = 0.5(0.7 + 0.7) = 0.7$$

		Taxon				Class
	A	B	C	D	E	1
Taxon A	–	X.X	0.5	0.7	0.7	*0.1*
Taxon B		–	0.1	0.7	0.7	*0.1*
Taxon C			–	0.7	0.7	0.3
Taxon D				–	0.35	0.7
Taxon E					–	0.7
Class 1						–

The samples that have been clustered together will be ignored in the course of further calculations (X.X), in Class 1 they will be marked by italics, as they cannot serve as lowest dissimilarity pair again.

Then, again the Taxa with the lowest distance are chosen, here B/C, grouped together in Class 2, an additional column is added and the mean aggregation index is calculated:

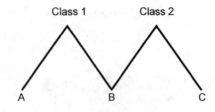

Class 1   Class 2

A   B   C

		Taxon				Class	
	A	B	C	D	E	1	2
Taxon A	–	X.X	0.5	0.7	0.7	*0.1*	*0.3*
Taxon B		–	X.X	0.7	0.7	*0.1*	–
Taxon C			–	0.7	0.7	0.3	*0.1*
Taxon D				–	0.35	0.7	0.7
Taxon E					–	0.7	0.7
Class 1						–	0.43
Class 2							–

$\mu(2/A) = 0.5(0.1 + 0.5) = 0.3$
$\mu(2/B) =$ must be ignored, having been clustered together twice, which is the maximum
$\mu(2/C) = 0.1$
$\mu(2/D) = 0.5(0.7 + 0.7) = 0.7$
$\mu(2/E) = 0.5(0.7 + 0.7) = 0.7$
$\mu(2/1) = 0.5(0.1 + 0.75) = 0.43$

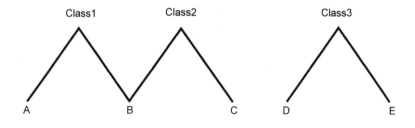

The taxa with the lowest distances are chosen, grouped together in Class 3, now D/E=0.35.

	Taxon				Class		
	A	C	D	E	1	2	3
Taxon A	–	0.5	0.7	0.7	0.1	0.3	0.7
Taxon C		–	0.7	0.7	0.3	0.1	0.7
Taxon D			–	X.X	0.7	0.7	0.35
Taxon E				–	0.7	0.7	0.35
Class 1					–	0.43	0.7
Class 2						–	0.7
Class 3							–

Now higher classifications are arranged by selecting the lowest dissimilarity, which is not yet been covered by any Classes—Class1/Class2=0.43

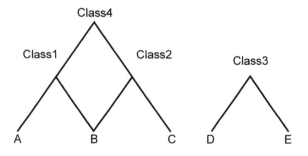

	Taxon				Class		
	A	C	D	E	1	2	3	4
Taxon A	–	0.5	0.7	0.7	*0.1*	*0.3*	0.7	0.2
Taxon C		–	0.7	0.7	*0.3*	*0.1*	0.7	0.2
Taxon D			–	X.X	0.7	0.7	*0.35*	0.7
Taxon E				–	0.7	0.7	*0.35*	0.7
Class 1					–	X.X	0.7	*0.43*
Class 2						–	0.7	*0.43*
Class 3							–	0.7
Class 4								–

For the establishment of Class5, a random connection is selected as there is a common dissimilarity between all classes.

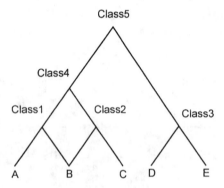

11.4.7   Split Decomposition

Split decomposition was developed by Bandelt and Dress [12] to analyze data or to produce instructive graphical outputs by transforming evolutionary distances into a sum of weakly compatible splits. The split decomposition method analyzes given distance data and finds deviations from the tree-like structure, namely splits, implied by homology, cases of convergence, or parallel evolutionary events [16,17].

A "split" of a graph $G=(V, E)$ is a grouping of taxa into two distinct sets (V_1, V_2) such that each species occurs in exactly one of the sets ($|V_1| \geq 2, |V_2| \geq 2$) with a common neighbor being presented in both subgroups. Therefore: $G=(V, E)$ is simple decomposed by the split V_1, V_2 into G_1 and G_2, where, for $i \in \{1, 2\}$, G_i is the subgraph of G induced by V_i with an additional vertex v, called marker, such that the neighborhood of v in G_i is the set of those vertices in V_i, which are adjacent in G to a vertex outside of V_i (see Fig. 11.6).

Additionally, a related composition is defined: $G_1=(V_1, E_1)$ and $G_2=(V_2, E_2)$ such that $V_1 \cap V2=\{v\}$. Then $G_1 \times G_2$ is the graph with vertex set $(V_1 \cup V_2) \setminus \{v\}$,

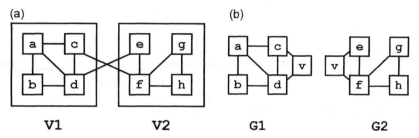

FIGURE 11.6   A graph with a split V1|V2 (*a*) and the two graphs G1 and G2 (*b*) obtained by simple split decomposition (modified after [16]).

and edges set $\{\{x, y\} \in E_1 : x \neq v \text{ and } y \neq v\} \cup \{\{x, y\} \in E_2 : x \neq v \text{ and } y \neq v\} \cup \{\{x, y\} : x \in NG_1(v) \text{ and } y \in NG_2(v)\}$. Obviously, if G is decomposable into G_1 and G_2, then $G = G_1 \times \ldots \times G_k$ instead of $((G_1 \times G_2)\ldots) \times G_k$.

The split decomposition of a graph is the recursive decomposition of the graph using simple decomposition until none of the obtained graphs can be decomposed further. The split decomposition tree of the graph G is the tree T in which each node h corresponds to a prime graph denoted by $G \times h$ obtained by the split decomposition. Furthermore, two nodes h and h' of T are adjacent if the corresponding graphs $G \times h$ and $G \times h'$ have a common marker, see Fig. 11.7.

SPLITS TREE [9,16,44] is the software to analyze phylogenetic networks using split decomposition, and the most recent version can also perform additional network reconstruction algorithms. Input data can be sequences, distances, quartets, trees, or splits, and the program can also perform statistical analyses. The splits and their isolation indices are part of the output. SPLITS TREE can be used to visualize complexity in phylogenetic data; it can indicate underlying biological processes but cannot prove existence of recombination and lateral gene transfer.

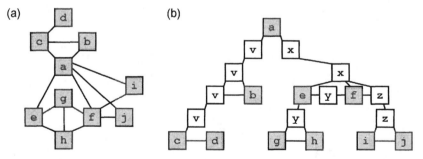

FIGURE 11.7   Example of a graph (*a*) and its split decomposition tree (*b*) with markers v, w, x, y, and z (modified after [16]).

11.5 SUMMARY

Although several methods for phylogenetic network reconstruction exist, models and analyses to detect lateral gene transfer, allopolyploidy, hybridization, as well as mechanisms operating at a microevolutionary level are still at an early stage of development. Difficulties for optimal network reconstructions are due to complex evolutionary processes as well as the differentiation between evolutionary signals and population genetic noise [51]. Even though these problems have not yet been solved, network approaches are appropriate methods for within-species phylogeny reconstructions as they incorporate population processes in the construction and refinement of haplotype relationships. More refined methods are needed to better reflect evolutionary mechanisms, which need to be transformed into mathematical and statistical forms to be incorporated in network reconstruction algorithms. However, to better detect population genetic processes of meiotic and sexual recombination as well as lineage sorting, not only the algorithms and software need to be refined, but biologists also need to adapt new software developments. First attempts have been made to provide statistical confidence assessments for different resolutions of data sets (e.g., parametric and nonparametric bootstrap methods on network reconstruction in the Bootscanning Package, [58]). Currently, there is no good general method for deriving networks, in the sense of an all-purpose tool for generating diagrams with reticulations; however, there is increasing interest in the development as well as in the application of intraspecific phylogenetics using network approaches to depict genealogical relationships [56]. As there also has not been any detailed comparative study of the various network methods, to compare and contrast their success rates in terms of false negatives and false positives, there is still much that needs to be done.

11.6 EXERCISES

1. Consider the following data matrix:

   ```
   Christopher Robin    ACGGAAATCGAA
   Winnie the Puh       ACGGATCGTTAT
   Tigger               ACGGATATCGAC
   Eeyore               GTACCAATCGAA
   Piglet               GTACCACGTTAG
   ```

   Apply the Galled Tree algorithm upon this data set using Christopher Robin as outgroup

2. Convert the following data matrix into a distance matrix and calculate a UP-GMA tree:

   ```
   1    ACTTCTATCATTGA
   2    ACTCCTATCATTGC
   3    ACTCCTATCTCTGC
   4    ACCCCTGTCACTAT
   5    ACTTCGGTCACTAG
   6    ACTTGTATGTATAG
   ```

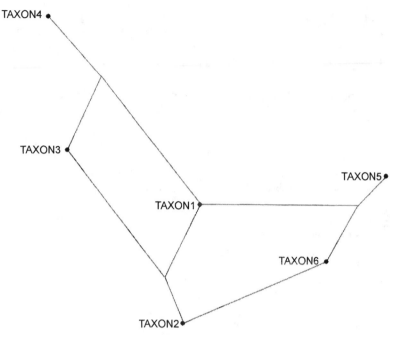

FIGURE 11.8   Example network.

   (a) Which network reconstruction methods can be applied from the distance matrix?

   (b) Apply the appropriate algorithm upon this dataset

3. Which different types of tree reconstruction methods are available and what are their disadvantages compared to network reconstruction methodologies?

4. Name four evolutionary processes leading to phylogenetic networks.

5. What type of network is displayed in Fig. 11.8 and what information does it contain?

REFERENCES

1. http://bioweb.pasteur.fr/seqanal/phylogeny/intro-uk.html.
2. http://corba.ebi.ac.uk/Biocatalog/Phylogeny.html.
3. http://evolution.genetics.washington.edu/phylip/software.html#methods.
4. http://www.rna.icmb.utexas.edu/linxs/1/phylogeny.html.
5. http://taxonomy.zoology.gla.ac.uk/rod/treeview.html.
6. http://www.csif.cs.ucdavis.edu/ gusfield/.
7. http://www.fluxus-engineering.com/sharenet.htm.
8. http://bioweb.pasteur.fr/docs/doc-gensoft/pyramids/.
9. http://www.splitstree.org/.

10. K. L. Adams, D. O. Daley, Y. L. Qiu, J. Whelan, and J. D. Palmer. Repeated, recent and diverse transfers of a mitochondrial gene to the nucleus in flowering plants. *Nature*, 408:354–357, 2000.

11. J. C. Aude, Y. Diaz-Lazcoz, J. J. Codani, and J. L. Risler. Applications of the pyramidal clustering method to biological objects. *Computational Chemistry*, 23:303–315, 1999.

12. H. J. Bandelt and A. W. Dress. Split decomposition: A new and useful approach to phylogenetic analysis of distance data. *Molecular Phylogenetics and Evolution*, 1:242–252, 1992.

13. H. J. Bandelt, P. Forster, and A. Röhl. Median-joining networks for inferring intraspecific phylogenies. *Molecular Biology and Evolution*, 16:37–48, 1999.

14. H. J. Bandelt, P. Forster, B. C. Sykes, and M. B. Richards. Mitochondrial portraits of human populations using median networks. *Genetics*, 141:743–753, 1995.

15. H. J. Bandelt, V. Macaulay, and M. Richards. Median networks: Speedy construction and greedy reduction, one simulation, and two case studies from human mtDNA. *Molecular Physiology and Evolution*, 16:8–28, 2000.

16. D. Bryant, D. Huson, T. Klöpper, and K. Nieselt-Struwe. Phylogenetic analysis of recombinant sequences. In *Proceedings of the 3rd International Workshop on Algorithms in Bioinformatics (WABI '03)*, volume 2812 of *LNBI*, pages 271–286. Springer, Berlin, 2003.

17. S. B. Carroll, J. K. Grenier, and S. D. Weatherbee. *From DNA to Diversity: Molecular Genetics and the Evolution of Animal Design*. Blackwell Science, Oxford, 2001.

18. I. Cassens, P. Mardulyn, and M. C. Milinkovitch. Evaluating intraspecific "network" construction methods using simulated sequence data: Do existing algorithms outperform the global maximum parsimony approach? *Systematic Biology*, 54:363–372, 2005.

19. M. A. Charleston and R. D. M. Page. *Spectrum*. Impkin Software, 1996–1997.

20. M. Clement, D. Posada, and K. A. Crandall. TCS: A computer program to estimate gene genealogies. *Molecular Ecology*, 9:1657–1659, 2000.

21. L. Mc Dade. Hybrids and phylogenetic systematics I. Patterns of character expression in hybrids and their implications for cladistic analysis. *Evolution*, 44:1685–1700, 1990.

22. L. Mc Dade. Hybrids and phylogenetic systematics II. The impact of hybrids on cladistic analysis. *Evolution*, 46:1329–1346, 1992.

23. E. Diday and P. Bertrand. An extension of hierarchical clustering: the pyramidal representation. In E. S. Gelsema and L. N. Kanal, editors, *Pattern Recognition in Practice II*, pp. 411–424. Elsevier Science Publishers, Amsterdam, 1986.

24. W. F. Doolittle. Phylogenetic classification and the universal tree. *Science*, 284:2124–2129, 1999.

25. J. Felsenstein. Evolutionary trees from DNA sequences: A maximum likelihood approach. *Journal of Molecular Evolution*, 17:368–376, 1981.

26. J. Felsenstein. Phylogenies and the comparative method. *American Naturalist*, 125:1–15, 1985.

27. J. Felsenstein. PHYLIP—Phylogeny Inference Package (Version 3.2). *Cladistics*, 5:164–166, 1989.

28. J. Felsenstein. PHYLIP: Phylogenetic Inference Package. University of Washington, Seattle, 1991.

29. D. J. Futuyama. *Evolutionary Biology*. Sinauer Associates, Sunderland, MA, 1998.

30. M. Y. Galperin and E. V. Koonin. Comparative genome analysis. *Methods of Biochemical Analysis*, 43:359–392, 2001.

31. P. A. Goloboff, S. Farris, and K. Nixon. *TNT (Tree analysis using New Technology) (BETA)*. Published by the authors, Tucumán, Argentina, 2000.

32. X. Gu. Maximum-likelihood approach for gene family evolution under functional divergence. *Molecular Biology and Evolution*, 18:453–464, 2001.

33. S. Guindon and O. Gascuel. A simple, fast and accurate method to estimate large phylogenies by maximum-likelihood. *Systematic Biology*, 52:696–704, 2003.

34. D. Gusfield. An overview of combinatorial methods for haplotype inference. In S. Istrail, M. S. Waterman, and A. G. Clark, editors, *Computational Methods for SNPs and Haplotype Inference*, volume 2983 of *Lecture Notes in Computer Science*, pp. 9–25. Springer, Berlin, 2004.

35. D. Gusfield. Optimal, efficient reconstruction of root-unknown phylogenetic networks with constrained and structured recombination. *Journal of Computer and Systems Sciences*, 70:381–398, 2005.

36. B. G. Hall. *Phylogenetic Trees Made Easy—A How-to Manual*. Sinauer Associates, Sunderland, MA, 2004.

37. W. Hennig. *Phylogenetic Systematics*. University of Illinois Press, Urbana, 1966.

38. D. M. Hillis. Phylogenetic analysis. *Current Biology*, 7:R129–R131, 1997.

39. D. M. Hillis, B. K. Mable, A. Larson, S. K. Davis, and E. A. Zimmer. Nucleic acids IV: Sequencing and cloning. In D. M. Hillis, C. Moritz, and B. K. Mable, editors, *Molecular Systematics*, pp. 321–381. Sinauer Associates, Sunderland, MA, 1996.

40. D. M. Hillis, C. Moritz, and B. K. Mable. *Molecular Systematics*. Sinauer Associates, 1996.

41. K. T. Huber, M. Langton, D. Penny, V. Moulton, and M. Hendy. Spectronet: A package for computing spectra and median networks. *Applied Bioinformatics*, 1:159–161, 2002.

42. J. P. Huelsenbeck, B. Rannala, and Z. Yang. Statistical tests of host-parasite co-speciation. *Evolution*, 51:410–419, 1997.

43. J. P. Huelsenbeck and F. Ronquist. MRBAYES: Bayesian inference of phylogenetic trees. *Bioinformatics*, 17:754–755, 2001.

44. D. H. Huson. SplitsTree: Analyzing and visualizing evolutionary data. *Bioinformatics*, 14:68–73, 1998.

45. A. G. Kluge and J. S. Farris. Quantitative phyletics and the evolution of anurans. *Systematic Zoology*, 18:1–32, 1969.

46. J. B. Kruskal. On the shortest spanning subtree and the traveling salesman problem. *Proceedings of the American Mathematical Society*, 7:48–50, 1956.

47. S. Kumar, K. Tamura, and M. Nei. MEGA: Molecular evolutionary genetics analysis, ver. 1.01. The Pennsylvania State University, University Park, PA, 1993.

48. P. Legendre and V. Makarenkov. Reconstruction of biogeographic and evolutionary networks using reticulograms. *Systematic Biology*, 51:199–216, 2002.

49. D. A. Libereles, D. R. Schreiber, S. Govindarajan, S. G. Chamberlin, and S. A. Benner. The adaptive evolution database (TAED). *Genome Biology*, 2:1–6, 2001.

50. C. R. Linder, B. M. E. Moret, L. Nakleh, A. Padolina, J. Sun, A. Tholse, W. Timme, and T. Warnow. An error metric for phylogenetic networks. Technical report TR-CS-2003-2026, University of New Mexico, Albuquerque, 2003.

51. C. R. Linder and L. H. Rieseberg. Reconstructing patterns of reticulate evolution in plants. *American Journal of Botany*, 91:1700–1708, 2004.

52. W. Madisson. A method for testing the correlated evolution of two binary characters: A gains or losses concentrated on certain branches of a phylogenetic tree? *Evolution*, 44:304–314, 1990.

53. E. P. Martins. Phylogenies and comparative data, a microevolutionary perspective. *Philosophical Transactions of the Royal Society of London B*, 349:85–91, 1995.

54. T. J. Merritt and J. M. Quattro. Evidence for a period of directional selection following gene duplication in a neurally expressed locus of triosephosphate isomerase. *Genetics*, 159:689–697, 2001.

55. C. Moritz and D. M. Hillis. Molecular systematics: Context and controversies. In D. M. Hillis, C. Moritz, and B. K. Mable, editors, *Molecular Systematics*. pp. 1–16. Sinauer Associates, Sunderland, MA, 1996.

56. D. Posada and K. A. Crandall. Intraspecific phylogenetics: Trees grafting into networks. *Trends in Ecology and Evolution*, 16:37–45, 2001.

57. A. Röhl. Phylogenetische netzwerke. Ph.D. thesis, Department of Mathematics, University of Hamburg, 1999.

58. M. O. Salminen, J. K. Carr, D. S. Burke, and F. E. Mc Cutchan. Identification of breakpoints in intergenotypic recombinants of HIV type 1 by bootscanning. *AIDS Research and Human Retroviruses*, 11:1423–1425, 1995.

59. S. Schneider, D. Roessli, and L. Excoffier. Arlequin ver.2.000: A software for population genetics data analysis. *Genetics and Biometry Laboratory*, University of Geneva, Switzerland, 2000.

60. D. S. Sikes and P. O. Lewis. PAUPRat: PAUP implementation of the parsimony ratchet. Technical report, Department of Ecology and Evolutionary Biology, University of Connecticut, Storrs, CT, 2001.

61. D. L. Swofford. Phylogenetic analysis using parsimony (PAUP) version 3.1.1. Smithsonian Institution, Washington, D.C., 1993.

62. D. L. Swofford, G. J. Olsen, P. J. Waddell, and D. M. Hillis. Phylogenetic inference. In D. M. Hillis, C. Moritz, and B. K. Mable, editors, *Molecular Systematics*. pp. 407–514. Sinauer Associates, Sunderland, MA, 1996.

63. A. R. Templeton. Nested clade analyses of phylogeographic data: Testing hypotheses about gene flow and population history. *Molecular Ecology*, 7:381–397, 1998.

64. A. R. Templeton. Statistical phylogeography: Methods of evaluating and minimizing inference errors. *Molecular Ecology*, 13:789–809, 2004.

65. A. R. Templeton, K. A. Crandall, and C. F. Sing. A cladistic analysis of phenotypic associations with haplotypes inferred from restriction endonuclease mapping and DNA sequence data. III. Cladogram estimation. *Genetics*, 132:619–633, 1992.

66. A. R. Templeton, E. Routman, and C. A. Phillips. Separating population structure from population history: A cladistic analysis of the geographical distribution of mitochondrial DNA haplotypes in the tiger salamander, *Ambystoma tigrinum*. *Genetics*, 140:767–782, 1995.

67. T. F. Turner, J. C. Trexler, J. L. Harris, and J. L. Haynes. Nested cladisitic analysis indicates population fragmentation shapes genetic diversity in a freshwater mussel. *Genetics*, 154:777–785, 2000.

68. K. Weising, H. Nybom, K. Wolff, and G. Kahl. *DNA Fingerprinting in Plants: Principles, Methods, and Applications*. CRC Press, Boca Raton, FL, 2005.

69. S. Z. Xu. Phylogenetic analysis under reticulate evolution. *Molecular Biology and Evolution*, 17:897–907, 2000.

70. H. Zuh, J. F. Klemie, S. Chang, P. Bertone, A. Casamayor, K. G. Klemic, D. Smith, M. Gerstein, M. A. Reed, and M. Snyder. Analysis of yeast protein kinases using protein chips. *Nature Genetics*, 26:283–289, 2000.

12

ECOLOGICAL NETWORKS

Ursula Gaedke

12.1 INTRODUCTION

Ecological networks typically represent *food webs*, which may be defined as networks of consumer-resource interactions between groups of organisms [40]. By these means, they describe who is present and who affects whom directly or indirectly by feeding interactions (for other interactions see below). Such information is essential for the understanding and management of the dynamics of the individual groups of organisms and of the entire ecosystem. In food webs, the vertices (nodes, points) are represented by individual species, certain life stages of one species, or by a number of species that were aggregated to form one group. A *biological species* is defined as the group of organisms that can, at least potentially, breed together in nature to produce fertile offspring [2]. When regarding the food web in a certain habitat, as it is mostly done, it is more appropriate to talk about populations rather than species. A *population* comprises all individuals of one species within a certain area (e.g., within a lake or on a leaf). To standardize across species and studies, the density (abundance) of a population is often provided, for example, as the number of individuals per square meter or per liter of water volume. To establish food webs it may, however, be more meaningful not to consider biological species that are units of reproduction but units of organisms that share the same predators and prey, that is, play similar trophic roles [7,50]. Hence, from the trophic point of view we may split or aggregate biological species into *trophic guilds*. A trophic guild consists of those species that share the same predators and prey [51]. By accounting simultaneously for the links to prey and predators this definition goes beyond the classical concept of trophic levels

Analysis of Biological Networks, Edited by Björn H. Junker and Falk Schreiber
Copyright © 2008 John Wiley & Sons, Inc.

that considers only the acquisition of resources and, thus, distinguishes, for example, primary producers (e.g., plants), herbivores, and carnivores. In some instances, a species may form a trophic guild. However, the diet of an individual predator species and/or the susceptibility of a prey species may change strongly during the individual ontogenetic growth from juveniles to adults. For example, a young fish larvae feeds on tiny zooplankton, whereas the adults may predate on other fish including the own offspring; and a newly hatched crocodile is at high risk of predation, whereas an adult one is not. Such ontogenetic differences between individuals of one species of different age in the feeding links may exceed the differences between biological species. This holds true in particular for species with pronounced ontogenetic growth (e.g., fish and other animals that produce a large number of small offspring that have to grow ten thousand fold or more to reach adult size), and for those undergoing metamorphosis, for example, dragon flies with predatory larvae living in lakes and adults foraging in the terrestrial realm, and for species with a large variability in their appearance (e.g., algal species occurring as single cells or large colonies). The concept of trophic guilds is also appropriate for small species, which can hardly be routinely distinguished at the biological species level. Its application also avoids that a vertex feeds on itself, which facilitates subsequent computations and the interpretation of the results (see below). Finally, using the concept of trophic guilds may reduce the effort required to establish all feeding links and to avoid the impression of unjustified accuracy.

However, the use of trophic guilds has also been criticized because their definition is at least to some extent subjective and mutually dependent or circular [29,30], rising similar problems as encountered in social science when trying to define social relations. For example, distinguishing numerous trophic guilds at the level of herbivores has consequences for the definition of the trophic guilds at the level of the primary producers (e.g., plants), because a smaller number of autotrophic species will belong to the same guild if feeding preferences of the herbivores are considered in more detail. As a consequence, new research motivating to split one trophic guild into two may have far reaching consequences for the definition of many other guilds within the food web. This motivated the search for more objective criteria of aggregation. For pelagic food webs (i.e., those in the open water body of lakes and oceans), body size is a useful criterion because many physiological and ecological properties of a pelagic organism are related to body size [37]. Hence, we may aim to have a similar number of trophic guilds within each size range (see Fig. 12.1). Furthermore, given a sufficient data basis, advanced statistical approaches may provide a more objective procedure to allocate species into trophic guilds [30] than the often used ad hoc methods. An additional point of concern when using the concept of trophic guilds [4] is that organisms may play similar roles in the food web to the extent that they have similar links to analogous (but not identical!) species. In former concepts, two species became members of the same trophic guild if and only if they had trophic links to (almost) the same set of other trophic groups. Stimulated by work in social sciences, this limitation was overcome by distinguishing between structural equivalence (same links) and regular equivalence (analogous but not identical links), see [26,29] and literature

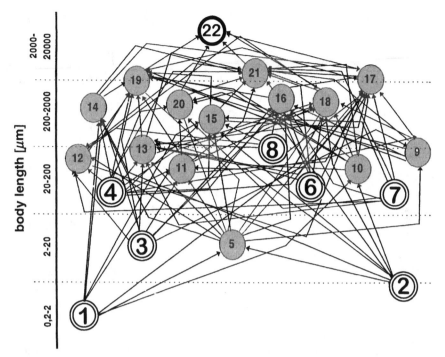

FIGURE 12.1   Example of a binary food web. The plankton food web consisting of more than 200 biological species in the open water body of large, deep Lake Constance was aggregated into 22 trophic guilds. The *y*-axis provides the body size of the (adult) organisms, which was used as an objective criterion to achieve an approximately even aggregation throughout the food web. This was not feasible for the smallest organisms (e.g., bacteria, #1, autotrophic picoplankton, #2) because here species or trophic guilds cannot be distinguished morphologically. Basal trophic guilds are represented by thick gray circles, intermediate ones by thin circles, and the top predator by a black thick circle.

cited therein. Using the concept of regular equivalence species are allocated into *isotrophic classes*, which have the same structural roles. They may or may not consume the same prey and may or may not share the same predators [29]. For example, small and larger herbivores preying upon different autotrophs and being consumed by different predators may play a similar trophic role although they may neither share any prey nor predator species. Hence, using the concept of structural equivalence, they would be regarded as totally different trophic guilds. In contrast, their coefficients of regular equivalence (*REGE coefficients*) may be high given their analogous role. The concept of regular equivalence is applicable to binary and quantitative food webs (for definition see below) [29]. To conclude, defining the trophic role of an organism remains a challenge in food web ecology since the first attempts by Elton [13]. This implies that the definition of the vertices in a food web model has to fit with the purpose for which the model is designed. Here, the expressions species, trophic guilds and compartments (typically used in the context of quantitative food webs) are used interchangeably for the vertices of the web.

The connections or links (edges, arcs, lines) in food webs are represented by trophic interactions, that is, the feeding of one species on other ones. When constructing a food web for a particular ecosystem, information on trophic interactions may be obtained by various means, which then, in turn, determine the aggregation of species into trophic guilds. In the natural environment, trophic interactions may be inferred from direct observations, the analysis of the stomach content (identify prey organisms visually or analyze genetic sequences), the analysis of the excrements, and by immunological and isotopic techniques. This information is typically supplemented by laboratory and field experiments (e.g., on the growth rate of a predator on a certain prey, reduction of prey densities at different predator densities), by studies on the morphology and the feeding behavior of a potential predator and by investigations of potential predator avoidance mechanisms of the prey. Inference on trophic interactions is sometimes also obtained from the analysis of time series on the densities of the potential predator and prey species. This may be done by visual inspection of the time series and by statistical approaches. Recently, preliminary attempts have been made to construct correlation networks (see also Chapter 13) from time series of population densities. In all cases, success has been limited to the best of my knowledge, which may be attributed to the strong noise overlaying the time series, time-lags in the response of predator abundance to prey availability, the complex diet and opportunistic feeding behavior of most consumers, and various indirect effects within the food web. Knowledge on feeding interactions may often by generalized across ecosystems, that is, it may be inferred from previous studies conducted elsewhere. However, many consumers are opportunistic in their feeding behavior and alter their diet depending on the supply with various resources. This can lead to a large spatio-temporal variability in the diet of one species. Hence, considerable uncertainty may remain about the existence or absence and the quantitative importance of a feeding link at a given time and location even for the best studied ecosystems.

Trophic interactions rule the flow of matter and energy in food webs and are important for the transfer of information in most ecosystems. However, it is increasingly understood that on top of trophic interactions other kinds of interactions between groups of organisms may influence their growth and loss processes and, thus, indirectly also the flow of matter and energy in the entire food web. Examples for such interactions include pollination, seed dispersal by animals, and kairomones. The latter are substances that are released unintentionally in very low concentrations by a predator and perceived by its prey that changes its behavior, morphology, and/or life cycle to reduce the predation risk at high predator densities. For example, a water flea can smell if fish are present and responds accordingly by changing its morphology and life cycle and hiding in deep dark water layers during the day. Such observations contributed to the development of a new branch in ecology, called chemical ecology.

Food webs and ecological networks, in general, are typically directed. In food webs, plants, bacteria, fungi, and detrivores (i.e., animals that consume dead organic matter and the attached microorganisms) represent the source vertices as they form

the energetic basis of the entire community whereas top predators are the end vertices. Circular relationships where, for example, A eats B, B eats C, and C eats A
do not exist among adults at least in pelagic food webs. However, pronounced ontogenetic growth may provoke circular feeding relationships. This may be due to
cannibalism within one species where large individuals prey on small ones. Such
cannibalism dampens the fluctuations in the population dynamics of this species as
the mortality of the juveniles is high at high population densities. Circular feeding
relationships may also occur among different species, which both exhibit pronounced
ontogenetic growth (Fig. 12.2). For example, adult jelly fishes predate on copepods
(small crustaceans) and may strongly reduce their densities. However, adult copepods, in turn, may predate on (or destroy) juvenile jelly fishes, that is, the offspring
of their predator. This results in a positive feed back mechanism that may have striking effects for the dynamics of both species. If, for example, copepods are first in
establishing high densities in spring, densities of jelly fishes will remain low due
to the high mortality of their juveniles. Otherwise, if the jelly fishes come first,
copepods will not achieve high densities (Fig. 12.2). Hence, seemingly minor differences in the initial conditions (here densities) may have a major impact on the
subsequent development of the system (and the nutrition of fish feeding on copepods

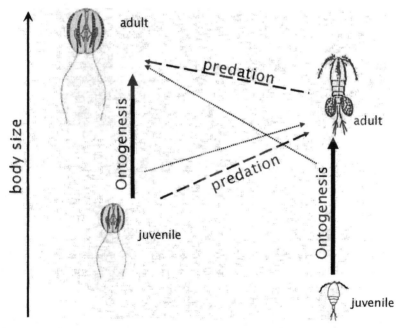

FIGURE 12.2 Example of circular feeding relationships between two biological species,
which both exhibit pronounced ontogenetic growth (i.e., eggs and hatchlings are much smaller
than the adults). Adult copepods (small crustaceans, right) prey upon juvenile jelly fish
(Ctenophora, left) whereas adult jelly fish prey upon the copepods, that is, the predators of
their juveniles.

but not jelly fish), which is one characteristic of chaotic behavior. Overall, such circular feeding relationships represent a network motif (see also Chapter 5) but are typically less numerous and quantitatively less important than unidirectional ones, see Ref. [34] and literature cited therein. From a functional and dynamic perspective, circular feeding relationships have to be clearly distinguished from the recycling of matter within the food web via dead organic matter and remineralization of nutrients. For example, all consumers release (egest) some part of their food because it is indigestible and they take up certain chemical compounds in excess (e.g., nitrogen, phosphorus). These substances are mostly recycled within the ecosystem, for example, bacteria decompose organic matter and plants take up nutrients released by animals. Such processes are of outstanding importance for ecosystem functioning but a differentiation between predation and egestion appears necessary from most ecological points of view, that is, if others than a book-keeping of fluxes is intended. Egestion, for example, exerts no dynamic feedback control on the releasing compartment whereas predation normally does. In addition, the composition of the egesta from a large range of different organisms may be similar. This enables an almost unlimited *omnivory* of the bacteria/detrivores, which are in food web models lumped together in one or a few compartments. This stands in contrast to predation, as individual predators are generally restricted to certain kinds of organisms.

Natural food webs may comprise hundreds of species and tens of thousands of feeding links even if they cover only one particular type of habitat. This complexity forces us to abstract from the situation and to develop verbal, graphical, and mathematical models, which portray different features of the same natural system. As a consequence, constructing ecological networks has a long tradition starting with Ref.[13] and different kinds of abstraction and of food web modeling have been developed since. Here, we will distinguish between

1. Binary food webs ("Who eats whom?")
2. Quantitative trophic food webs ("Who eats whom (how much)?", "Who recycles (how much)?")
3. Ecological information networks ("Who influences whom (how much)?")

These approaches proceed in the given order along a gradient of increasing requirements for data and knowledge. They will be described and compared with respect to their theoretical foundation, data requirements, operational problems, and the information they provide on food web structure and function (for details see also Ref. [16], for a review Ref. [18]). Binary food webs only consider which species are present and who eats whom but not the density of the organisms and the quantity of flows between the different species. They form the basis for the so-called *food web theory* that aims to recognize generalities and scaling laws in the structure of natural food webs, which are then compared to networks found in other disciplines [41]. Quantitative food webs take into account the magnitude of flows among different species and often also between species and their abiotic environment, that is, they may account for recycling.

They provide comprehensive descriptions of the fluxes and cycling of matter (e.g., energy, nutrients, etc.) and of the trophic structure when evaluated by network analysis. The amount of matter and energy that is passed along an individual flux varies by several orders of magnitude within a food web. Hence, the realism of a food web is greatly improved by quantifying the fluxes. However, for the time being this complicates the evaluation of the food web structure and the comparison with networks found by other disciplines, which is a major goal of the present book. Hence, the following chapter gives more emphasis to the binary than to the quantitative webs although this does not reflect their relevance for answering most ecological questions. Binary and quantitative food web models have two short-comings in common. They are both static in the sense that they represent a snapshot or a spatio-temporal average of a food web at a certain location and time, and they are restricted to trophic interactions. These disadvantages may, in principal, be overcome by developing dynamic models of quantitative food webs, which also include other kinds of information exchange than trophic interactions. However, given the complexity of this task and the amount of data and knowledge required to set up such ecological information models, they are typically restricted to a subset of processes within a food web, which are regarded as particularly important for overall system dynamics or the specific question under consideration. The latter determines whether the model is developed and parameterized for a particular ecosystem or to study the dynamics of certain processes such as predator–prey cycles in principal without referring to particular species or habitats.

To set up an ecological network always requires the definition of system boundaries since establishing a detailed food web or information network of the entire planet is out of scope. System boundaries are typically chosen such that interactions within them are stronger than across them. Difficulties may arise for all systems and approaches but they are usually most severe for binary webs because here all links are equally weighted.

12.2 BINARY FOOD WEBS

12.2.1 Introduction and Definitions

Binary food webs (or topological or descriptive webs) only consider the existence or absence of a feeding link between the component species but not its magnitude or interaction strength, which gave rise to the name "binary." This information is typically provided in the so-called *community matrix* in which predators are usually allocated to the rows and prey to the columns. The values of the matrix are either 1 or 0 representing the presence or absence of a feeding relationship between the respective pair of predator and prey. This matrix corresponds to the adjacency matrix described in Chapter 2. Binary webs provide a static description of "who eats whom."

12.2.2 Descriptors of the Network

A large number of metrics have been developed to describe the structure of binary food webs, which were often closely linked to graph theory (see Chapter 2) [50]. They include the number of species/trophic guilds, S, and the number of links, L, representing the vertices and edges of the web. The species are classified as either top predators (T, having no predators but prey on other species), intermediate species (I, preying on others and being themselves preyed upon), or basal species (B), which feed on no other species but are fed upon by I and T. Basal species comprise *autotrophs* deriving their energy from photosynthesis, and often also bacteria and other *osmotrophs* that derive their energy from dissolved organic substances, and *detrivores* that consume dead organic matter. They form the basis of the food web as they are preyed upon by other species and typically rely themselves on inorganic or organic carbon sources and inorganic nutrients. The existence of true top predators may be questioned as they are at least attacked by parasites and pathogens or may be subject to cannibalism, which is often ignored in food web studies [22]. To quantify the trophic structure of the binary web, the proportions of top, intermediate, and basal species, and the ratio of the number of prey species to the number of predator species, $(T + I)/(I + B)$, is computed. The mean chain length represents the average number of links connecting top to basal species. The maximum chain length is defined accordingly. The values of the mean chain length obtained from binary webs are higher than those obtained from quantitative webs because binary webs give equal weight to all links [47]. Quantitative webs account for the fact that with each trophic transfer a large part (at least 66%, but often 90% and more) of the energy is lost by egestion and respiration (see below for details). As a consequence, energy fluxes at higher trophic levels and along long food chains are much smaller than at the basis of the food web and along short food chains. For example, bears or wild pigs gain most of their energy by eating plants rather than by eating herbivores, which have eaten plants. This is not accounted for by binary webs but see [47] to reduce the problem. The degree of connectedness within a binary web is measured by the *linkage density*, $D = L/S$ and the ratio of the number of observed links to the number of all possible links (*directed connectance* $C = L/S^2$ or, excluding cannibalism, $C = L/(S \times (S - 1))$, or, excluding cannibalism and cyclic feeding relationships, $C = 2L/(S \times (S - 1))$. The connectance represents the fraction of all possible links that are realized in the specific food web. D is a function of C and S (e.g., [32]). More recently, stimulated by ongoing research in social science, emphasis has also been given to a potential clustering or compartmentation within the webs where neighbors of a vertex are more likely to be connected to each other than in a random graph [28] (for details on network clustering see also Chapter 6). In a compartmented web, relatively isolated sub webs exist that are relatively strongly connected within themselves and less with the other sub webs. In an extreme case, a food web may consist of several hardly interconnected linear food chains. In theory, compartmentation increases the stability of networks. One formula used as an index of compartmentation is

$$C_1 = 1/(S(S - 1)) \times \sum_{i=1}^{S} \sum_{j=1}^{S} p_{ij}, \qquad i \neq j,$$

with p_{ij} being the number of species interacting both with species i and j, divided by the number of species interacting either with i or j (see [34] and literature cited therein). Related information is provided by the link distribution frequency defined as the frequency of species S_L which have L links either to a predator or a prey [33]. According to expectations, a stronger tendency for compartmentation was found in food webs that comprised species from different habitats (e.g., aquatic and terrestrial) than in food webs originating from more homogenous habitats. However, as frequently encountered in ecology, the situation is not as clear and trivial as that as, for example, food webs in one (sub-)habitat may perceive a substantial subsidy of matter and energy from others (e.g., leaves of trees falling into a creek) (see [2] and literature cited therein).

12.2.3   Operational Problems

Binary webs have the seemingly advantage to provide comprehensive information on the structure of food webs from many different habitats. However, a closer look reveals that there are numerous operational problems associated with this approach, which may greatly reduce its reliability and overall relevance [7,20,34,40]. Especially during the first decades of research, the catalogues of established binary webs that were used to search for generalizations, contained mostly webs that were established for purposes other than those of food web theoreticians. Hence, they were often focused on some part of the entire food web (e.g., the endangered or commercially interesting species or those which were easy to assess) and did not deliver an adequate description of the entire web. Such an unequal precision in the representation of a food web may strongly affect the above-mentioned metrics. For example, the taxonomic resolution at the higher trophic levels was often larger than at the lower ones. As a consequence, some studies were undertaken that specifically addressed this problem and resulted in some more resolved food webs [6,31,39,48] (for recent compilations of binary food webs see also http://www.foodwebs.org/index.html). Other problems encountered when establishing binary food webs are the in- or exclusion of less known or quantitative very minor feeding links [50] and of species that are not regular present in the habitat (e.g., migratory birds, transient species). The latter strongly depends on the definition of the system boundaries. In binary webs, the spatiotemporal variability in the food web structure may be inferred from changes in the species list. However, many species occur year round or throughout the habitat but in very different densities. If a species is not encountered in a certain sample, we only know that its density was below the — often variable — detection limit, which complicates spatiotemporal and cross-system comparisons [6,48]. For further discussion and suggestions for improvements see also [7,20].

12.2.4   Aims and Results

The motivation to study food web structure is manifold. First, knowledge of the major feeding links is essential for any kind of management, either of individual species or the entire ecosystem. However, as the quantitative importance of the numerous feeding links differs greatly, binary webs have little to offer for the solution of applied ecological problems. Hence, they play at most a marginal role in applied ecology. Regarding basic science, a long lasting goal has been to identify potential regularities in the structure of natural food webs of different habitats. Binary webs are of focal interest for those searching for phenomenological regularities in food web structures and their dependence on the type of habitat, its characteristics (e.g., disturbance regime, productivity), the size of the web and the history of assembly [8,20,22]. Based on a large compilation of binary food webs from very different habitats several patterns were suggested, such as a constant ratio of predator to prey species, constant proportions of links between B, I, and T and constant linkage density, D, which implies a hyperbolic decrease of C with S; and infrequent occurrence of three-species loops, for example, [7]. Furthermore, food chains in two-dimensional habitats (e.g., grasslands) appeared to be shorter than those in three-dimensional habitats (e.g., open water bodies, forests with well-established canopy) [32,34]. These patterns, some of which have also been termed *scaling laws* [7,22], are not discussed here in detail, as the underlying binary webs were often subject to the operational problems mentioned above. That is, they may or may not be artifacts and some are supported by more recent detailed food web studies whereas others are not [10,22,34]. The interest in such regularities first arose in the context of the discussions on relationships between food web stability and complexity, with complexity being often expressed by the size and connectance of the web [38,49,50]. A key question was, and given the dramatic loss of biodiversity to some extent still is, "Are species-rich food webs more or less stable than species-poor webs?" In this context, the relationship between S and C was analyzed by comparing numerous empirical binary webs. There is no final answer yet as the empirical evidence is contradicting (C may increase, decrease, or remain constant with increasing S [20], which is partly due to the operational problems discussed above). These studies also revealed that concepts such as food web stability, complexity, size, and interaction are neither trivial to define nor to measure, and that considering only the existence or absence of feeding links may fall short in several respects (for some details see below, [2,15] and literature cited therein).

Interest in the structure of binary food webs was further promoted by studies of networks of other systems including those in physics, life sciences, economy, and social science, which is sometimes also termed *complex system research* [41]. They led to an exchange of analysis techniques [29] and attempts to compare network properties across disciplines. One way to search for patterns in the binary food web structure is to compute the above-mentioned metrics for empirical webs and for artificial webs, which were assembled at random following some rules to maintain a certain degree of realism (e.g., having the same S and C as the empirical ones) [11]. Statistically significant deviations between the patterns found in the empirical and the random webs indicate that biological interactions or other mechanisms not considered when creating the random webs influence the food web structure. This approach, sometimes termed *null model analysis* [2], resembles the familiar testing of null hypotheses in a statistical context and is more widely used in ecological research. The observed patterns are compared with what may be expected purely by chance without the consequences of any biological interactions. However, the latter is often subject to debate [34]. It has been attempted to find some simple rules for constructing food webs that reproduce major properties of real food webs. First, the *cascade model* was suggested [8] where the species get numbers from 1 to S. They only prey on species with a lower number than themselves and do so with the probability d/S with d being the density of links per species. d allows to fit the model to the data. With $d = 4$, some structural properties of the earlier, more aggregated webs were reproduced but deficiencies exist when using the recent, more resolved webs [10]. A similarly constructed model, the *niche model* [46], also puts the species into a certain order describing their *niche*, and the prey window of a consumers has a randomly chosen center with a lower niche value than that of the consumer. Comparing the performance of the cascade and the niche model in reproducing 12 food web properties revealed a better fit of the niche model [46]. It has, however, to be kept in mind that such static binary models cannot provide a mechanistic explanation for the potential existence of general patterns in natural food webs [10,41]. Both models order species along an abstract gradient and the niche model assigns a certain niche to each species. Such an ordering may be based on body size for pelagic food webs from large open water bodies, as predators are here typically larger than their prey. However, there is also an upper boundary to the difference between predator and prey for energetic reasons resulting in a certain prey size window relative to the size of the predator.

12.2.5 Conclusion

There were and still are substantial operational problems when establishing and comparing binary webs from different habitats and researchers, which implies that in particular early results on generalities in food web structures may be out-dated by analyses of more accurate food web studies. Nevertheless, they form the basis of ongoing research aiming to identify generalizations in food web structure and to compare them to other kinds of networks in biological, social, and other science. The analysis of binary food webs has a great power to summarize binary structural patterns of complex food webs at the expense of contributing little directly to the understanding of ecosystem functioning. Reasons for this include that binary web ignore the pronounced differences in the relative importance of the different feeding links [35], hardly consider the often pronounced spatio-temporal variability, and do not include other interactions than feeding links. This explains why important ecological concepts may not be based on binary webs and why they have rarely been used to solve applied problems.

12.3 QUANTITATIVE TROPHIC FOOD WEBS

12.3.1 Introduction, Definitions, and Database

Binary food webs may be converted into trophic food webs (also called *flow* or *bioenergetic webs*) by quantifying the biomass of the trophic guilds (typically called *compartments* in this context) and the individual links (typically called *fluxes* or *flows* in this context), that is, the diet composition of each consumer compartment has to be known quantitatively. In addition, the exchange with the nonliving environment is often quantified as well (e.g., the uptake of carbon dioxide by plants, the release of carbon dioxide and dead organic carbon by the organisms, and the uptake of organic substances by bacteria, fungi, and detrivores). Most often, this quantification is done in units of carbon as this represents the most suitable surrogate for energy that is typically the most decisive resource for consumers (but see below). The quantity of the individual living compartments (i.e., its biomass) is often determined by counting the number of individuals of each species (or life stage of each species, or higher taxonomic unit) per unit area (or volume of water), measuring the weight of each individual and computing the product. The latter represents the biomass per unit area (volume), expressed, for example, in $mgCm^{-2}$. Depending on the group of organisms more automated procedures for estimating biomass may exist (e.g., measuring the chlorophyll concentration of phytoplankton). Measuring the fluxes among the living compartments and their exchange with the nonliving environment is with few exceptions more demanding than establishing the compartmental biomasses. For each living compartment, quantitative information is required about the food uptake (I, ingestion), release of organic matter (E, e.g., excretion of nondigestible parts of the food), respiration (R), and production (P, comprising somatic growth and reproduction) per unit biomass and time. For each living compartment, the equation of mass-balance has to be fulfilled: $I = E + R + P$, that is, ingestion has to cover the needs for respiration and production after accounting for the nonassimilated part of the food. A typical unit for these fluxes is $mgCm^{-2}d^{-1}$. Flux measurements are nowadays obtained by a large variety of technologies depending on the ecosystem, organism, and flux under consideration. Photosynthesis providing the energetic basis of plant growth and, thus, of most ecosystems, may be inferred, for example, from the measured uptake of carbon dioxide or release of oxygen, the activity of the electron transport system within the photosynthesis apparatus and the net growth of plants if this is observable in the field. Bacterial production, representing often the second basal energy input into most food webs, may be estimated from the uptake of radioactively labeled substances. In addition to measurements conducted in the field, laboratory experiments are routinely used to obtain flux estimates. They may comprise measurements of the growth, respiration, and reproduction rates of the component species as a function of food concentration and temperature, and the maximum food uptake as a function of prey availability and quality. Very importantly, the efficiencies by which ingested food is assimilated, $(R + P)/I$, and invested into the production of new biomass, $P/(R + P)$, should be quantified as far as possible. When setting up the trophic food web model, they may not surpass certain threshold values as a part of the food is always nondigestible (this fraction tends to be higher for plants than animals), and the maintenance of the metabolism always demands energy (i. e., respiration that is higher for vertebrates, and birds and mammals in particular, than for unicellulars and noninvertebrates). Mass-balance conditions have to be fulfilled for all individual compartments as well as for the entire system. For example, in autochthonous ecosystems that do not receive subsidies from other ecosystems, primary production must be sufficient to allow for a potential increase in biomass and at the same time compensate all losses from the ecosystem, for example, by respiration and sedimentation. Furthermore, the release of organic material by the different compartments has to cover the demands of bacteria and detrivores. This leads to a set of equations that is on the one hand highly overdetermined, that is, the number of equations is much larger than the number of free parameters. On the other hand, all estimates of diet compositions, biomasses, and the amount of fluxes are subject to often considerable uncertainty and spatiotemporal variability. Several programs have been developed, which try to cope with these facts. For example, the software developed by [21] is based on an inverse approach [27,42] that keeps the balanced quantitative food web models closest to their initial estimates (for alternatives see, e.g., [5,14,23] and literature cited therein). If the data base is sufficiently comprehensive, the internal consistency of the different sources of information used to quantify the network can be controlled and some fluxes may be estimated, which are particularly inaccessible to measurements.

12.3.2   Multiple Commodities

Most trophic models were quantified in units of carbon representing the best surrogate for energy and energetic constraints are thought to drive most food web processes. However, it is increasingly recognized that food quality may also play a decisive role on top of quantity. For example, if plants grow under severely nutrient depleted conditions, their internal ratio between nutrients and carbon declines strongly. Such flexibility in the stoichiometric composition is only found in plants and bacteria but not in animals, which, in addition, have higher nutrient/carbon ratios than plants. As a consequence of these two facts, growth of herbivores may be limited by nutrients rather than carbon/energy when they have to rely exclusively on strongly nutrient depleted plants, for example, [17]. This is one example, how food quality may determine growth of herbivores. Other examples include a lack of polyunsaturated fatty acids or sterols, for example, [45]. These findings motivated the recent development of trophic models in two commodities (e.g., carbon and the most limiting nutrient) [9,17,25,44] and of programs that can mass-balance food webs in two commodities simultaneously, for example, [21]. This increases the realism of the network model and implies additional constraints on the mass-balance conditions if fluxes in the second commodity and/or ratios between the two commodities in the compartmental biomasses have also been measured. As already pointed out in the context of binary webs, it should be acknowledged that the large spatiotemporal variability and adaptability inherent in almost all ecological processes and entities cannot be fully reflected in any food web model. Hence, a considerable degree of uncertainty in all flux estimates remains even for well-studied systems. One way to account for these problems and to avoid the impression of unjustified accuracy is to stronger aggregate the component organisms. That is, the number of compartments in a trophic food web model is typically smaller than the number of trophic guilds in a binary web (see Fig. 12.2 and 12.3, [30]).

12.3.3   Descriptors of the Network and Information to be Gained

Various measures have been developed within the framework of *network analysis* [14] to describe and compare trophic food webs in time and space and among habitats.

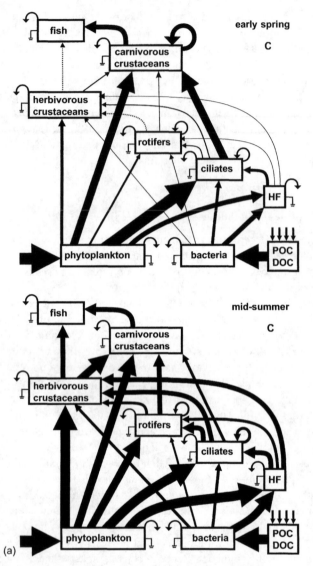

FIGURE 12.3 Example of a quantitative food web model and its seasonal variability (top: early spring, bottom: mid-summer, 1991) in the open water body of large, deep Lake Constance. The network was simultaneously mass-balanced in units of carbon (surrogate for energy) and of phosphorus, representing the most limiting nutrient. The width of the arrows represents the magnitude of flows in units of carbon or phosphorus on a logarithmic scale. ⊥ symbolizes respiration and ↷ the release of carbon or phosphorus by egestion, which enters the pool of dead organic matter (POC—particulate organic carbon, DOC—dissolved organic carbon, POP—particulate organic phosphorus, DOP—dissolved organic phosphorus). Respiration and egestion are not drawn at scale. Flux diversity increases from spring to summer. The food web models are similar in units of carbon and phosphorus during early spring when plenty of nutrients (here phosphorus) are available and, thus, also phytoplankton has a relative high phosphorus content.

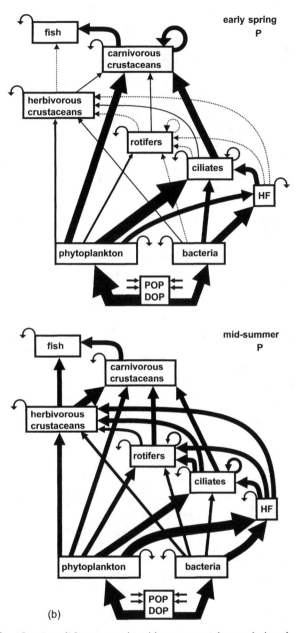

FIGURE 12.3 *(Continued)* In contrast in mid summer, carbon and phosphorus webs differ due to the nutrient depletion of phytoplankton. Then, the carbon (energy) and nutrient supply of the consumers is less tightly coupled; they derive most of their energy from phytoplankton but substantial amounts of phosphorus from the phosphorus-rich bacteria and other animals. Note that bacteria and phytoplankton are competitors in terms of phosphorus but not in terms of carbon, which leads to a major difference in food web structure and functioning in respect to carbon and phosphorus (data from [17,24]).

They are partly similar to those used for binary food web analysis but complemented with others, sometimes derived from thermodynamic considerations. They quantify the cycling and remineralization of nutrients, intercompartmental dependencies ("who gets directly or indirectly how much energy or nutrients from whom?"), and the trophic structure. The latter provides information for example, on "how much biomass and energy is available at which trophic level?" "how efficient is energy channeled from basal to top species?" "how important are which parts of the food web for these processes?" "which organisms are at which trophic level/trophic position?" A measure for the size (or productivity) of the food web is not the number of species but the total system throughput, which is the sum of all fluxes within the network. The average path length measures the mean number of trophic transfers that a unit of biomass goes through from its entry into the food web (e.g., by primary or bacterial production) until it leaves it (typically by respiration) and represents a weighted average of the food "chain" length. As pointed out above, this average path length obtained from quantitative food webs is generally smaller than the corresponding value obtained from a binary web because it is weighted by the quantitative importance of each flux. In addition, mass-balanced trophic food webs can be used to follow the pathway of a unit of biomass through the food web as time progresses. The computation of these compartmental and system resident times provides information on the velocity at which the unit of biomass is channeled through and lost from the system and on the potential accumulation in certain compartments. This may be relevant to understand nutrient cycling and the behavior of toxic substances. The computation of the various measures is facilitated by software packages such as NETWRK [14,43] and ECOPATH [3,4] (and ECOSIM, http://www.ecopath.org, for summary and activities); for a comparison of both packages see [5,23].

12.3.4   Conclusion

To conclude, quantitative food web models provide comprehensive descriptions of fluxes and the cycling of carbon and nutrients in an ecosystem, especially when they are mass-balanced and evaluated by network analysis. They answer questions such as "who eats whom how much?" but also "who respires how much?" and "who excretes/remineralises how much?" By accounting for the large variability in the relative importance of the different fluxes they provide a more realistic picture than binary webs. This motivates ongoing attempts to include quantitative scales into theoretical network analyses [35,41]. However, even quantitative food web models still have limitation in explaining probable causalities in food web dynamics, as they are primarily static and the interaction strength between two compartments does not only depend on the quantity of the fluxes among them. Such tasks may be approached using dynamic information networks.

12.4   ECOLOGICAL INFORMATION NETWORKS

The ecological networks presented so far were primarily static and restricted to trophic interactions (including remineralization). As most ecological processes are highly variable in time and space, numerous other models have been developed, which are dynamic and account for spatial structure. They may either be based on a set of coupled differential equations that describe the interactions between the different species and with abiotic forcing factors. This type of model is most appropriate if a large number of individuals behaves in a similar way, for example, unicellular algae in an open water body. If, in contrast, the individual life history and distinct neighborhood relationships to other organisms largely determine species interactions and other ecological processes, individual-based, spatially explicit modeling approaches should be used [19]. They are often used in terrestrial plant vegetation ecology, for example, to describe the growth of individual trees depending on the trees in the neighborhood. Identifying the most relevant processes that have to be incorporated into the model to answer a certain question, and finding sufficiently realistic mathematical expressions and parameters for them is very challenging given the complexity and adaptability of ecological processes and networks, especially when compared to the usual amount of funding devoted to ecological research. The model design has to be strictly focused on answering specific questions, and trade-offs are required between generality, realism, our ability to track the model behavior, and the effort required to set up and test the model. To date, most dynamic food web models related to particular ecosystems have either focused on parts of the entire food web (e.g., plant — herbivore interactions) or strongly aggregated the component species into a few categories. An exception is the European Regional Seas Ecosystem Model (ERSEM) [1,12] but also this model cannot overcome the inherent problem that the many-layered complexity of natural ecosystems can never be depicted in a single mathematical model. In order to be useful, a model has to be an abstraction from reality. It is beyond the scope of this section to present these various types of dynamic models in detail. In the food web models considered so far, species interactions and the flow of information were linked to trophic interactions, that is, the feeding of one species on the other ones. Direct predation and competition for resources are often very important factors for regulating population dynamics, especially in pelagic food webs. However, they may be modulated by other interactions that are characterized by a very small flow of energy as compared to their relevance for food web dynamics. In the previous section, it has already been discussed that quantitative minor parts in the diet such as polyunsaturated fatty acids, vitamins, and sterols may determine the growth rate of the consumer. Further examples where a marginal flow of matter may strongly influence food web dynamics include pollination, dispersal of seeds, kairomones, or pheromones. The latter promote communication within one biological species, for example, males may release pheromones to attract females. That is, the interaction strength that represents the pair wise *per capita* effect of one species on another one, is neither constant for all links nor proportionally to the quantity of flow along a certain link. In some networks based on such nontrophic interactions, partly similar investigations have been performed as described above for the binary food webs. For example, in mutualistic networks between terrestrial plants and their pollinators relationships between species richness and the number of interactions and the connectance as a function of the habitat type were established [36].

All ecological networks considered so far were static in respect to their species inventory and the kinds of interactions. This may be adequate when regarding shorter time intervals. However, local extinction and invasion of new species are common to most food webs and, on a longer time scale, existing species may change their properties, new species may evolve, and other may go extinct. These processes will alter the network structure and, thus, demand consideration to understand its emergence, for example, by using assembly and evolutionary models [10,41]. To conclude, a comprehensive understanding and forecasting of the dynamics of ecological networks demands a dynamic representation accounting for the high spatiotemporal variability of natural food webs and for other flows of information on top of trophic interactions.

12.5 SUMMARY

Ecological networks typically represent food webs that may be defined as networks of consumer–resource interactions between groups of organisms. By these means, they describe who is present and who affects whom directly or indirectly by feeding interactions. In food webs, the vertices (nodes, points) are represented by individual species, certain life stages of one species, or by an aggregation of species (trophic guild). The connections or links (edges, arcs, lines) are represented by trophic interactions, that is, the feeding of one species on other ones. Depending on the question under consideration and the availability of data, the highly complex natural food webs are depicted by various types of network models such as binary food webs ("Who eats whom?"), quantitative trophic food webs ("Who eats whom how much?" Who recycles how much?), and ecological information networks ("Who influences whom (how much)?"). Comparing binary webs from different habitats and researchers is still hampered by operational problems. Nevertheless, they are used to identify generalizations in food web structures that are then related to other kinds of networks in biological, social, and other sciences. Binary food webs are well suited to summarize qualitative structural patterns of complex food webs at the expense of contributing little directly to the understanding of ecosystem functioning, mostly because they ignore the pronounced differences in the relative importance of the different feeding links. Quantitative food web models provide comprehensive descriptions of fluxes and the cycling of carbon (energy) and nutrients in an ecosystem, especially when they are mass-balanced and evaluated by network analysis. By accounting for the large variability in the relative importance of the different fluxes, they provide a more realistic picture than binary webs and have been preferred in some recent studies although they are more data demanding. Nevertheless, quantitative food web models still have limitation in explaining probable causalities in food web dynamics as they are primarily static like binary webs, and the interaction strength between two compartments does not only depend on the quantity of the fluxes among them. Such tasks may be approached using dynamic information networks, which facilitate a comprehensive understanding and forecasting of the dynamics of ecological networks. However, their implementation again demands an even higher effort.

TABLE 12.1   Adjacency Matrix for a Hypothetical Food Web for Exercise 1

Species	1	2	3	4
1	0	0	1	0
2	0	0	1	1
3	0	0	0	1
4	0	0	0	1

12.6   EXERCISES

1. Construct a graphical representation of the binary food web defined by the community (adjacency) matrix (Table 12.1) and compute S, L, D, and C. Identify basal species and top predators.

2. Sketch quantitative trophic food webs in units of carbon and phosphorus based on the following information: The web consists of primary producers, herbivores, and omnivores. The latter consume primary producers and the herbivores. The herbivores consume 75% of the primary production in terms of C and the omnivores the remaining 25%. Primary production amounts to $100 \, mg \, m^{-2} d^{-1}$. Each consumer converts 33% of the food ingestion into new production, 33% is egested, and 33% is respired. How much herbivorous production is available for the omnivore to ingest in units of C? Construct a food web model in units of P by first assuming that the ratio between carbon and phosphorus of 100:1 for the primary producer and of 40:1 for the animals. How much phosphorus ingests the omnivore? How much phosphorus is released by the two consumers? Assume now a ratio between carbon and phosphorus 200:1 for the primary producer and of 40:1 for the animals (i.e., the plants are nutrient-depleted, see Fig. 12.3 "summer situation"). Before you construct a model, evaluate whether the herbivore is now limited by C or P? What is the maximum production in units of C and P the herbivore can now achieve? What are the consequences for the omnivore?

3. Describe the most important differences between binary food webs, quantitative trophic food webs, and ecological information networks and provide their major respective advantages and limitations.

REFERENCES

1. J. W. Baretta, W. Ebenhöh, and P. Ruardij. The european regional seas ecosystem model, a complex marine ecosystem model. *Netherlands Journal of Sea Research*, 33:233–246, 1995.

2. M. Begon, C. R. Townsend, and J. L. Harper. *Ecology. From Individuals to Ecosystems.* Blackwell Publishing Ltd, 2006.

3. V. Christensen and D. Pauly. Ecopath II — a software for balancing steady-state ecosystem models and calculating network characteristics. *Ecological Modelling*, 61:169–185, 1992.

4. V. Christensen and D. Pauly, editors. *Trophic models of aquatic ecosystems. Conference Proceedings*, volume 26, International Centre for Living Aquatic Resource Management, 1993.

5. R. R. Christian, D. Baird, J. Luczkovich, J. C. Johnson, U. M. Scharler, and R. E. Ulanowicz. Role of network analysis in comparative ecosystem ecology of estuaries. *Aquatic Food Webs*, pp. 25–40. Oxford University Press, Oxford, UK, 2005.

6. G. P. Closs and P. S. Lake. Spatial and temporal variation in the structure of an intermittent-stream food web. *Ecological Monographs*, 64:1–21, 1994.

7. J. E. Cohen, R. A. Beaver, S. H. Cousins, D. L. DeAngelis, L. Goldwasser, K. L. Heong, R. D. Holt, A. J. Kohn, J. H. Lawton, N. Martinez, R. O'Malley, L. M. Page, B. C. Patten, S. L. Pimm, G. A. Polis, M. Rejmanek, T. W. Schoener, K. Schoenly, W. G. Sprules, J. M. Teal, R. E. Ulanowicz, P. H. Warren, H. M. Wilbur, and P. Yodzis. Improving food webs. *Ecology*, 74:252–258, 1993.

8. J. E. F. Cohen, F. Briand, and C. M. Newman. *Community Food Webs*, volume 20 of *Biomathematics*. Springer Verlag, Berlin, 1990.

9. R. Constanza and B. Hannon. *Network Analysis in Marine Ecology. Methods and Applications*, volume 32 of *Coastal and Estuarine Studies*, chapter Dealing with the 'mixed units' problem in ecosystem network analysis. Springer-Verlag, Berlin, 1989.

10. B. Drossel and A. J. McKane. *Handbook of Graphs and Networks*, chapter Modelling food webs, pp. 218–247. Wiley-VCH, Weinheim, Germany, 2003.

11. J. A. Dunne, R. J. Williams, and N. D. Martinez. Food-web structure and network theory: The role of connectance and size. *Proceedings of National Academy Sciences USA*, 99:12917–12922, 2002.

12. W. Ebenhöh, C. Kohlmeier, and P. J. Radford. The benthic biological submodel in the european regional seas ecosystem model. *Netherlands Journal of Sea Research*, 33:423–452, 1995.

13. C. S. Elton. *Animal Ecology*. Sidgwick and Jackson, London, *Research*, 1927.

14. K. H. Mann F. Wulff, F. G. Field, editor. *Network Analysis in Marine Ecology. Methods and Application*, volume 32 of *Coastal and Estuarine Studies*. Springer-Verlag, New York, 1989.

15. G. F. Fussmann and G. Heber. Food web complexity and chaotic population dynamics. *Ecology Letters*, 5:394–401, 2002.

16. U. Gaedke. A comparison of whole community and ecosystem approaches to study the structure, function, and regulation of pelagic food webs. *Journal of Plankton Research*, 17:1273–1305, 1995.

17. U. Gaedke, S. Hochstädter, and D. Straile. Interplay between energy limitation and nutritional deficiency: Empirical data and food web models. *Ecological Monographs*, 72:251–270, 2002.

18. U. Gaedke and D. Straile. Food webs and chains. *Encyclopedia of Environmental Science*. Kluwer Academic Publishers, Dordrecht, 1998.

19. V. Grimm and S. F. Railsback. *Individual-based Modeling and Ecology*. Princeton University Press, 2005.

20. S. J. Hall and D. G. Raffaelli. Food webs: Theory and reality. *Advances in Ecological Research*, 24:187–239, 1993.

21. D. Hart, L. Stone, A. Stern, D. Straile, and U. Gaedke. Methods for balancing ecosystem flux charts: New techniques and software. *Environmental Modelling & Assessment*, 2:23–28, 1997.

22. K. E. Havens. Scale and structure in natural food webs. *Science*, 1107–1109, 1992.

23. J. J. Heymans and D. Baird. Network analysis of the northern benguela ecosystem by means of NETWRK and ECOPATH. *Ecological Modelling*, 97–119, 2002.

24. S. Hochstädter. *Erstellung und Analyse von Phosphor-Nahrungsnetzen im pelagischen Kreislauf des Bodensees. Dissertation.* Hartung-Gorre Verlag, University of Constance, Germany, 1997.

25. A. G. Jackson and P. M. Eldridge. Food web analysis of a planktonic system off southern california. *Progress in Oceanography*, 30:223–251, 1992.

26. J. C. Johnson, S. P. Borgatti, J. J. Luczkovich, and M. G. Everett. Network role analysis in the study of food webs: An application of regular role coloration. *Journal of Social Structure*, 2:e3, 2001.

27. O. Klepper and J. P. G. van de Kramer. The use of mass balances to test and improve the estimates of carbon fluxes in an ecosystem. *Mathematical Bioscience*, 85:229–234, 1987.

28. A. E. Krause, A. K. Frank, D. M. Mason, R. E. Ulanowicz, and W. W. Taylor. Compartments revealed in food-web structure. *Nature*, 426:282–285, 2003.

29. J. J. Luczkovich, S.P. Borgatti, J.C. Johnson, and M.G. Everett. Defining and measuring trophic role similarity in food webs using regular coloration. *Journal of Theoretical Biology*, 220:303–321, 2003.

30. J. J. Luczkovich, G. P. Ward, R. R. Christian, J. C. Johnson, D. Baird, H. Neckels, and W. Rizzo. Determining the trophic guilds of fishes and macroinvertebrates in a seagrass food web. *Estuaries*, 25:1143–1164, 2002.

31. N. D. Martinez. Artifacts or attributes? Effects of resolution on the little rock lake food web. *Ecological Monographs*, 61:367–392, 1991.

32. N. D. Martinez. Constant connectance in community food webs. *American Naturalist*, 140:1208–1218, 1992.

33. J. M. Montoya and R. V. Solé. Topological properties of food webs: from real data to community assembly models. *Oikos*, 102:614–622, 2003.

34. P. Morin. *Community Ecology.* Blackwell Science, Oxford, UK, 1999.

35. A.-M. Neutel, J. A. P. Heesterbeek, and P. C. de Ruiter. Stability in real food webs: Real links in long loops. *Science*, 296:1120–1123, 2002.

36. J. M. Olesen and P. Jordano. Geographic patterns in plant-pollinator mutualistic networks. *Ecology*, 83:2416–2424, 2002.

37. C. Pahl-Wostl. Food webs and ecological networks across temporal and spatial scales. *Oikos*, 66:415–432, 1993.

38. S. L. Pimm, J. H. Lawton, and J. E. Cohen. Food web patterns and their consequences. *Nature*, 350:669–674, 1991.

39. G. A. Polis. Complex desert food webs : an empirical critique of food web theory. *American Naturalist*, 138:123–155, 1991.

40. G. A. Polis and K. O. Winemiller, editors. *Food Webs, Integration of Patterns and Dynamics.* Chapman & Hall, Boca Raton, FL, 1995.

41. S. R. Proulx, D. E.L. Promislow, and P. C. Phillips. Network thinking in ecology and evolution. *Trends in Ecology and Evolution*, 20:345–353, 2005.

42. L. Stone and T. Berman. Positive feedback in aquatic ecosystems: The case of the microbial loop. *Bulletin of Mathematical Biology*, 55:919–936, 1993.

43. R. E. Ulanowicz. *NETWRK 4: A Package of Computer Algorithms to Analyze Ecological Flow Networks*. University of Maryland Chesapeake Biological Laboratory, Solomons, Maryland, 1987.

44. A. F. Vezina and T. Platt. Food web dynamics in the ocean. I. Best-estimates of flow networks using inverse methods. *Marine Ecology Progress Series*, 42:269–287, 1988.

45. A. Wacker and E. von Elert. Polyunsaturated fatty acids: evidence for non-substitutable biochemical resources in *Daphnia galeata*. *Ecology*, 82:2507–2520, 2001.

46. R. J. Williams and N. D. Martinez. Simple rules yield complex food webs. *Nature*, 404:180–183, 2000.

47. R. J. Williams and N. D. Martinez. Limits to trophic levels and omnivory in complex food webs: Theory and data. *American Naturalist*, 163:458–468, 2004.

48. K. O. Winemiller. Spatial and temporal variation in tropical fish trophic networks. *Ecological Monography*, 60:331–367, 1990.

49. P. Yodzis. The indeterminancy of ecological interactions as perceived through perturbation experiments. *Ecology*, 69:508–513, 1988.

50. P. Yodzis. *Species Diversity in Ecological Communities*. chapter Environment and trophodiversity, pp. 26–38. University of Chicago Press, Chicago, 1993.

51. P. Yodzis and K. O. Winemiller. In search of operational trophospecies in a tropical aquatic food web. *Oikos*, 87:327–340, 1999.

13

CORRELATION NETWORKS

Dirk Steinhauser, Leonard Krall, Carsten Müssig,
Dirk Büssis, and Björn Usadel

13.1 INTRODUCTION

Over the past few years biological science has experienced substantial technological advances that have led to the rediscovery of systems biology [22–24,32]. These advances were ignited with the technological ability to completely sequence the genome from virtually any organism [30,39]. Although these initial sequencing efforts focused mainly on gene discovery and genome structure analysis, they triggered the development of various multiplex high-throughput assays, such as GC–MS (gas chromatography coupled with mass spectrometry) based metabolomics [25], microarray based transcriptomics [27], or MS and enzyme assay based proteomics [48]. Today these analytical technologies permit a simultaneous monitoring (profiling) of all of the components of the cellular inventory: genes, transcripts, proteins, and metabolites. With genome information and profiling technologies now readily accessible, the mining and exploitation of data derived from such multiparallel "omics" technologies open up the possibility to gain comprehensive insight into understanding biological systems and their complexity.

However, it is clear that knowledge of qualitative and quantitative data of the cellular inventory is necessary, but clearly not sufficient to address and understand system

Analysis of Biological Networks, Edited by Björn H. Junker and Falk Schreiber
Copyright © 2008 John Wiley & Sons, Inc.

responses. Cellular processes are determined by a large number of functionally diverse, differently active, and frequently multifunctional sets of the cellular elements [43]. These elements interact selectively and in many cases nonlinearly to execute specific cellular functions [22].

The understanding of biological complexity through the modeling of cellular systems permits one to shift from a component-centric focus to integrative and system level investigations. Whereas systems biology is not consistently defined, it represents an analytical approach to unravel interrelations among and between the cellular elements in biological systems [17,52]. Such interrelationships can then be approached experimentally and/or described by statistical measures.

In this chapter, we describe and discuss the crucial steps of measuring and interpreting statistical interrelations between cellular elements by correlations and correlation networks, as well as describing and discussing correlations and correlations networks in the context of interrelations between heterogeneous elements of the cellular inventory. The potentials and constraints, challenges, as well as pitfalls of these approaches will be discussed. We attempt to show the usefulness of linking computational science and experimental biology through component-centric driven and systems level investigations. While we mainly focus on transcriptional correlations in this chapter, basically these guides are applicable for all cellular elements.

13.2 GENERAL REMARKS

The following sections deal with correlation networks, their generation, and their interpretation. First, correlation values can be looked upon from two perspectives, namely from the probability point of view (based on strength) or from the (real) strength point of view. In the former viewpoint, one is interested if a correlation is seen by mere coincidence or if there really is a connection between the two variables, albeit being possibly weak. This is due to the fact that the significance of a correlation coefficient is not only related to its strength but also to the number of samples examined. Thus, through examining a large number of samples, weak correlations can become significant. The second vantage point just considers everything from the strength of the interaction (see Section 13.3.4).

Regardless of the point of view, usually these networks are transferred into discretized networks for standard network analyses. However, this crucial step might introduce several mistakes, since the threshold used for discretization might not have been optimally chosen, thus leading to distortions of the network and thus distortion in the interpretation of the information. Moreover, correlation networks are *a priori* undirected. Therefore it is difficult to attach flow directions to these networks. Finally, correlation is not synonymous to causality. Therefore one must be careful when interpreting these networks!

With these caveats in mind, we can now proceed to the next sections.

13.3 BASIC NOTATION

Here we discuss the analysis of correlations and correlation networks based on various statistical as well as mathematical definitions, computational algorithms, and finally biological interpretation of the results obtained. This section aims to address some basic terminology, notations, and equations of correlations and correlation networks used in this manuscript. Interested readers with a basic knowledge of statistical analyses as well as profiling technologies and their use may skip parts or even the entire section. We present a brief overview of the mathematical basis underlying correlation networks and present only the necessary concepts. Interested readers can get more detailed descriptions by referring the cited references and the references therein as well as elsewhere in this book. Most of the terms described are based on data and data structures normally generated and handled in biological labs. Therefore, the notations described may be slightly different and may be wielded differently than from other areas of natural science.

13.3.1 Data, Unit, Variable, and Observation

The essential fundamentals for correlations and the resulting correlation networks are data. Data represent the substantial raw material for any method of statistical analyses. In this chapter, we use the term *data* to describe values that are analyzed and/or even modified by statistical analyses. An *observation*, a datum, is considered as a value for some object on some variable. An assembly of data, that is, the collection of all data, for a particular study is referred to as a *data set*. Such data sets are often visualized as tables. An exemplary data set is given in Fig. 13.1.

In literature, various terms can be found to indicate objects of a data set. In this chapter, we use the neutral term *unit*, which consists of some variables each with observation(s). "Heat" represents a unit in Fig. 13.1. In contrast, a *variable* is a characteristic or attribute or entity of interest about a unit that can take on different values. In Fig. 13.1, "b0003" is a variable.

		Unit	
Gene\profile	Control	Heat	Cold
b0001	0.8	-8.2	4.1
b0002	1.3	NaN	-4.5
b0003	8.2	-3.3	-5.7
b0004	-2.4	0.1	NA
b0005	-1.2	4.1	0.1
b0006	NA	3.2	-0.2
Variable			Observation

FIGURE 13.1 Simple illustration of an exemplary partial data set. Its notations have been labeled. (Legend: NA = not available, NaN = not a number).

Data sets generated in biological science are often represented as variable-unit-tables. In such tables the units are represented by the column header and the variables as 1st cells of each row as illustrated in Fig. 13.1. Such representation can be transposed by turning rows into columns and vice versa.

13.3.2 Sample, Profiles, and Replica Set

The basic sources for data in molecular and biological science are samples. In principle, the *sample* describes the biological material used for analyses, for example, by multiplex high-throughput methods. The sample represents a particular origin, such as a part of an organism or its entirety, which can be specified by various attributes. These sample attributes can be related by the treatment, the condition of growth, the age, the time of sampling, and so forth, which are often standardized [4,7]. In most of the cases samples are processed according to the designed method before they are used for analyses. The processing and analysis methods are manifold and are discussed elsewhere [1,27,41].

For construction of correlation networks a set of diverse variables ($n \geq 8$) have to be measured. Multiplex high-throughput technologies enable scientists to perform such simultaneous measurements of diverse variables. Basically, the *profile* represents the output of such a technology for a sample. The general design of profiles depends on the method and technology platform used. The primary profile, which is generally unprocessed and reflects the raw measurements, is simply the readout of the technology platform. Such primary profiles can be images (Fig. 13.2a), for example, generated by expression profiling technologies for transcript abundance measurements, or chromatograms derived from various protein and metabolite profiling technologies etc. (Fig. 13.2b).

Primary profiles are normally converted to secondary profiles. In this chapter, we consider the term *profiles* as *secondary profiles* that represent the text-based output of a profiling technology platform and may be preprocessed by particular statistical

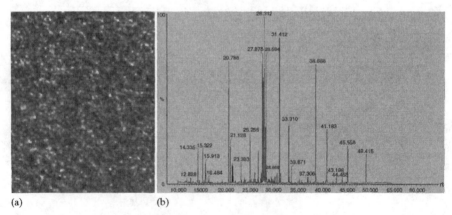

(a)   (b)

FIGURE 13.2 Two examples illustrate types of primary profiles for an expression (*a*) image and a metabolite (*b*) chromatogram profiling technology platform.

Gene\attribute	Signal	Detection	Description
b0001	6.7	Absent	Isomerase
b0002	23.2	Marginal	Metal transporter
b0003	0.8	Absent	Cytochrome P450
b0004	53.7	Present	Unknow protein
b0005	1285.1	Present	Hypothetical protein
b0006	2	Absent	Metallothionein-like protein

FIGURE 13.3 Simple illustration (based on Affymetrix microarray technology platform) of a partial, normalized profile with variables (gene, e.g., b0003) and various attributes reflecting the relative adjusted signals ("signal"), the reliability of detection ("detection") and the description of the respective variable.

algorithms. In general, such profiles consist of a number of diverse variables (up to many thousands) each with their respective observations. These observations describe semi-/quantitative or relative values of the abundance (amount) of the variables. Secondary profiles can contain further attributes, for example, values or scores, which may be used by further algorithms for normalization.

Such *normalized profiles* are derived from secondary profiles by the application of statistical algorithm and/or logical rules to adjust observations in terms of technical and experimental, or even known artificial variations. Thus, normalized profiles consist of adjusted or transformed observations for each variable and may contain further attributes that reflect the quality and/or the reliability of the individual observations (Fig. 13.3).

13.3.3 Measures of Association

For construction of correlation networks, associations between variables derived from quantitative or qualitative data have to be measured. Such *associations* characterize at least the strength of relation between two variables. They can be measured by a multitude of different coefficient types, which can be basically classified into similarity and dissimilarity measures.

Similarity measures monotonically reflect the extent of similarity between variables $s_{ij}, i \in I, j \in J$: the larger the similarity s_{ij}, the more similar are i and j. Similarities between variables can be obtained as primary data or secondary data derived from variable-to-unit data sets. Matrices (see below) based on similarity measures are called similarity matrices $S \equiv (s_{ij})$. Similarity data may be nonpositive ($s_{ij} < 0$) and asymmetrical ($s_{ij} \neq s_{ji}$).

In contrast, *dissimilarity* matrices $D \equiv (d_{ij})$ reflect dissimilarities d_{ij} between variables $i, j \in I$: the larger the dissimilarity d_{ij}, the less similar the variables i and j are. Dissimilarities are supposed to be non-negative ($d_{ij} \geq 0$). The resulting matrix D is symmetrical ($d_{ij} = d_{ji}$) with diagonal dissimilarities of zero values ($d_{ii} = 0$). Dissimilarities can be observed empirically through conversion of similarities or as secondary data derived from variable-to-unit data sets. The Euclidean distance, the distance between two points or vectors, is probably one of the most widespread dissimilarity measures used in the natural sciences [28].

Interestingly, a direct relationship between similarities and dissimilarities may occur, which cannot be assumed always to be relevant. Despite this, dissimilarities can be obtained by transforming similarities by equations such as $d_{ij} = \sqrt{(1 - s_{ij})}$, $d_{ij} = 1 - s_{ij}$, $d_{ij} = 1 - |s_{ij}|$, or $d_{ij} = \exp(-s_{ij})$ [28]. Moreover, dissimilarities can be scaled for example, by dividing all dissimilarity d_{ij} by the maximum dissimilarity $\max(d_{ij})$ of D.

13.3.4   Simple Correlation Measures

Correlation coefficients [10,38] belong to the group of similarity measures and describe at least the magnitude of the relation between two variables (*association*). In contrast, some coefficients can describe both magnitude and direction (relationship). Correlation coefficients can take on values in range of -1 to $+1$.

The *magnitude* of a correlation estimates the strength of the relation: the strength of the tendency of variables to move in the same (or opposite) direction or how strong they covary across the set of underlying paired observations. The larger the absolute correlation, the stronger the variables are associated.

The *direction* of the correlation describes how the variables are related. If the correlation is positive, the two variables have a positive relationship (move in the same direction) or the correlation can be negative (and move in the opposite direction). A positive relationship means that as one variable increases so does another one. In contrast, a negative relationship reflects that the other variable decreases.

The most prominent correlation measure is the *Pearson product moment correlation* (ρ or r, Equation 13.1). The Pearson correlation describes the linear relationship between two variables that are on an interval or ratio scale. Its values range from -1 to $+1$. For some applications Pearson's r requires bivariate (two-dimensional) normal distributed observations. The Pearson correlation can be strongly affected by bivariate outliers, which may increase or decrease the magnitude of relationship.

The following equation computes the Pearson product moment correlation r of variable X and Y. $\overline{X}$ and $\overline{Y}$ are the mean of X and Y, respectively, and n the number of paired observations.

$$r(X, Y) = \frac{\sum_{i=1}^{n} (X_i - \overline{X})(Y_i - \overline{Y})}{\sqrt{\sum_{i=1}^{n} (X_i - \overline{X})^2} \sqrt{\sum_{i=1}^{n} (Y_i - \overline{Y})^2}} \tag{13.1}$$

A special case of the Pearson's r is the nonparametric (i.e., distribution-free) *Spearman rank-order correlation* (r_s, Equation 13.2). It requires fewer assumptions compared to its relative. The Spearman correlation is estimated on ranked observations of two variables and ranges from -1 to $+1$. It can describe linear as well as nonlinear relationships, if the observations are continuously increasing or decreasing. According to its nonparametric nature, Spearman correlation is more robust to bivariate outliers than its parametric counterpart and does not necessarily required bivariate normal distributed observations.

The following equation computes the Spearman rank-order correlation r_s of variable X and Y. The number of paired observations is n. The Spearman correlation can be computed by Equation 13.2 or by Equation 13.1 on ranked observations.

$$r_s(X, Y) = 1 - \frac{6 \sum_{i=1}^{n} (\text{rank}(X) - \text{rank}(Y))^2}{[n(n^2 - 1)]} \tag{13.2}$$

In contrast to the above-mentioned correlation coefficients, a variety of other coefficients can be found in the literature, such as Kendall's τ [38], each with their own "advantages" and "disadvantages." Most are special cases or based on Pearson product moment correlation, such as the James–Stein Shrinkage Pearson correlation [11].

13.3.5 Complex Correlation and Association Measures

With the ever increasing number of publicly available profiles derived from multiplex high-throughput technologies and the increasing computational power, various other, more complex, measures of dependencies have been applied. For instance, the *mutual information* (I, *MI*), derived from information theory [37], provides a general measure of dependencies. This measure allows for the deciphering of association between variables on general criterion beyond linear or slightly nonlinear dependencies. It quantifies the reduction in the uncertainty of one random variable given knowledge about another random variable. The mutual information can be computed in different ways [44]. Often it is based on the Shannon (cross-)entropy (Equation 13.3). Improving computation, adjustment for finite sample effect, and significance testing has been suggested [44].

The following equation computes the mutual information (I) based on Shannon entropy for X and $Y(H(X), H(Y))$ and their joint entropy $(H(X, Y))$. M_x, M_y represent the possible states of X and Y, respectively. P is the probability of a particular state M_i.

$$I(X, Y) = H(X) + H(Y) - H(X, Y) \quad \text{with} \tag{13.3}$$

$$H(X) = -\sum_{i=1}^{M_x} p(x_i) \log p(x_i) \quad \text{and}$$

$$H(X, Y) = -\sum_{i=1}^{M_x} \sum_{j=1}^{M_y} p(x_i, y_j) \log p(x_i, y_j)$$

Partial correlation coefficients [10,38] represent another way of describing dependencies between variables. Such coefficients have been introduced to bypass

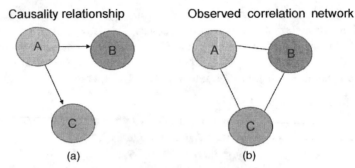

FIGURE 13.4 Illustration of causal relationships between variables (A, B) and (A, C) (*a*) and the resulting network (*b*) derived by simple correlation analysis.

problems derived from simple correlation analysis. For instance, one drawback of dependency analysis by simple correlations can be the observation of strong relationships which might result from secondary effects or due to the action of another variable. This can play an important role in networks derived from a multiplex high-throughput assay with hundreds of variables. For a better illustration, one can assume a simple model consisting of three variables A, B, and C. In this model, variables A and B are highly correlated by a causal relationship. Furthermore, variable A might also be causally correlated to the variable C. If the resulting correlations of (A, B) and (A, C) are therefore strong, we will also observe a strong correlation of (B, C). This might jeopardize downstream network analysis by putting an edge (a connection, see below) between B and C because there will be more weight placed to the nodes B and C than there actually is (Fig. 13.4).

A method to circumvent such a problem is the use of partial correlation coefficients. This would measure the correlation between any pair of variables if a third (or more) specified variable(s) has been held constant. For instance, in the case above it would measure the association $r_{BC.A}$ between variables (B, C) if variable A would be constant (controlled or removed from the model). This is also called partial correlation of first order, since only one variable is controlled. In this simple case, the partial correlation can be computed by the equation.

The following equation computes the partial correlation of first order based on Pearson product moment correlation.

$$r_{BC.A} = \frac{r_{BC} - r_{BA}r_{CA}}{\sqrt{\left(1 - r_{BA}^2\right)\left(1 - r_{CA}^2\right)}} \tag{13.4}$$

Higher order partial correlations of order n (n variables controlled) can be obtained by a similar formula based on the partial correlations of order $n - 1$. Obviously, these calculations get more complicated when dealing with higher order partial correlation. Therefore, these are usually calculated using the inverse of the correlation matrix, which is related to the partial correlation. However, the inverse can usually not be solved if there are more variables than units. To circumvent this problem a pseudo-inverse for partial correlation estimation combined with graphical Gaussian model has been introduced [36].

13.3.6 Probability, Confidence, and Power

Beyond measuring the relationship of variables the correlation obtained can be used for further statistical testing or comparison. Often such further statistical analyses, for example, confidence, power, or comparison of correlations, require transformation of the correlations obtained.

The basic *transformation* and *back-transformation* of a correlation are computed by using *Fisher-z-transformation* (Equation 13.5) or their improvements [18]. Such transformations may be necessary for the comparison of obtained correlations, especially if they are derived from not fully assigned data matrices [38].

$$z = \frac{1}{2} \ln \frac{1+r}{1-r} \quad \text{or} \quad z = \arctan h\,(r) \tag{13.5}$$

Moreover, correlations can be examined as to whether there are differences from a null hypothesis (H_0) or an alternative hypothesis (H_A) by statistical testing. Because of the extensive definitions required, and the specific test statistics (e.g., Equation 13.6 for Pearson product moment correlation) in relation to the coefficient in question, it will not be discussed in detail here [5,10,38].

$$t_s = r \sqrt{\frac{n-2}{1-r^2}} \quad \text{with} \quad n-2 \quad \text{degree of freedom (df)} \tag{13.6}$$

Basically, correlations can be converted into a test value by using the respective test statistic, which can than be compared against a table of probabilities or the respective probability distribution. The later would yield the probability of alpha. Such *probability* estimates the chance to observe a correlation as a coincidence of random sampling. It ranges from 0 to 1. Usually a value of <0.05 (95%), referred as a significance level, will be accepted as significant. Rejecting a null hypothesis that is true is referred to as type I error or false positive.

Although population correlations can be estimated by a point, that is, the sample correlation, it can also be estimated by an interval between two points. The interval between these two points specifies a defined level of confidence for the location of the estimated true population correlation. Such interval is termed as *confidence interval* or *confidence level*.

A further parameter in statistical testing is referred to as *power* or *power of test*. The power is computed as $1 - \beta$ with β as the probability of not rejecting a false null hypothesis (Type II error, false negative). The power explains the fact how likely one is to find significant correlations given that the alternative hypothesis is also true. Its values ranges from 1 to 0, with values closer to 1 reflecting less chance to make a type II error.

13.3.7 Matrices

Generation and analyzing of correlation-based probability networks basically require the use and conversion of different types of matrices (e.g., rectangular arrays of numbers) by computer programs. The basic matrix type is referred to as *data matrix B*, which is similar to variable-to-unit data sets (see above) of profiling data. Whereas such data sets can contain additional attributes, the data matrix contains only the observations used for computation of the relationships between variables. The observations can mirror quantitative values, ranks or Boolean values, for example, qualitative values. The data matrix may be preprocessed in different ways and/or by different statistical algorithms, for example, variable or units can be normalized to unit variance etc. Such matrices are usually asymmetric with n number of variables and m number of units ($n \times m$ matrix).

Computation of all pairwise relationships between the variables will result in a squared and symmetrical matrix with n variables ($n \times n$ matrix), termed a *relationship matrix R*. Depending on the coefficient type used it can either be a similarity or a dissimilarity matrix (see above). Usually, the upper or lower triangle may be computed and represented based on the fact that relationship matrices are symmetrical ($r_{ij} = r_{ji}$). Such matrices can be converted into a *probability matrix P*, a relationship matrix, by using the appropriate test statistic for the coefficient in use (see above).

The application of graph analyzing methods may require the conversion of relationship matrices. For instance, for clustering analyses a similarity-based relationship matrix will be usually converted into a distance range (see above). Beyond this, the main conversion step for graph/network analyses are matrix discretization into an ($n \times n$) *adjacency matrix A*. Relationship matrices contain numerical values for each pairwise comparison $R_{ij}, i \in I, j \in J$. Based on arbitrary or probability cutoffs, the relationship matrix-element R_{ij} will be usually converted into a binary adjacency matrix-element A_{ij}. For instance, one assumes a probability matrix containing significant levels in range of 0–1 and a cutoff 95% confidence (alpha of 0.05) to accept only "significant" relations. Each variable-to-variable relationship (or more precisely, probability) with values less than 0.05 are set as 1 ($A_{ij} = 1$, i.e., we draw an edge between them), all others are set to 0 ($A_{ij} = 0$, we do not draw an edge). Adjacency matrices derived from relationship matrices are usually undirected *graphs G*.

13.4 CONSTRUCTION AND ANALYSES OF CORRELATION NETWORKS

The construction and analysis of correlation-based probability networks is comprised of various steps through a series of data processing and conversion methods (Fig. 13.5). In the following section, we will describe and discuss the crucial steps

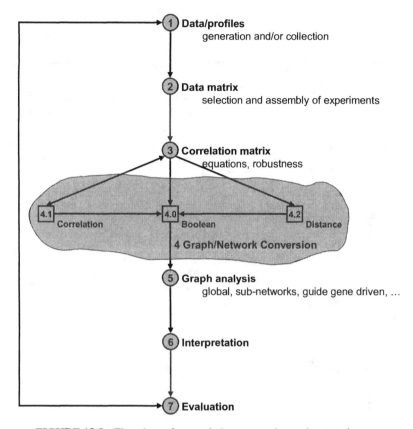

FIGURE 13.5   Flowchart of network data processing and conversion steps.

of this process as well as a few of the major pitfalls. Because of the myriad possible modifications, this section is more intended to be a rough guideline instead of a precise recipe.

13.4.1   Data and Profiles

The first step for correlation network analysis is the generation and/or collection of profile data. This is usually derived from multiplex high-throughput technologies. As mentioned above, profiles can be generated to address a particular biological phenomenon. Both the biological phenomenon under examination and the precise biological question will influence the experimental design as well as the selection of the technology platform (the readout). Such biological questions are usually related to the unraveling of significant changes of variables, such as gene expression or metabolite pools, among different units, that is, treatments or conditions. The generated profiles are often limited to a certain number (less than 10) because of the costs of high-quality profiling studies.

An alternative source for profiles is represented by the ever-growing number of public repositories for profiling experiments, especially for storage and analysis of gene expression or metabolite data [14,33]. These ever growing resources provide access to thousands of profiles. Therefore they represent substantial data resources to tackle different questions through correlation/correlation network analyses or other cross-experimental investigations. One drawback of publicly available data sets is the presence of insufficiently described experiments, even though the MIAME (gene expression microarray) [7] or MIAMET (metabolites) [4] standards demand at a minimal set of information that must be supplied. In cases where the minimal information sets are not exhaustive enough, organism specific resources have begun to be developed, for example, MIAME-Plant [55].

13.4.2   Data Set and Matrix

With a set of profiles in hand, one can select and assemble data sets and ultimately convert them into data matrices. Basically, two main strategies can be utilized. Which should be used is based on the question that needs to be answered.

Therefore, the first question that one needs to ask is if one is searching for a generalized trend or if one is looking for a special (context-dependent) effect. Generalized trends are best perceived in a complex matrix. Complex matrices comprise profiles that differ with respect to genotypes, environmental conditions, experimental treatments, and developmental stages, and great care must be taken not to add bias toward one kind of experimental condition, for example, stress related profiles. To aid in the selection, the experimental description may help, depending on the details available in regard to the limitations mentioned above. However, inadequately described profiles could lead to a bias due to consistent but nonanno-tated experimental bias or additional biological effects. Thus, interpretation of the generalized matrix should still be exercised with care. Alternatively, the profiles can be preanalyzed by clustering to define groups of similar profiles. Out of each group, one or more profiles could then be randomly selected. Nevertheless, an unknown factor may still underlie the clustering and therefore would bias the profile selection.

Complex matrices can be easily established (e.g., using publicly available expression profiles), and due to the large number of available profiles it is unproblematic to create several data matrices in parallel for testing the associations observed in a specific data matrix. A combination of numerous, different experiments in data matrices allows for the selection for both the removal of outliers and also conditional changes of transcript levels. Only constitutive associations will be identified, because the combination of different profiles masks associations, which occur only under specific conditions (i.e., in a subset of underlying profiles). Another quality of complex matrices is that associations with an unknown common factor become unlikely.

In the other case, where one wants to study context dependent effects, one chooses or assembles data sets that contain experiments of a given condition or from the effect under examination. These context-specific matrices are generated using profiles that have something in common (e.g., stress condition, specific tissue, or developmental stage). The limited variability of profiles underlying context-specific matrices produces both advantages and problems. A main advantage is that numerous genes may be expressed in a specific context only; therefore a complex matrix will not provide any associations (as discussed above). For example, a special type of stress such as limited nutrient supply could be studied in greater detail, such as in the case where the effect of sulfate limitation was investigated by using correlation networks based on artificially introduced directionality [31]. Using this approach, one may gain deeper insights into the regulation network underlying the special stress or condition applied. However, again great care must be taken in the evaluation and interpretation of the data in regard to the fact that effects that are not directly related to the specific context are included in the matrix, but that are nevertheless influencing the correlations. Since often in such conditional matrices the number of arrays becomes limiting, it might be tempting to take many arrays from a time course or a concentration series. Even though this is *per se* not a drawback, this can lead to severe interpretation problems. For example, in a time course experiment, the variable time is usually not part of the matrix, but both time and time-coupled variables such as day–night cycle, nutrient supply, medium changes, growth effects, and so forth can play a significant role in consorted gene expression and can therefore drive correlations between genes. However, the mutual correlation between these genes is only an effect of their correlation to the driving variable (time) and in no way causal. Because these additional variables are not part of the data matrix, they cannot be removed by sophisticated methods like partial correlations, which might otherwise alleviate the driving effect of these variables. Thus, one might see a tight correlation network between several genes, and one may attribute this network to the primary effect, for example, the stress applied, but in effect one sees only the results of a secondary, unconsidered variable, such as growth.

Regardless of the data studied—global or conditional—it is of vital importance to normalize the data consistently and remove variables, for example, genes and/or units (such as experiments) from the data matrices that do not fulfill minimum quality criteria. In terms of normalization, various methods have been developed to consistently normalize, for instance, two and one color microarrays. However, not all of these normalization strategies are equally well suited for correlation analyses and have both their own merits as well as disadvantages. Probably the most common normalization strategy for Affymetrix$^{®}$ type microarrays today, the so-called MAS5/GCOS1 method, might give a better indication of the actual expression value as compared to quantitative real-time PCR [35] but reaches higher standard deviations in replicates, whereas the RMA method [20] gives better reproducibility but requires the normalization of all considered arrays in one batch, but which might finalize the data matrix, because adding experiments would require recalculation of all expression values. On the contrary, the MAS5 method flags some genes as not expressed/not validly measured that might both be considered either an advantage or disadvantage. This is a plus in terms of additional information, but, however, it requires handling of such genes and eventually removal of genes to often flagged, since one might otherwise be observing effects due to random noise. Here RMA gives for all these values a small, but nonzero expression value. Thus, one is able to utilize these data points, even though they most likely will not add too much effect to the correlation matrix. To resolve these points, [16] used bacterial operons to validate the different normalization techniques for correlation analysis and came to the conclusion that a combination of different methods works best. For metabolite or protein quantification measurements, no normalization strategy with similar power has been applied thus far to the authors' knowledge. The maturity of the normalization for transcripts may be attributed to the large number of transcripts that can be measured at one time and the refinement this technique has already reached.

Nevertheless, regardless of study, power, or maturity of normalization, one should not compare apples to oranges. All microarray normalization strategies essentially assume that only a small, negligible fraction of transcripts change between experiments, which is a prerequisite clearly violated if comparing different tissues or organs. This can sometimes lead to effects that then could distort the correlation matrices derived from such data matrices. If one is not sure about such experiments, it might be useful to detect and subsequently delete outlying experiments using a technique such as deleting residuals as shown in [34] before proceeding to the correlation matrix calculation.

13.4.3   Correlation Matrix

After having generated an appropriate data matrix, the next step is converting this variable-by-unit matrix into a symmetrical variable-by-variable correlation matrix. Often, the Pearson correlation coefficient is used due to the speed with which it can be calculated and the familiarity that "wet bench" scientists have with it. However, both of these reasons should generally not be a major consideration but rather the generation of the correlation matrix should be determined, again, on the question that needs to be answered.

Unfortunately, even though many tools exist, which aid in converting a data matrix into a correlation matrix, most of these programs are not able to cope with large data sets and often fail due to the size of the resulting correlation matrix in memory. Recently, due to the advent of inexpensive 64 bit technology in the form of both AMD and now Intel chips paired with a 64 Bit Linux system, it is now feasible to simply increase the available RAM of a computer so that even huge matrices can be calculated in memory. If such systems are not available, the user may still depend on self-made tools or could download precomputed data sets that are publicly available, such as through CSB.DB [42], Genevestigator [54], Expression Angler [47], and the Arabidopsis Co-expression Tool [21].

13.4.4   Network Matrix

The next step to bring the correlation matrix into actual network form is discretization of the correlation matrix, if an unweighted network is the target. Therefore a cut-off parameter has to be set to judge if a correlation is to be considered or not. This can be done by converting correlation coefficients into p-values and then using the usual arbitrary p-values of 0.05 (95% confidence) or 0.01 (99% confidence) as a cutoff. The conversion is done by standard statistical transformations (Equations 13.5 and 13.6, [42] and Web sites described therein for a longer discussion) or one could use Fisher's conversion to generate z values and ask if a true zero correlation would be included in the confidence interval. Because a larger number of units (i.e., experiments) greatly influence the results, processing a large enough number of experiments (say 50) can give extremely low p-values even after conservative correcting for multiple testing. Therefore a second filter, based on the strength of the correlation, could be used. For example, if one shows that a correlation is significant using the aforementioned formulas, one could then ask if the correlation coefficient is greater than 0.7. This seemingly arbitrary cutoff is derived from regression. Here, if one variable is declared independent and the other one declared a dependent variable that could be linearly regressed against the former, assuming that one could explain 50% of the variance of the dependent variable by the independent, one would thus achieve an R^2 of 0.50, which is numerically very close to an r of 0.7 ($0.7 * 0.7 = 0.49$), even though correlation does not distinguish between independent and dependent variables. This cutoff of 0.7 might be useful if one is not only interested in strong effects between the two variables, but rather small and possibly spurious effects as well.

After having thus discretized the correlation matrix, one arrives at the adjacency matrix A that is an immediate representation of the underlying network.

13.4.5   Correlation Network Analysis

For generalized as well as specific network analyses, the reader is best referred to the other chapters of this book. Here only a basic outline will be given, taking into special consideration the fragility and dependence of the correlation network on the underlying correlation matrix. Because of this dependency, analysis of correlation networks is just in its infancy.

Basically, there are two possible ways to analyze correlation-based networks. The first option takes all the nodes of the network (i.e., the topology) into account. One may analyze this structure or compare it between other networks (global approach). The second option begins by selecting a particular "guide gene." This guide gene may represent a node or any component in the network. The guide gene is used to predict biological function by looking at the network from its particular perspective. This is achieved by identifying connected nearest nodes and, in addition, by identifying other parts of the network, which show a similar behavior. A comprehensive analysis of a biological network would utilize both approaches to extract and predict important biological function.

One commonly applied analysis of correlation-based networks is related to the networks connectivity. Usually, the numbers of nodes n with connectivity k are plotted against their respective connectivity k. Such analyses revealed that connectivity often follows a power-law distribution (where $n(k) \sim k - y$). This means that only a few nodes are highly connected [2,51]. This observation seems to be a general and global parameter of transcriptional correlation networks as revealed in six different organisms as diverse as single prokaryotic cells to complex eukaryotic species and humans [2]. Interestingly, even the exponent was observed to be on a similar range for the selected species. However, such distributions can only be observed after careful selection of the cutoff value (see above) chosen for the correlation in order to construct the adjacency matrix. Random and biologically meaningless networks are usually produced if the cutoff value is inappropriately selected to low.

Probably the most easily performed network analysis is to compare the connectivity of the correlation network with some "true" or "given" network properties before delving into more involved network analysis methods. Indeed, this is a necessary and important step since many more sophisticated analysis methods will most likely fail if the underlying network contains too "fuzzy" information. But even this nearly trivial analysis can reveal meaningful and interesting information, and it lays the cornerstone for all subsequent analyses. Using this method it has been shown that significantly correlated yeast genes encode for proteins that are several times more likely to interact than randomly picked proteins [3,15]. Interestingly as well, it is often observed that highly connected genes, more often than not, represent essential genes for an organism [19]. Other analyses, using the specific genes underlying the nodes, compared connectivity with the presence of an orthologus gene from among other organisms. Again, a correlation between connectivity and conservation could be observed [2,9]. One should keep in mind, though, that many well correlated genes in the network do not encode for interacting proteins.

However, even with the inherent noise in the data, it is still possible to not only perform these analyses on a global scale, but also within given subnetworks, which can be delineated by biological function. This can then lead not only to identification of global properties, but to actually pinpoint biological modules that might have more important functions than others within the gene network. For example, one can plot the number of nodes from within a module against the connectivity exclusively against module members [9]. Another way to investigate this is to remove or randomize those experiments that lead to the definition of these modules, and thus vary the networks. This will reveal context dependent properties within the network. Nevertheless, many network submodules are highly stable even when looking in such diverse data sets such as cell cycle, DNA damage, and environmental perturbations [9].

Along the same lines, [40] showed that operon encoded bacterial genes, which by the definition of an operon should be transcribed together, have significantly higher correlations than nonoperon encoded genes. However, the best correlated genes were not always encoded by the same operon, and correlation information alone was not sufficient to detect operon encoded genes. Even though these studies are encouraging, they also bring up the question for what "truth" the experimenter is really looking. Here, the experimental design gets utmost importance again, especially considering that usually static correlation networks only represent the overlay of several "true" dynamic networks at best. Therefore it seems to be a good strategy to take this fuzziness into account and to concentrate on more robust network properties. This might be done by moving away from the general network analysis toward a more local analysis that might be based on "guide genes." In this case, the focus is concentrated in the immediate proximity of the guide genes and the genes that may affect this immediate area of the network (see Section 13.5.1). Often this is combined with an overrepresentation analysis of functional classes. If one finds such an overrepresentation and as well as some new genes of hitherto unknown function, one might assume that they have a similar role. This approach seems to work very well for genes of the ribosome, or plant cell wall biosynthesis genes [49]. This kind of analysis is therefore very similar to other function prediction machine learning techniques such as k-nearest neighbors or correlation based clustering (e.g., [28]).

13.4.6   Interpretation and Validation

As stated in the preceding section, it is important for correlation networks (or for any networks for that matter) to be able to validate the analysis results utilizing independent data (sets) that were not included in the modeling of the network. In this case, other already known networks can be used for validation (as exemplified above) or for interpretation. Nevertheless an experimental validation based upon the results of the analysis is surely the best and most straightforward corroboration. Moreover, using this strategy one might iterate through the network analysis steps and thus gradually improve the network and its analysis. An example is given by [26] who identified potential target genes, based on a guide gene driven approach and then validated these experimentally by an independent experiment (see Section 13.5.3). Unfortunately, this way is often the most cumbersome, expensive, or simply not practical. Therefore, oftentimes the data are grouped into biological classes using either GO [12] or MapMan [50] terms. Using these biological classes, it becomes possible to attribute some biological function to the variables that is both machine readable and can be visualized or used for grouping, thus making it amendable for interpretation by the researcher.

For further cases of interpretation, the reader can also follow the biological examples given below.

13.5   BIOLOGICAL USE OF CORRELATION NETWORKS

13.5.1   The Global Analysis Approach

The global analysis approach for the biological interpretation of correlation networks serves to give a broad overview of the state of an organism, be it on the metabolic level, the proteomic level, the transcript level, or a combination of these. As discussed previously, based on the experiments chosen for the underlying matrix, the output network can be a frame-capture image of a dynamic process.

In Ref. [31] the authors demonstrated that coresponse networks can be used efficiently to perform network analysis. To this aim, they used a mixed metabolite and gene network, constructed from Arabidopsis sulfur starvation data. For further analysis, they decreased the network size, concentrating only on those submodules interesting for them. Interestingly, they observed that their network followed a power-law distribution, and also that the genes having the highest centrality were coming from the biological classes of nucleotide metabolism, protein destination, and intracellular transport. Unfortunately, no comparison was made to the size of these biological classes within their network.

A similar approach was taken by Ref. [9]. Using publicly available yeast microarray data, they demonstrated a modular structure of the correlation networks. They also showed that within some modules, a correlation between centrality and essentiality could be observed.

In Ref. [45], the authors proposed a more advanced form of analysis by using a conserved correlation network, which was inferred from different organisms, and again they found a power-law distribution of the obtained network. After formatting their network, they could identify highly connected components within the network, which were significantly enriched in biological processes. On a similar trail [2] compared coexpression networks from several model organisms and reported that for the studied organisms, connectivity followed a power law. Moreover like the other studies they inferred that high degree of connectivity is often associated with gene essentiality as well as with gene/function conservation. Furthermore, by comparing the networks they could show that functionally related genes are frequently coexpressed in multiple organisms. However, the relative importance of the coexpressed genes amongst the studied organisms varies.

These holistic analyses provide insight into the organization of the biological processes in an organism and can identify critical connections. Network analyses which emanate from and are targeted on specific physiological pathways represent an alternative approach which may hold greater potential to both identify further components involved in the pathway and to identify the functional context of it. In the context of gene expression network analysis, the guide gene approach represents such a targeted approach.

13.5.2 The Guide Gene Approach[1]

As mentioned in the previous section, guide genes are tools one may use to probe a correlation network. This allows one to focus on the connections in the immediate proximity of the guide gene, and on the genes in this area, providing direct access to transcript coresponse analyses. This approach not only focuses on a localized region in the complete network but may also delineate those areas within the network that may be associated with the process in which the guide gene has a role (see footnote). The term *guide gene* reflects the importance of a careful selection of these genes since the quality of results critically depends on their specificity. The proper selection of guide genes allows the addressing of various biological questions, that is, What could be the function of this unknown gene? Could this protein function in a large protein complex? and so forth. Initially, potential guide genes could be selected based on biological knowledge, such as components of signaling or biosynthetic pathways, stimulus-response genes, components of a protein complex, a known subcellular localization, or regulatory factors. Guide genes could also be derived from preliminary experimental data, and transcript coresponse analysis may help with further guide gene selection refinement. For example, if only a few target genes of a transcription

---

[1]While we generally speak of transcript coresponse analyses here, most examples can be directly transformed for similar analyses based on metabolite or protein samples.

factor have been identified, then the expression of these target genes should be associated with the transcription factor. Thus, the transcription factor and the target genes could be used in an intersection gene query (see below) to screen for further potential target genes. Interestingly, because the coresponses mirror networks, associated genes also mirror factors that in turn regulate expression of the transcription factors. In contrast to direct approaches, transcript coresponse analysis has, in principle, the potential to identify upstream regulatory factors.

Either a single gene can be used in a screen for associated genes (single gene query) or two or more genes can be used (intersection or multiple gene query). Single gene queries may result in the identification of several hundred or even thousands of genes. Thus, single guide genes may not provide enough specificity. Specificity could be conferred by a second or additional guide genes. The additional guide gene(s) must meet certain requirements, however. Transcript levels of guide genes must show association to each other (see Section 13.3.3 for statistical details), otherwise no associated genes would be identified in common. Guide genes also must show variability (i.e., nonperfect association), because otherwise a single gene query would give identical results. The guide genes should be involved in the same pathway (which could also mean a specific intersection of two pathways) otherwise the associated genes will not reflect a common regulation.

The suitability of guide genes for identification of associated genes critically depends on the data matrix. The data matrix must meet several requirements:

- Possess a high number of variables that meet quality parameters in most profiles (discussed in Section 13.3.7; the experiments underlying a data matrix must allow detection of associated genes).
- Include tissues in which the guide gene is expressed.
- Provide sufficient variability (i.e., experiments are different and include an adequate number of profiles) (low variability may not separate unspecific associations; however if the data matrix includes very different conditions only constitutive and highly robust associations can be identified).
- Does not include profiles of mutants or transgenics impaired in the pathway or response of interest (organisms that do not show an intact response or pathway of interest cannot reveal associations with that response or pathway).

If from a biological point of view, or due to experimental findings, a guide gene appears suitable, it may still perform poorly in an association analysis. The most common reasons are if there is weak expression of the guide gene (i.e., detectable transcript levels in only a limited number of profiles), the experiments are based on biologically artificial conditions such as treatment of tissues with synthetic hormones, or the overexpression of a gene (whereas the data matrix reflects intact physiological networks) or finally, the guide gene and/or associated genes are also involved in and/or regulated by other pathways, resulting in weak or unspecific associations.

Suitability of guide genes can be tested using known pathway-involved genes. For example, if a receptor is required for the activity of certain genes, these genes should be identified in a coresponse analysis using the receptor gene as guide gene. Likewise, if a transcription factor is required for the activity of certain genes, these target genes should show association to the transcription factor. However, initial evaluation of guide genes using published data harbors the risk to confer bias, since published data are based on specific experimental conditions (as discussed in Section 13.4.1).

Depending on the guide gene(s), the data matrix and the statistical parameters (see above) screens for associated genes could identify only a few genes or thousands of genes. Both a low number of associations and a high number of associated genes may be undesired. A low number of associated genes may not reveal sufficient insight into associated networks, may not allow the detecting of potential genes for further analysis, and could reflect the use of an improper data matrix. Conversely, a high number of associated genes likely mirrors different regulatory pathways, and associated genes likely are under control of a common (unknown) regulatory factor. Thus, numerous associated genes may not mirror specific regulatory events. An intersection gene query with additional guide genes may improve the result.

Importantly, associations do not allow discrimination between primary and secondary events. Possible approaches to dissect between primary and secondary effects include analyzing a time series (e.g., adding an essential gene by an inducible system in a knock out mutant background), or the detailed analysis of mutants or transgenic organisms that show altered levels of a transcript, stimulus, or metabolite of interest. The usefulness of these experiments for association studies is dubious, however. Association studies of time series experiments may barely provide novel information. In fact, transient effects may result in an increase of weak association measures (see Section 13.4.2). Association studies of mutants and transgenic organisms in which the pathway of interest is disrupted hold the problem that physiological networks (and associations) are disrupted as well. Thus, direct approaches are required for dissecting primary and secondary events.

13.5.3   A Simple Coregulation Test: Photosynthesis

Photosynthesis is a complex process involving the concerted action of numerous proteins with various protein complexes [8]. Light is collected by two large protein complexes, the light-harvesting complexes I and II (LHC I and LHC II), and the energy is transmitted in the form of excitons to two reaction centers (P700 and P680). Here electron transport is driven by these excitons and mediated through other protein complexes. The electrons stem from one of the photosystems (PS II) involving a water-splitting complex-producing molecular oxygen. The electron transport causes a proton gradient over membranes, which is used by ATPases to phosphorylate ADP and thereby providing ATP for the Calvin cycle and other physiologic processes. The electrons finally reduce redox-proteins. Some of the redox-proteins reduce sulfur bridges of enzymes of the Calvin cycle and hereby activate them. Other redox-proteins provide reducing equivalents for sustaining the Calvin cycle.

To further increase the complexity, genes for proteins involved in photosynthesis are expressed in the chloroplast itself as well as in the nucleus. For some proteins, gene expression for subunits even occurs in both the nucleus and the chloroplast.

The best-known example is RubisCO, where the small subunit is expressed in the nucleus and the large subunit in the chloroplast. Furthermore, many photosynthetic proteins depend upon cofactors. Chlorophyll a/b binding (cab) proteins obviously need chlorophyll for light harvesting. This means that expression of genes involved in tetrapyrrol biosynthesis have to be tightly coordinated with genes coding for cab proteins.

Based on all these reasons, gene expression for proteins involved in photosynthesis can be regarded as a highly coregulated network. Expression for some of the genes is known to depend on light. For others, the precise means of regulation is still unknown. Further, the exact nature how chloroplastic and nuclear photosynthesis gene expression is coordinated is unknown. The following exemplary correlation analysis was undertaken to confirm the strong coregulation of the expression of genes for proteins involved in photosynthesis.

To analyze coregulation of photosynthetic genes the CSB.DB tool [42] was used. Because one wishes to observe a general overview of a complex process, the set of experiments on which the coexpression analysis is based should be as broad as possible, as a bias could distort the results (see Section 13.4.2). Therefore the data matrix nasc0271 was chosen. This data matrix contains results of 51 experiments carried out under various experimental conditions. Nasc0271 consists of validly measured expression levels of 9694 genes. The ranking of the pair wise comparison was based on the nonparametric Spearman coefficient. Because positive coresponse of photosynthetic genes was the aim of the analysis, the output of positive significant coresponding genes with Bonferroni correction [6] was used.

To carry out the coresponse analysis, a gene of interest was chosen. In this case we chose a gene coding for a chlorophyll a/b binding protein CP26 (*lhcb5*, At4g10340). CP26 is associated with the light-harvesting complex of photosystem II [53]. Two groups of chl a/b binding protein are usually distinguished within the LHC II. As the major LHC II proteins of the outer antennae bind more than 50% of the total chlorophyll, their role is most likely collecting light energy [46]. CP26 belongs to the minor LHC II proteins. Minor LHC II proteins are supposed to be more involved in energy distribution and photoprotection [13]. These proposed functions make CP26 an interesting candidate gene to analyze coresponse behavior.

The output of the analysis revealed 787 significantly coresponding genes for *lhcb5*. Values of Spearman's Rho rank correlation for the best coresponding genes were very high. The best coresponding gene coding for the chloroplastic NADP-GAP DH (At3g26650) showed a Spearman value of 0.9843. This means that changes of NADP-GAP DH gene expression were almost identical to those of *lhcb5* in 51 expression profiles using various experimental conditions.

The aim of the analysis was to identify how strong photosynthetic genes corespond over 51 expression profiles. To achieve this aim photosynthetic genes have to be identified and separated from other genes. We therefore manually reannotated the 50 closest coresponding genes (Spearman 0.9843 down to 0.8729) and assigned the genes into bins. We used four bins directly involving photosynthesis, namely genes coding for photosystem II proteins, for photosystem I proteins, for protein involved in electron transport, and for proteins of the Calvin cycle. The fifth bin contains genes

TABLE 13.1 Bins of the 50 Closest Coresponding Genes to CP26

Bin	Count	Genes
PS II	14	At1g06680, At1g29930, At1g44575, At1g62520, At1g67740, At1g79040, At2g06520, At2g30570, At2g34430, At3g21055, At3g47470, At4g21280, At5g01530, At5g54270
PS I	13	At1g08380, At1g30380, At1g31330, At1g52230, At1g55670, At3g16140, At3g54890, At3g61470, At4g02770, At4g12800, At4g28750, At5g64040, AtCg00340
Electron transport	5	At1g20340, At1g60950, At2g26500, AtCg00480, AtCg00720
Calvin cycle	5	At1g12900, At2g39730, At3g26550, At5g38420, At5g61410
Plastidic ribosome	4	At3g27830, At3g54210, At4g34620, At5g30510
Other	9	At1g08380, At1g23400, At1g52220, At1g72610, At3g49260, At3g56940, At4g01150, At4g30950, At5g35630

coding for plastidic ribosomal subunits. The rest of the genes were assigned into the bin other (Table 13.1).

Overall there was a very strong coresponse of *lhcb5* with genes coding for photosynthesis proteins, as these represented 37 out of the 50 highest coresponding genes. Interesting to note is that the number of coresponding genes coding for proteins of PS II and PS I was almost identical. Five genes each coding for proteins involved in electron transport and Calvin cycle were amongst the 50 best coresponding genes. Interestingly, two out of the best coresponding genes were from the Calvin cycle bin, namely the aforementioned gene for NADP-GAP DH (At3g26650, Spearman 0.9843, Rank 1) and a gene coding for the small subunit of RubisCO (At5g38420, Spearman 0.9641, Rank 3).

That four genes coding for plastidic ribosomal subunits were in the group of the 50 best coresponding genes reflects nicely that plastidic gene expression is closely concerted with nuclear expression of photosynthesis genes.

Closer analysis of bin "others" revealed that seven out of the nine genes code for chloroplastic proteins. Four are annotated as expressed proteins (At1g08380, At1g52220, At1g23400, and At4g01150). One gene codes for a putative ZIP protein, located in the thylakoid membrane (At3g56940). The other two genes code for well known plastidic proteins involved in fatty acid desaturation (FAD6, At4g30950) and plastidic glutamine synthesis (GS, At5g35630). There were only two genes not fitting into this photosynthesis/plastidic environment, a gene coding for a calmodulin binding protein (At3g49260) and the gene coding for GER1 (At1g72610).

In a second approach, a neighborhood gene analysis was carried out with all 787 highly significantly coresponding genes. All the genes were assigned to the MapMan [50] bins using the CSB.DB tool [42].

TABLE 13.2   Neighborhood Gene Analysis. The 787 Coresponding Genes were Assigned into the MapMan Bins.

MapMan Bin name	%
Photosynthesis	22.0
Tetrapyrrole synthesis	18.9
Redox regulation	5.9
Others	53.3

Table 13.2 shows that by far largest groups of coresponding genes consisted of genes involved in photosynthesis and tetrapyrrole synthesis. This was an expected outcome. Particularly that genes involved in tetrapyrrol synthesis coresponded with *lhcb5* shows the expected link between genes for chlorophyll-producing proteins and chlorophyll-binding proteins. The third largest bin contained genes coding for proteins involved in redox regulation. This result was again expected, as redox regulation is involved in activating Calvin cycle enzymes.

In summary, starting from a single gene coding for an LHC II protein, we could show that coresponse analysis yielded an expected pattern of genes involved in photosynthesis and related processes. Through closer analysis of the 50 best coresponding genes, we identified only two genes that did not match the postulated photosynthesis/chloroplast environment. Expanding the analysis, the 787 highly significantly coresponding genes revealed that the analysis reflected the known framework of genes involved in photosynthesis. Neighborhood search revealed that most of the coresponding genes were indeed involved in photosynthesis, tetrapyrrol synthesis, and redox regulation.

13.5.4   A Complex Coregulation Test: Brassinosteroids

In the following section, we discuss critical aspects of transcript coresponse analysis of a more complex biological system by describing a coresponse screen for brassinosteroid (BR)-related genes. Similar approaches are conceivable for other signaling pathways. BRs are essential for plant growth and development [29]. The primary direct action of BR is the modification of gene expression. Therefore, identification of BR-regulated genes provides insights into their mode of action and physiological effects. Both direct approaches (i.e., expression profiling experiments using mutants or BR-treated plants) and transcript coresponse analysis hold the potential to identify pathway-involved genes. However, transcript coresponse takes the analysis a step further in comparison to direct approaches because BR-related non-BR-responsive genes can be identified. These associated genes provide insight into the functional environment of guide gene(s) used. Both BR-biosynthesis and BR-signaling genes represent specific guide genes that may be used for the identification of BR-associated genes [26]. However, due to their weak expression, BR-biosynthetic genes are barely detected in numerous expression profiles, whereas several BR-signaling genes were reliably detected in many profiles.

Seven components of BR-signaling (*BRI1*, *BRL1*, *BRL3*, *BAK1*, *BIN2*, *BES1*, and *BZR1*) were tested using CSB.DB and the data matrix nasc0271 (described above, see Section 13.5.3), as well as the complex matrices nasc0272 and nasc0273. These complex matrices comprised data for approximately 45% of known BR-responsive genes (see [26] for details). Only the *BRI1*, *BAK1*, and *BIN2* genes identified several known BR-responsive genes. *BRI1* and *BAK1* outperformed *BIN2* in single gene queries, and the *BRI1/BAK1* intersection gene query also outperformed *BRI1/BIN2* and *BAK1/BIN2* intersection gene queries. The BRI1 receptor is the major BR-receptor required for most responses. Loss of function mutations result in extreme dwarfism and nearly complete BR-insensitivity in Arabidopsis. BAK1 is a coreceptor with BRI1. The *BRI1/BAK1* intersection gene query produced a recovery rate of 34.7% of known BR-related genes and demonstrated that an intersection query using *BRI1* and *BAK1* was optimal to identify pathway-involved genes in the nasc0271 data matrix.

Suspected pathway-involved genes (identified using coresponse analysis) can then be tested experimentally. In the BR-example, publicly available expression profiles and additional experiments were used to test BR-responsiveness of all genes showing significantly correlated expression with the *BRI1* and *BAK1* genes. These profiles and experiments analyzed transcript levels in BR-mutants (i.e., BR-deficient and BR-insensitive mutants), BR-treated plants, and plants treated with inhibitors of BR-biosynthesis. 24% of the associated genes turned out to be BR-responsive in at least two experiments [26].

The finding that only a subset of known BR-responsive genes is identified via associations with *BRI1* and *BAK1* likely reflects that published expression profiles represent specific experimental conditions. On the contrary, after experimental confirmation, 76% of all associated genes appeared not to be under direct control of BR. These associated genes mirror the "functional environment" of the guide genes used for analysis. Virtually, no pathway acts independently of other pathways, but instead is embedded in a context. This context is difficult to identify through direct approaches, but becomes accessible by transcript coresponse analysis. Thus, transcript coresponse analysis can both identify pathway-involved genes and provide insights into underlying biological associations and cross talk.

13.6 SUMMARY

In this chapter, we have attempted to describe the basic notation and mathematics behind generating a correlation matrix, and have discussed the critical parameters that must be considered before data are included in the matrix. We have also tried to stress the importance of a proper discretization and filtering of the correlation matrix to ensure that biologically meaningful results are generated so that the generated network is the best reflection of the underlying data. We have tried to indicate the strengths and weaknesses in analyzing correlation networks both at the global and local levels. We have as well tried to show some of the results that may be garnered through looking at the complete network at once (global analysis) and as well have stressed a way to analyze the network at a local level with a hypothesis-driven guide gene approach. Finally, using two biological examples of the guide gene approach, we have attempted to aid the reader in a biological interpretation of the output.

The authors would like to stress again that the analysis of correlation-based networks is critically dependent on the underlying data that makes the matrix, and the specific biological question that one asks. In the end, the specific tool(s) used to probe the network for meaningful output is also question dependent, and the resulting output must be critically evaluated.

13.7 EXERCISES

The following exercise section is intended to demonstrate basic statistical steps regarding generation and analyses of correlation-based probability networks. It will mainly focus on statistical points. For biological interpretation, the critical reader is refered to the Section 13.5. The examples in the Section 13.5 may serve as a basis for repeating the analyses by using different publicly available web resources as cited in Section 13.4.3.

1. One of the crucial steps in correlation-based network analyses represents the choice of association measure. The first example will briefly illustrate the different results that may be observed depending on the coefficient of association. This choice has advantages and disadvantages depending on the researcher's question.

   A data matrix with five variables (var.1–var.5) and 10 units (exp.1–exp.10) is given in Table 13.3.

TABLE 13.3 Artificial $m \times n$ Data Set

Variables / Units	Experiment (exp.)				
	1	2	3	4	5
var.1	1.11	0.08	0.11	0.78	−0.33
var.2	0.83	−0.25	−0.43	0.60	−0.82
var.3	0.26	−1.37	−1.25	−0.54	−1.59
var.4	−0.65	−0.07	0.17	−0.66	0.59
var.5	0.12	−0.10	0.24	0.23	−0.02

Variables / Units	Experiment (exp.)				
	6	7	8	9	10
var.1	−0.67	−0.15	0.56	−0.35	1.95
var.2	−1.56	−0.91	−0.03	−0.92	1.36
var.3	−2.00	−1.23	−0.81	−1.73	0.60
var.4	0.84	0.22	−0.54	0.97	−0.98
var.5	0.32	0.14	−0.16	−0.10	2.33

(a) Generate a (line) plot for each of the variables by plotting the units (x-axis) against their respective observations (y-axis) to get a first impression of the data matrix. You may add lines to aid interpretation.

(b) Compute the Pearson correlation of var.1 versus all other variables (five results). In Microsoft (MS) Excel you can use the function "pearson"; in the statistical software environment R, you can use the function "cor" or "cor.test."

(c) Compute the Spearman correlation of var.1 versus all other variables (five results) as Pearson correlation on ranked data. In MS Excel, you can use the function "rank" to rank the observations and than run the "pearson" function. In R you can use the function "cor" or "cor.test" by changing the method parameter to "spearman."

(d) Compute the Euclidean distance of var.1 versus all other variables (five results). In R you can use the function "dist." In MS Excel you have to calculate it by using the following formula:

$$d_e(X, Y) = \sqrt{\sum_{i=1}^{n} (X_i - Y_i)^2}$$

that is, calculate the square difference of var.1 and, for example, var.2 for each unit, sum over all units, and calculate the square root of this sum.

(e) Compare the calculated similarity and dissimilarity coefficients for each variable. Generate scatterplot(s) of var.1 (x-axis) versus all other variables to aid interpretation.

2. A Pearson correlation between two variables of 0.7 is observed on the basis of 3, 10, 50, and 100 paired observations n. Compute the degree of freedom (df), the t statistic (t_S, acc. Equation 13.6) and the p-value. In MS Excel, the p-value can be computed by the function "tvert" and using the parameter t_S, df, and 2 for a two-sided test. In R you can use the function "pt" (Note: A one-sided value will be returned).

REFERENCES

1. Affymetrix. Technical instructions for technology platform, e. g. Statistical Algorithms Description Document. http://www.affymetrix.com/index.affx, 2002.

2. S. Bergmann, J. Ihmels, and N. Barkai. Similarities and differences in genomewide expression data of six organisms. *PLoS Biology*, 2:85–93, 2004.

3. N. Bhardwaj and H. Lu. Correlation between gene expression profiles and protein-protein interactions within and across genomes. *Bioinformatics*, 21:2730–2738, 2005.

4. R. J. Bino, R. D. Hall, O. Fiehn, J. Kopka, K. Saito, J. Draper, B. J. Nikolau, P. Mendes, U. Roessner-Tunali, M. H. Beale, R. N. Trethewey, B. M. Lange, E. Syrkin Wurtele, and

L. W. Sumner. Potential of metabolomics as a functional genomics tool. *Trends in Plant Science*, 9:418–425, 2004.

5. D. G. Bonett and T. A. Wright. Sample size requirements for estimating Pearson, Kendall and Spearman correlations. *Psychometrika*, 65:23–28, 2000.

6. C. E. Bonferroni. Il calcolo delle assicurazioni su gruppi di teste. In *Studi in Onore del Professore Salvatore Ortu Carboni*. pp. 13–60. Rome, 1935.

7. A. Brazma, P. Hingamp, J. Quackenbush, G. Sherlock, P. Spellman, C. Stoeckert, J. Aach, W. Ansorge, C. A. Ball, H. C. Causton, T. Gaasterland, P. Glenisson, F. C. Holstege, I. F. Kim, V. Markowitz, J. C. Matese, H. Parkinson, A. Robinson, U. Sarkans, S. Schulze-Kremer, J. Stewart, R. Taylor, J. Vilo, and M. Vingron. Minimum information about a microarray experiment (MIAME)—toward standards for microarray data. *Nature Genetics*, 29:365–371, 2001.

8. B. B. Buchannan, W. Gruissem, and R. L. Jones. *Biochemistry & Molecular Biology of Plants*. American Society of Plant Physiologists (ASPP), Maryland, 2000.

9. M. R. Carlson, B. Zhang, Z. Fang, P. S. Mischel, S. Horvath, and S. F. Nelson. Gene connectivity, function, and sequence conservation: predictions from modular yeast coexpression networks. *BMC Genomics*, 3:7–40, 2006.

10. P. Y. Chen and P. M. Popovich. *Correlation: Parametric and Nonparametric Measures*. Quantitative Applications in the Social Science. Sage Publications, Thousand Oaks, CA, 2002.

11. V. Cherepinsky, J. Feng, M. Rejali, and B. Mishra. Shrinkage-based similarity metric for cluster analysis of microarray data. *Proceedings of the National Academy of Sciences*, 100:9668–9673, 2003.

12. Gene Ontology Consortium. The Gene Ontology (GO) project in 2006. *Nucleic Acids Research*, 34:D322–326, 2006.

13. L. Dall'Osto, S. Caffarri, and R. Bassi. A mechanism of nonphotochemical energy dissipation, independent from PsbS, revealed by a conformational change in the antenna protein CP26. *Plant Cell*, 17:1217–1232, 2005.

14. R. Edgar, M. Domrachev, and A. E. Lash. Gene Expression Omnibus: NCBI gene expression and hybridization array data repository. *Nucleic Acids Research*, 30:207–210, 2002.

15. H. Ge, Z. Liu, G. M. Church, and M. Vidal. Correlation between transcriptome and interactome mapping data from *Saccharomyces cerevisiae*. *Nature Genetics*, 29:482–486, 2001.

16. B. Harr and C. Schlotterer. Comparison of algorithms for the analysis of Affymetrix microarray data as evaluated by co-expression of genes in known operons. *Nucleic Acids Research*, 34:e8, 2006.

17. L. Hood and R. M. Perlmutter. The impact of systems approaches on biological problems in drug discovery. *Nature Biotechnology*, 22:1215–1217, 2004.

18. H. Hotelling. New light on the correlation coefficient and its transformations. *Journal of the Royal Statistical Society*, 15:193–232, 1953.

19. J. Ihmels, R. Levy, and N. Barkai. Principles of transcriptional control in the metabolic network of Saccharomyces cerevisiae. *Nature Biotechnology*, 22:86–92, 2004.

20. R. A. Irizarry, B. M. Bolstad, F. Collin, L. M. Cope, B. Hobbs, and T. P. Speed. Summaries of Affymetrix GeneChip probe level data. *Nucleic Acids Research*, 31:e15, 2003.

21. C. H. Jen, I. W. Manfield, I. Michalopoulos, J. W. Pinney, W. G. Willats, P. M. Gilmartin, and D. R. Westhead. The Arabidopsis co-expression tool (ACT): A www-based tool and database for microarray-based gene expression analysis. *Plant Journal*, 46:336–348, 2006.

22. H. Kitano. Computational systems biology. *Nature*, 420:206–210, 2002.

23. H. Kitano. Looking beyond the details: A rise in system-oriented approaches in genetics and molecular biology. *Current Genetics*, 41:1–10, 2002.

24. H. Kitano. Systems biology: A brief overview. *Science*, 295:1662–1664, 2002.

25. J. Kopka, A. Fernie, W. Weckwerth, Y. Gibon, and M. Stitt. Metabolite profiling in plant biology: Platforms and destinations. *Genome Biology*, 5:109, 2004.

26. J. Lisso, D. Steinhauser, T. Altmann, J. Kopka, and C. Müssig. Identification of brassinoid-related genes by means of transcript co-response analyses. *Nucleic Acids Research*, 33:2685–2696, 2005.

27. D. J. Lockhart and E. A. Winzeler. Genomics, gene expression and DNA arrays. *Nature*, 405:827–836, 2000.

28. B. Mirkin. *Mathematical Classification and Clustering*, volume 11 of *Nonconvex Optimisation and Its Application*. Kluwer Academic Publishers, Norwell, 1996.

29. C. Müssig. Brassinosteroid-promoted growth. *Plant Biology*, 7:1–8, 2005.

30. NCBI. http://www.ncbi.nlm.nih.gov/Genomes/index.html.

31. V. J. Nikiforova, C. O. Daub, H. Hesse, L. Willmitzer, and R. Hoefgen. Integrative gene-metabolite network with implemented causality deciphers informational fluxes of sulphur stress response. *Journal of Experimental Botany*, 56:1887–1896, 2005.

32. Z. N. Oltvai and A.-L. Barabási. Life's complexity pyramid. *Science*, 298(5594): 763–764, 2002.

33. H. Parkinson, U. Sarkans, M. Shojatalab, N. Abeygunawardena, S. Contrino, R. Coulson, A. Farne, G. G. Lara, E. Holloway, M. Kapushesky, P. Lilja, G. Mukherjee, A. Oezcimen, T. Rayner, P. Rocca-Serra, A. Sharma, S. Sansone, and A. Brazma. ArrayExpress—a public repository for microarray gene expression data at the EBI. *Nucleic Acids Research*, 33:D553–555, 2005.

34. S. Persson, H. Wei, J. Milne, G. P. Page, and C. R. Somerville. Identification of genes required for cellulose synthesis by regression analysis of public microarray data sets. *Proceedings of the National Academy of Sciences*, 102:8633–8638, 2005.

35. L. X. Qin, R. P. Beyer, F. N. Hudson, N. J. Linford, D. E. Morris, and K. F. Kerr. Evaluation of methods for oligonucleotide array data via quantitative real-time PCR. *BMC Bioinformatics*, 7:23, 2006.

36. J. Schäfer and K. Strimmer. An empirical Bayes approach to inferring large-scale gene association networks. *Bioinformatics*, 21:754–764, 2005.

37. C. E. Shannon. A mathematical theory of communication. *The Bell System Technical Journal*, 27:379–423, ibid 623–656, 1948.

38. R. R. Sokal and F. J. Rohlf. *Biometry: The Principles and Practice of Statistics in Biological Research*. W. H. Freeman and Company, New York, 1995.

39. C. Somerville and S. Sommerville. Plant functional genomics. *Science*, 285:380–383, 1999.

40. D. Steinhauser, B. H. Junker, A. Luedemann, J. Selbig, and J. Kopka. Hypothesis-driven approach to predict transcriptional units from gene expression data. *Bioinformatics*, 20:1928–1939, 2004.

41. D. Steinhauser and J. Kopka. Methods, applications, and concepts of metabolite profiling: Primary metabolism. In *Plant Systems Biology*, Vol. 97 of Experientia Supplementum, Springer Verlag, Heidelberg, 2006.

42. D. Steinhauser, B. Usadel, A. Luedemann, O. Thimm, and J. Kopka. CSB.DB: A comprehensive systems-biology database. *Bioinformatics*, 20:3647–3651, 2004.

43. G. Stephanopoulos, H. Alper, and J. Moxley. Exploiting biological complexity for strain improvements through systems biology. *Nature Biotechnology*, 22:1261–1267, 2004.

44. R. Steuer, J. Kurths, C. O. Daub, J. Weise, and J. Selbig. The mutual information: Detecting and evaluating dependencies between variables. *Bioinformatics*, 18:S231–240, 2002.

45. J. M. Stuart, E. Segal, D. Koller, and S. K. Kim. A gene-coexpression network for global discovery of conserved genetic modules. *Science*, 302:249–255, 2003.

46. H. Teramoto, T. Ono, and J. Minagawa. Identification of Lhcb gene family encoding the light-harvesting chlorophyll-a/b proteins of photosystem II in *Chlamydomonas reinhardtii*. *Plant Cell Physiology*, 42:849–856, 2001.

47. K. Toufighi, S. M. Brady, R. Austin, E. Ly, and N. J. Provart. The botany array resource: e-Northerns, Expression Angling, and promoter analyses. *Plant Journal*, 43:153–163, 2005.

48. M. Tyers and M. Mann. From genomics to proteomics. *Nature*, 422:193–197, 2003.

49. B. Usadel, A. M. Kuschinsky, D. Steinhauser, and M. Pauly. Transcriptional co-response analysis as a tool to identify new components of the wall biosynthetic machinery. *Plant Biosystems*, 139:69–73, 2005.

50. B. Usadel, A. Nagel, O. Thimm, H. Redestig, O. E. Blaesing, N. Palacios-Rojas, J. Selbig, J. Hannemann, M. C. Piques, D. Steinhauser, W. R. Scheible, Y. Gibon, R. Morcuende, D. Weicht, S. Meyer, and M. Stitt. Extension of the visualization tool MapMan to allow statistical analysis of arrays, display of corresponding genes, and comparison with known responses. *Plant Physiology*, 138:1195–1204, 2005.

51. W. Weckwerth, M. E. Loureiro, K. Wenzel, and O. Fiehn. Differential metabolic networks unravel the effects of silent plant phenotypes. *Proceedings of the National Academy of Sciences*, 101:7809–7814, 2004.

52. A. D. Weston and L. Hood. Systems biology, proteomics, and the future of health care: Toward predictive, preventative, and personalized medicine. *Journal of Proteome Research*, 3:179–196, 2004.

53. A. E. Yakushevska, W. Keegstra, E. J. Boekema, J. P. Dekker, J. Andersson, S. Jansson, A. V. Ruban, and P. Horton. The structure of photosystem II in Arabidopsis: Localization of the CP26 and CP29 antenna complexes. *Biochemistry*, 42:608–613, 2003.

54. P. Zimmermann, M. Hirsch-Hoffmann, L. Hennig, and W. Gruissem. GENEVESTIGATOR. Arabidopsis Microarray Database and Analysis Toolbox. *Plant Physiology*, 136:2621–2632, 2004.

55. P. Zimmermann, B. Schildknecht, D. Craigon, M. Garcia-Hernandez, W. Gruissem, S. May, G. Mukherjee, H. Parkinson, S. Rhee, U. Wagner, and L. Hennig. MIAME/Plant—adding value to plant microarrray experiments. *Plant Methods*, 2:1, 2006.

INDEX

Analysis of Biological Networks, Edited by Björn H. Junker and Falk Schreiber
Copyright © 2008 John Wiley & Sons, Inc.